jacaranda *plus*

Next generation teaching and learning

Access all formats of your online Jacaranda resources in three easy steps!

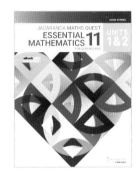

To access your resources:

1 ● go to **www.jacplus.com.au**
2 ● log in to your existing account, or create a new account
3 ● enter your unique registration code(s).

Note
- Only one JacPLUS account is required to register all your Jacaranda digital products.
- By registering the code(s) within your JacPLUS bookshelf, you are agreeing to purchase the resource(s). Please view the terms and conditions when registering.

REGISTRATION CODE

Electronic versions of this title are available online; these include eBookPLUS and PDFs. Your unique registration codes for this title are:

5FHHWJHVQ7J

5JJEPR76MUV

5NNXSZ6E3U5

Each code above provides access for one user to the eBookPLUS and PDFs.

Two subsequent users of this code will be able to access only a Lite version of the eBook. You can upgrade from the Lite version to the full-access eBookPLUS at any time by selecting the 'Upgrade' button on your JacPLUS bookshelf.

NEED HELP?

If you would like to discuss specific digital licensing options or request digital trials, or if you require any other assistance, email support@jacplus.com.au or telephone 1800 JAC PLUS (1800 522 7587).

T0360077

A Wiley Brand

JACARANDA MATHS QUEST

ESSENTIAL MATHEMATICS 11

FOR QUEENSLAND

UNITS 1&2

jacaranda
A Wiley Brand

JACARANDA MATHS QUEST
ESSENTIAL MATHEMATICS 11
FOR QUEENSLAND

UNITS 1&2

MARK BARNES | PAULINE HOLLAND

CONTRIBUTING AUTHORS

Shirly Griffith | Libby Kempton

First published 2019 by
John Wiley & Sons Australia, Ltd
42 McDougall Street, Milton, Qld 4064

Typeset in 11/14 pt Times LT Std

ISBN: 978-0-7303-6702-4

Front cover image: © antishock/Shutterstock

Illustrated by various artists, diacriTech and Wiley Composition Services

Typeset in India by diacriTech

Printed in Singapore by
Markono Print Media Pte Ltd

A catalogue record for this
book is available from the
National Library of Australia

10 9 8 7 6 5 4 3 2

CONTENTS

11 Data collection   442

PRACTICE ASSESSMENT 4

Unit 2 examination

APPENDIX Maths skills workbook   478

ABOUT THIS RESOURCE

Jacaranda Maths Quest 11 Essential Mathematics Units 1 & 2 for Queensland is expertly tailored to address comprehensively the intent and structure of the new syllabus. The *Jacaranda Maths Quest for Queensland* series provides easy-to-follow text and is supported by a bank of resources for both teachers and students. At Jacaranda we believe that every student should experience success and build confidence, while those who want to be challenged are supported as they progress to more difficult concepts and questions.

Preparing students for exam success

Chapter openers provide students with their learning sequence and syllabus outcomes.

A variety of online resources are available in the eBookPLUS; these include video eLessons, interactivities, SkillSHEETS and weblinks.

Each subtopic concludes with a carefully graded exercise which provides opportunities for success and challenge.

Every chapter concludes with exam practice questions classified as Simple familiar, Complex familiar and Complex unfamiliar.

Two complete sets of practice assessments modelled on QCAA guidelines — a set for student revision and a quarantined set for teachers — are included. Exemplary responses and worked solutions are provided for teachers.

Chapter questions and activities are aligned with Marzano and Kendall's taxonomy of cognitive process — retrieval, comprehension, analysis and knowledge utilisation.

Features of the *Maths Quest* series

Questions and topics are sequenced from lower to higher levels of complexity; ideas and concepts are logically developed and questions are carefully graded, allowing every student to achieve success.

An extensive glossary of mathematical terms is provided in print and as a hover-over feature in the eBookPLUS.

Fully worked examples in the Think/Write format provide guidance and are linked to questions.

Free fully worked solutions are provided, enabling students to get help where they need it, whether at home or in the classroom — help at the point of learning is critical. Answers are provided at the end of each chapter in the print and offline PDF.

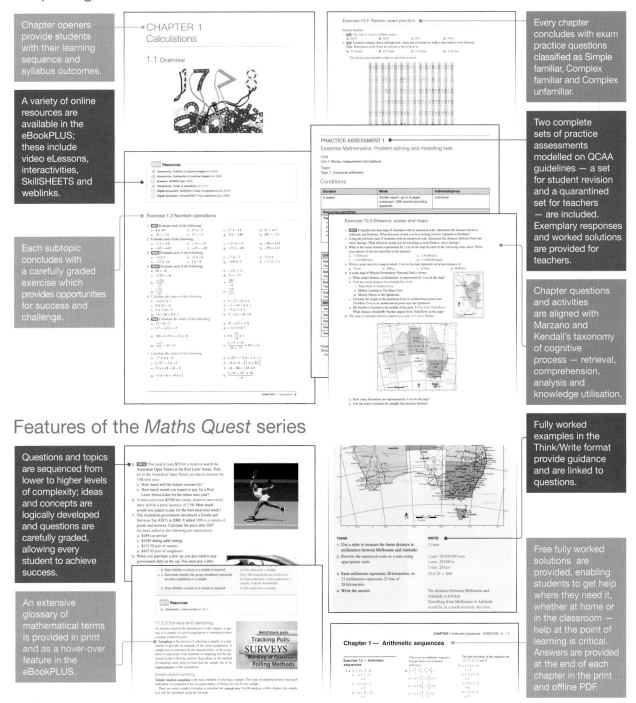

eBookPLUS features

Fully worked solutions for every question

Digital documents: downloadable SkillSHEETS to support skill development and SpreadSHEETS to explore mathematical relationships and concepts

A downloadable PDF of the entire chapter of the print text

Interactivities and video eLessons placed at the point of learning to enhance understanding and correct common misconceptions

In the Prelims section of your eBooKPLUS

A downloadable PDF of the entire solutions manual, containing worked solutions for every question in the text

A set of four practice assessments: a problem solving and modelling task and three examination-style assessments

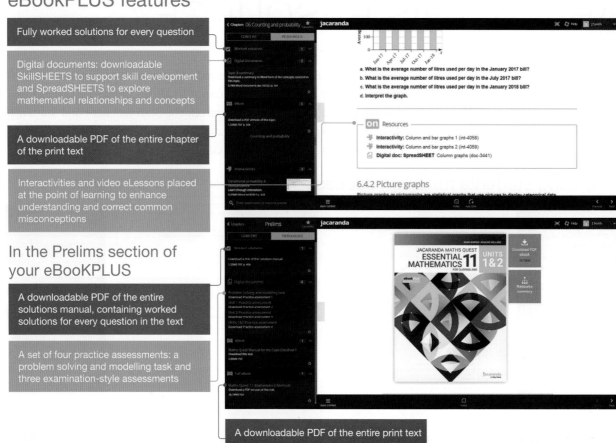

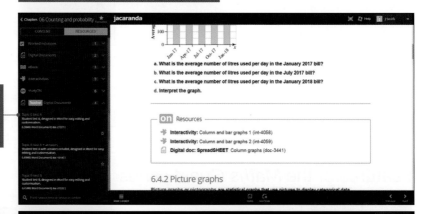

A downloadable PDF of the entire print text

Additional resources for teachers available in the eGuidePLUS

In the Resources tab of every chapter there are two topic tests in downloadable, customisable Word format with worked solutions.

In the Prelims section of the eGuidePLUS

Work programs are provided to assist with classroom planning.

Practice assessments: in addition to the four provided in the eBookPLUS, teachers have access to a further four quarantined assessments. Modelled on QCAA guidelines, the problem solving and modelling tasks are provided with exemplary responses while the examination-style assessments include annotated worked solutions. They are downloadable in Word format to allow teachers to customise as they need.

About eBookPLUS

jacaranda *plus*

This book features eBookPLUS: an electronic version of the entire textbook and supporting digital resources. It is available for you online at the JacarandaPLUS website (**www.jacplus.com.au**).

Join **thousands** of other students and teachers in discovering the **next generation** in **teaching** and **learning solutions**...

Using JacarandaPLUS

To access your eBookPLUS resources, simply log on to **www.jacplus.com.au** using your existing JacarandaPLUS login and enter the registration code. If you are new to JacarandaPLUS, follow the three easy steps below.

Step 1. Create a user account

The first time you use the JacarandaPLUS system, you will need to create a user account. Go to the JacarandaPLUS home page (**www.jacplus.com.au**), click on the button to create a new account and follow the instructions on screen. You can then use your nominated email address and password to log in to the JacarandaPLUS system.

Step 2. Enter your registration code

Once you have logged in, enter your unique registration code for this book, which is printed on the inside front cover of your textbook. The title of your textbook will appear in your bookshelf. Click on the link to open your eBookPLUS.

Step 3. Access your eBookPLUS resources

Your eBookPLUS and supporting resources are provided in a chapter-by-chapter format. Simply select the desired chapter from the table of contents. Digital resources are accessed within each chapter via the resources tab.

Once you have created your account, you can use the same email address and password in the future to register any JacarandaPLUS titles you own.

Using eBookPLUS references

eBookPLUS logos are used throughout the printed books to inform you that a digital resource is available to complement the content you are studying.

Searchlight IDs (e.g. **INT-0001**) give you instant access to digital resources. Once you are logged in, simply enter the Searchlight ID for that resource and it will open immediately.

Minimum requirements

JacarandaPLUS requires you to use a supported internet browser and version, otherwise you will not be able to access your resources or view all features and upgrades. The complete list of JacPLUS minimum system requirements can be found at **http://jacplus.desk.com**.

Troubleshooting

- Go to **www.jacplus.com.au** and click on the Help link.
- Visit the JacarandaPLUS Support Centre at **http://jacplus.desk.com** to access a range of step-by-step user guides, ask questions or search for information.
- Contact John Wiley & Sons Australia, Ltd. Email: support@jacplus.com.au Phone: 1800 JAC PLUS (1800 522 7587)

ACKNOWLEDGEMENTS

The authors and publisher would like to thank the following copyright holders, organisations and individuals for their assistance and for permission to reproduce copyright material in this book.

Images

• Alamy Australia Pty Ltd: **328**/Folio Images; **379**/Art Directors & TRIP • Alamy Stock Photo: **9** (bottom)/NDP; **19** (top)/GoAustralia; **20**/CrowdSpark; **41** (top)/GL Archive; **371**/david hancock • American Economic Association: **155** (bottom)/Copyright American Economic Association; reproduced with permission of the Journal of Economic Perspectives • Australian Bureau of Statistics: **24** (bottom), **154**, **239** (top), **443** • Copyright Agency Limited: **147**/Step right up and do it yourself, *The Age*, John Collett, 27/03/13 • Creative Commons: **27** (bottom), **61** (bottom), **236**, **238**, **274** (bottom), **398**, **399** (bottom); **194**/Department of Home Affairs; **215**/Based on information from the Bureau of Meteorology; **216** (bottom), **217** (bottom), **217** (centre), **217** (top), **218** (centre)/Based on information taken from the Bureau of Meteorology; **216** (centre), **225** (bottom)/Based on information taken from the Bureau of Meteorology; **218** (bottom), **226** (top), **291**/Bureau of Meteorology; **218** (top), **242** (centre)/Based on the information taken from the Bureau of Meteorology; **222** (bottom), **222** (top), **223** (bottom), **223** (centre), **223** (top), **224** (centre), **224** (top)/© Commonwealth of Australia; **224** (bottom)/Based on information taken from the Bureau of Meteorology; **225** (centre)/Based on information from the Bureau of Meteorology; **228** (bottom)/Based on information taken from the Australian Bureau of Statistics; **229** (top)/Based on information taken from the Australian Government Department of Social Services; **230**/Australian Institute of Health and Welfare; **236** (bottom)/© Commonwealth of Australia.; **237** (bottom), **243** (centre)/Based on information taken from the Australian Bureau of Statistics; **242** (bottom)/Based on information taken from exchange rates between $AU and $US between 13 February and 15 March 2013; **243** (bottom)/Based on information taken from Australian Bureau of Statistics; **246**/Based on information taken from 2006 census Townsville B01 selected person characteristics by sex count of persons based on usual place of residence; **247**/Based on information taken from the Australian Bureau of Meteorology; **269**/Australian Bureau of Statistics; **270**/Department of Mapping and Surveying 1980, Queensland resources atlas, 2nd rev ed. Courtesy Dept of Lands; **284**, **285** (top)/© Copyright, Commonwealth of Australia; **285** (bottom), **286** (top)/Australian Government — Department of Infrastructure and Transport; **291**/© Australian Prudential Regulation Authority; **384**/© 2010, Commonwealth of Australia • Excel: **277** (a), **277** (b), **277** (c), **277** (d), **277** (e), **278** (a), **278** (b), **278** (c), **278** (d), **279** (a), **279** (b), **279** (c), **280** (a), **280** (b), **280** (c), **280** (d), **280** (e), **281** (a), **281** (b), **281** (c), **281** (d), **282** (a), **282** (b), **282** (c), **282** (d), **283** (a), **283** (a), **283** (b), **283** (c), **283** (d), **283** (e), **284**, **287**, **288** (a), **288** (b), **288** (c), **288** (d), **292** (a), **292** (b), **292** (c), **292** (d), **292** (e), **292** (f), **292** (g), **292** (h), **292** (i), **445** (a), **445** (b), **446**/Microsoft • Getty Images: **312** (bottom)/PeopleImages; **366**/Uppercut RF • Getty Images Australia: **13** (bottom)/Andrew D. Bernstein; **22** (bottom)/Andrew D. Bernstein/Contributor; **22** (centre)/AFP; **76** (centre)/Billy Stickland/Allsport; **83** (centre)/Mark Dadswell/Staff; **111**/Ian Hitchcock; **112**/Chris Hyde; **113** (top)/Mark Nolan; **166**/Auscape/Contributor; **408** (centre)/William West • iStockphoto: **315** (bottom)/@ Anja Koppitsch royalty free — extended use; **346**/© René Mansi royalty free — extended use; **361**/royalty free — extended use • John Wiley & Sons Australia: **13** (top)/Taken by Kari-Ann Tapp; **74**, **113** (q)12; **127**/Renee Bryon; **480**/Kari-Ann Tapp • MAPgraphics: **392**, **397**, **399** (top), **400**, **401** • Mark Zandi: **156**/The Economic Impact of the American Recovery and Reinvestment Act, Mark Zandi, Chief Economist, Moody's Economy.com, January 21, 2009. http://www.economy.com/markzandi/documents/economic_stimulus_house_plan_012109.pdf • Microsoft Excel: **89** (a), **89** (b), **89** (c), **89** (d), **89** (e), **90** (a), **90** (b), **90** (c), **90** (d), **90** (e), **90** (f), **91** (a), **91** (b)/Microsoft Corporation • Public Domain: **23** (bottom), **387**; **233**/Based on information taken from Statistics Canada, Pictographs. • Shutterstock: **1**/Anton Mezinov; **4**, **108** (top)/Ariwasabi; **6** (bottom), **136** (top)/Stuart Jenner; **6** (centre)/Andy Dean Photography; **6** (top)/Tupungato; **9** (top)/gpointstudio; **10** (bottom), **135** (top), **313**, **450**/michaeljung; **10** (top)/oliveromg; **14** (bottom)/manfredxy; **14** (top)/Kathie Nichols; **18** (bottom)/sandystifler; **18** (bottom)/Byron W. Moore; **18** (centre)/Huguette Roe; **18** (top)/Golden Pixels LLC; **21** (centre)/Mega Pixel; **21** (centre)/urbanbuzz; **21** (top), **138**/Elena Elisseeva; **23** (centre)/photogolfer; **23** (top), **118** (top), **120** (top)/Leonard Zhukovsky; **27** (centre)/Radu Bercan; **27** (top)/Anton Balazh; **28**/Ekaterina Kamenetsky; **29** (bottom)/Racheal Grazias; **29** (top)/ONYXprj; **33**/Lotus_studio; **34** (bottom)/Shane White; **34** (top), **48** (top right), **82** (centre)/M. Unal Ozmen; **35**, **77** (centre)/Pakhnyushchy; **36** (bottom)/sabri deniz kizil; **36** (top)/Mellimage; **38** (bottom)/43372; **38** (centre)/Roobcio; **38** (top)/Yoko Design; **39**/SS1001; **40** (bottom)/Alexandra Lande; **40** (top)/Olinchuk; **41** (bottom)/Vitaliy Hrabar; **42**/Bardocz Peter; **43**/zstock; **44**/Studio Barcelona; **45**/stockcreations; **46** (bottom)/bonchan; **46** (top)/Konstantin Faraktinov; **47** (bottom)/artellia; **47** (bottom)/AridOcean; **47** (centre)/Songsook; **47** (top), **117** (bottom)/ThomasLENNE; **48** (bottom)/ben bryant; **48** (bottom)/Ashley van Dyck; **48** (centre)/creatOR76; **48** (top)/Eladstudio; **48** (top left)/Unal Ozmen; **49** (bottom)/LoopAll; **49** (centre left)/Aprilphoto; **49** (centre right)/Sashkin; **49** (top)/Sorbis; **49** (top), **211** (top), **334**, **353**, **358**, **364**/Monkey Business Images; **50** (bottom)/Dan Kosmayer; **50** (centre)/aodaodaodaod; **50** (top)/happymay; **51**/Anteromite; **52**/Nonnakrit; **53**/photowind; **54**/cTermit; **55** (bottom)/Iryna Denysova; **55** (centre)/David Svetlik; **55** (top)/TanjaJovicic; **56** (bottom)/JM-Design; **56** (centre)/Golf_chalermchai;

CHAPTER 1
Calculations

1.1 Overview

CONTENT

In this chapter, students will learn to:
- solve practical problems requiring basic number operations
- apply arithmetic operations according to their correct order
- ascertain the reasonableness of answers to arithmetic calculations
- use leading-digit approximation to obtain estimates of calculations
- use a calculator for multi-step calculations
- check results of calculations for accuracy
- recognise the significance of place value after the decimal point
- evaluate decimal fractions to the required number of decimal places
- round up or round down numbers to the required number of decimal places
- apply approximation strategies for calculations.

Fully worked solutions for this chapter are available in the Resources section of your eBookPLUS at www.jacplus.com.au.

1.2 Number operations

1.2.1 Adding and subtracting integers

Integers are positive and negative whole numbers, including zero. We can use a number line to add or subtract integers.

To add a positive integer, move to the right on the number line.

To add a negative integer, move to the left on the number line. Follow the rules shown in the table below to add and subtract positive and negative numbers.

Adding a positive	$+ + = +$
Adding a negative	$+ - = -$
Subtracting a positive	$- + = -$
Subtracting a negative	$- - = +$

WORKED EXAMPLE 1

Evaluate each of the following:

a. $-3 + +2$ b. $-3 + -2$

THINK   **WRITE**

a. 1. Rewrite the two signs ($+ +$) as single sign ($+$).   a. $-3 + +2 = -3 + 2$

   2. Perform the calculation using a number line (adding means moving to the right).

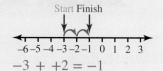

   3. Write the answer. $-3 + +2 = -1$

b. 1. Rewrite the two signs ($+ -$) as a single sign ($-$).   b. $-3 + -2 = -3 - 2$

   2. Perform the calculation using a number line (subtracting means moving to the left).

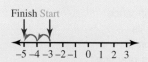

   3. Write the answer. $-3 - 2 = -5$

1.2.2 Multiplying or dividing integers

Follow the rules shown in the table below to multiply or divide positive and negative numbers.

When multiplying or dividing two numbers with the same sign, the answer is positive.	$+ \times + = +$ or $+ \div + = +$ $- \times - = +$ or $- \div - = +$
When multiplying or dividing two numbers with different signs, the answer is negative.	$+ \times - = -$ or $+ \div - = -$ $- \times + = -$ or $- \div + = -$

WORKED EXAMPLE 2

Evaluate each of the following:

a. 12×-3 b. $-9 \div -3$

THINK   **WRITE**

a. 1. The two numbers have different signs so the answer will be negative. $(+ \times - = -)$ a. Negative answer

    2. Write the answer. 12×-3
                                                          $= -36$

b. 1. The two numbers have the same signs so the answer will be positive. $(- \div - = +)$ b. Positive answer

    2. Write the answer. $-9 \div -3$
                                                          $= 3$

1.2.3 Order of operations

BIDMAS can help us to remember the correct order in which we should perform various operations. We do brackets first; then powers or indices; then division and multiplication (working from left to right); and finally addition and subtraction (working from left to right).

Brackets
Indices (powers and roots)
Division and **M**ultiplication (left to right)
Addition and **S**ubtraction (left to right)

If there are multiple operations within a bracket then BIDMAS should be applied inside the brackets first.

WORKED EXAMPLE 3

Calculate the values of the following:

a. $23 - 6 \times 4$ b. $(12 - 8) + 5^2 - (10 + 2^2)$ c. $\dfrac{3(4 + 8) + 4}{4 + 2(3^2 - 1)}$

THINK   **WRITE**

a. 1. Apply BIDMAS to determine the first step (in this instance multiplication).   a. $23 - 6 \times 4$
                                                  $= 23 - 24$

    2. Complete the next step in the calculation (in this instance subtraction) and write the answer. $= -1$

b. 1. Apply BIDMAS to determine the first step (perform the calculations in brackets first, then remove the brackets).   b. $(12 - 8) + 5^2 - (10 + 2^2)$
                                                  $= 4 + 5^2 - (10 + 4)$
                                                  $= 4 + 5^2 - 14$

    2. Complete the next steps in the calculation (resolve the powers, then carry out addition and subtraction from left to right). $= 4 + 25 - 14$
                                                  $= 29 - 14$

    3. Complete the final step and write the answer. $= 15$

c. 1. Apply BIDMAS to determine the first step (brackets).	c. $\dfrac{3(4+8)+4}{4+2(3^2-1)} = \dfrac{3 \times 12 + 4}{4 + 2(9-1)}$
2. Complete the next step in the calculation (multiplication).	$= \dfrac{3 \times 12 + 4}{4 + 2 \times 8}$
	$= \dfrac{36 + 4}{4 + 16}$
3. Perform the calculations on the numerator and denominator separately.	$= \dfrac{40}{20}$
4. Complete the division and write the answer.	$= 2$

1.2.4 Reasonableness

When calculating mathematical answers, it is always important to understand the question so you have an idea of what a *reasonable* answer would be.

When you complete arithmetic calculations, either manually or using your calculator, you can make mistakes. Checking the reasonableness of answers can indicate possible mistakes in your working.

WORKED EXAMPLE 4

Cathy goes to the shop to purchase an outfit for dinner. She buys a $145 dress and a $180 jacket, and gets $30 off the combined price by purchasing the two together. From the shoe shop, she buys a pair of shoes for $120 and a pair of stockings for $15.

How much did Cathy spend in total? Check your answer for reasonableness.

THINK	WRITE
1. Read the question carefully to understand what it is about.	The question relates to adding up Cathy's total shopping bill including discounts.
2. Have an idea of what sort of answer you expect.	Cathy purchased three items, each over $100 dollars, so the answer should be over $300.
3. Work out what needs to be added and subtracted.	$145 + $180 − $30 + $120 + $15
4. Calculate and write the answer.	= $430 spent
5. Does the answer seem reasonable?	This answer is above $300, so it seems reasonable. Since the price of two items end with 5 and the rest end with 0, the answer should end in 0.

Exercise 1.2 Number operations

1. **WE1** Evaluate each of the following:
   - a. $4 + +8$
   - b. $15 + -7$
   - c. $37 + -14$
   - d. $21 - +17$
   - e. $18 - -10$
   - f. $27 - -3$
   - g. $124 - +86$
   - h. $267 - -33$

2. Evaluate each of the following:
   - a. $-12 - +6$
   - b. $-18 - -8$
   - c. $-27 + -15$
   - d. $-56 + +54$
   - e. $-63 - +45$
   - f. $-45 - -65$
   - g. $-23 + -49$
   - h. $-79 + +43$

3. **WE2a** Evaluate each of the following:
   - a. -4×5
   - b. 12×-6
   - c. -7×-7
   - d. 15×4
   - e. -25×-6
   - f. 8×-9
   - g. -400×5
   - h. -12×-11

4. **WE2b** Evaluate each of the following:
   - a. $48 \div -6$
   - b. $-121 \div 11$
   - c. $-120 \div -6$
   - d. $33 \div -11$
   - e. $\dfrac{-150}{25}$
   - f. $\dfrac{100}{-4}$
   - g. $\dfrac{-72}{-8}$
   - h. $\dfrac{110}{-11}$

5. Calculate the values of the following:
   - a. $-4 \times 2 + 1$
   - b. $8 \div (2 - 4) + 4$
   - c. $9 \times (8 - 3)$
   - d. $-3 - 40 \div 8 + 2$
   - e. $4 + 12 \times -5$
   - f. $-5 \times 12 + 2$
   - g. $-6 - 36 \div 9 + 3$
   - h. $12 \div (2 - 4) - 6$

6. **WE3** Calculate the values of the following:
   - a. $12 - 6 \div 3$
   - b. $45 \div (27 \div -3)$
   - c. $(17 - +7) \div -5$
   - d. $-12 + 8 \times 7$
   - e. $100 \div (-50 \div -2) + 10$
   - f. $9 + \dfrac{24}{-6} \times 3$
   - g. $\dfrac{-15}{-3} - 20 \div 4$
   - h. $\dfrac{(-12 + 4)}{36 \div 18} + 80 \div -8$

7. Calculate the values of the following:
   - a. $-7 + 4 \times -4$
   - b. $(-63 \div -7) \times -3 + -2$
   - c. $(-5)^2 - 3 \times -5$
   - d. $-6 \times -8 - \left[3 + (-6)^2\right]$
   - e. $52 \div (-9 - 4) - 8$
   - f. $-6 - 64 \div -16 + 8$
   - g. $-3 \times -6 \div -9 + 12$
   - h. $\dfrac{(-48 \div 8)^2 \times 36}{-4}$

8. **WE4** Fred went to the sports department to get some clothes for the gym. He purchased a $120 tracksuit and a pair of $150 runners as a package and got $35 off the combined price. Fred also purchased a pair of running shorts for $30 and a singlet top for $25.

   How much did he spend at the sports department? Check your answer for reasonableness.

9. Erika goes shopping for some party food. She buys 6 packets of chips at $2.50 each and 2 boxes of cola at $8.50 each, but since she purchased 2 she gets a $2 discount. She also buys 5 packets of biscuits at $1 each and 3 dips at $2 each. How much did Erika spend in total? Check your answer for reasonableness.

10. Bob went to the warehouse to get some building supplies.

    He bought three 4 m long pieces of timber at $4.50 per metre, 2 packets of nails at $5.50 each and 3 tubes of liquid nails at $4 each. Because Bob is a regular customer he received a $15 discount off his total purchase.

    How much did Bob spend in total? Check your answer for reasonableness.

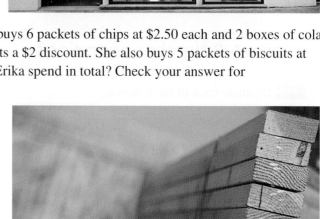

11. Taiki headed north on a bike ride initially travelling 25 km. He then turned around and travelled 15 km south before stopping for a drink. After his drink, Taiki continued to ride south for another 20 km, before again turning around and travelling north for a further 25 km.
    a. How many kilometres did Taiki cover on his ride?
    b. How far north did he finish from where he started?

12. Students were given the following question to evaluate.
    $$4 + 8 \div (-2)^2 - 7 \times 2$$
    a. A number of different answers were obtained, including -8, -12 and -17. Which one of these is correct?
    b. Using only brackets, change the question in two ways so that the other answers would be correct.

13. Calculate the number required to make the following equation true.
    $$13 \times (15 - 1) = 180 + ?$$

14. Calculate the number required to make the following equation true.
    $$\left(10^2 + 12 \div 3\right) \div (-8) = ?$$

1.3 Estimation

An **estimate** is not the same as a guess, because it is based on information.

Estimations can be useful in many situations; for example, when we are working with calculators or calculating the total of a shopping bill. By mentally estimating an approximate answer, we are more likely to notice if we make a data entry error.

To estimate the answer to a mathematical problem, round the numbers and calculate an approximate answer.

1.3.1 Rounding

Numbers can be rounded to different degrees of accuracy. For example, they can be rounded to 1, 2, 3 or more decimal places. The more decimal places a number has, the more accurate it is. However, more decimal places are not always necessary or relevant.

Rounding to the nearest 10

To round to the nearest 10, think about which multiple of 10 the number is closest to. For example, 34 rounded to the nearest 10 is 30, because 34 is closer to 30 than 40.

$34 \approx 30$

(*Note:* The $\approx$ symbol represents 'is approximately equal to'.)

Rounding to the nearest 100

To round to the nearest 100, think about which multiple of 100 the number is closest to.

Rounding to the first digit

When rounding to the leading (first) digit, the second digit needs to be looked at.
- If the second digit is 0, 1, 2, 3 or 4, the first digit stays the same and all the following digits are replaced with zeros.
- If the second digit is 5, 6, 7, 8 or 9, the first digit is increased by 1 (rounded up) and all the following digits are replaced with zeros.

For example, if 2345 is rounded to the first digit, the result is 2000 because 2345 is closer to 2000 than it is to 3000.

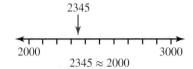

$2345 \approx 2000$

WORKED EXAMPLE 5

Round each of the following as directed.
a. 563 to the nearest 10
b. 12 786 to the nearest 100
c. 7523 to the leading digit

THINK	WRITE
a. 1. To round to the nearest 10, look at the units place value.	a. 56**3**
2. Is the units place value less than 5? Since 3 is less than 5, we round down. Replace the 3 with a 0 and write the answer.	≈ 560
b. 1. To round to the nearest 100, look at the tens place value.	b. 12 7**86**
2. Is the tens place value greater than or equal to 5? Since 8 is greater than or equal to 5, we round up. Increase the 7 to 8 and replace the 8 and 6 with zeros.	12 786 ↓↓↓ 800
3. Write the answer.	12 786 $\approx$ 12 800

▶

c. 1. When rounding to the leading digit, look at the second digit.	c. 7523
2. Since 5 is greater or equal to 5, we round up. Increase the first digit (7) by 1 and replace the remaining digits with zeros.	7523 ↓↓↓↓ 8000
3. Write the answer.	$7523 \approx 8000$

1.3.2 Estimation

Estimation is useful for checking that your answer is reasonable when performing calculations. This can help avoid simple mistakes, like typing an incorrect number into a calculator.

WORKED EXAMPLE 6

Estimate 66 123 × 749 by rounding each number to the leading (first) digit and then completing the calculation.

THINK	WRITE
1. Round the first number (66 123) by focusing on the second digit. Since the second digit (6) is 5 or greater, increase the leading digit (6) by one and replace the remaining digits with zeros.	$66\,123 \approx 70\,000$
2. Round the second number (749) by focusing on the second digit. Since the second digit (4) is less than 5, leave the leading digit and replace the remaining digits with zeros.	$749 \approx 700$
3. Multiply the rounded numbers by multiplying the leading digits ($7 \times 7 = 49$) and then add the number of zeros $4 + 2 = 6$.	$70\,000 \times 700$ $7 \times 7 = 49$ $70\,000 \times 700 = 49\,000\,000$
4. Write the estimated answer and compare to the actual answer.	$66\,123 \times 749 \approx 49\,000\,000$ This compares well with the actual answer of 49 526 127, found using a calculator.

Resources

Digital document: SkillSHEET Rounding to the first (leading) digit (doc-6418)

Interactivity: Rounding (int-3932)

Interactivity: Rounding to the first digit (int-3731)

Interactivity: Checking by estimating (int-3983)

Exercise 1.3 Estimation

1. **WE5** Round each of the following as directed.
   a. 934 to the nearest 10
   b. 12 963 to the nearest 100
   c. 85 945 to the leading digit

2. Round the following.
   a. 347 to the nearest 10
   b. 86 557 to the nearest 100
   c. 65 321 to the leading digit

3. Round each of the following numbers to the nearest 10.
   a. 47
   b. 82
   c. 129
   d. 162
   e. 250
   f. 2463
   g. 4836
   h. 7

4. Round each of the following numbers to the nearest 100.
   a. 43
   b. 87
   c. 142
   d. 177
   e. 3285
   f. 56 346
   g. 86 621
   h. 213 951

5. Round each of the following numbers to the nearest 1000.
   a. 512
   b. 3250
   c. 1324
   d. 6300
   e. 7500
   f. 13 487
   g. 435 721
   h. 728 433

6. Round each of the following numbers to the leading digit.
   a. 12
   b. 23
   c. 45
   d. 153
   e. 1388
   f. 16 845
   g. 23 598
   h. 492 385

7. **WE6** Estimate 74 852 × 489 by rounding each number to its first digit.

8. Estimate 87 342 ÷ 449 by rounding each number to its first digit.

9. Estimate the answer to the following expressions by rounding each number to the leading digit and then completing the calculation.
   a. 482 + 867
   b. 123 + 758
   c. 1671 − 945
   d. 2932 − 1455
   e. 88 × 543
   f. 57 × 2632
   g. 69 523 ÷ 1333
   h. 3600 ÷ 856

10. In your own words, explain how to round using leading digit approximation.

11. Give three examples of situations where it is suitable to use an estimate or a rounded value instead of an exact value.

12. Emily purchased a number of items at the supermarket that cost $1.75, $5.99, $3.45, $5.65, $8.95, $2.35 and $7.45. She was worried she didn't have enough money, so she used leading digit approximation to do a quick check of how much her purchases would cost. What approximation did she come up with?

13. The crowds at the Melbourne Cricket Ground for each of the five days of the Boxing Day Cricket test were: 88 214, 64 934, 55 349, 47 567 and 38 431. Using leading digit approximation, what was the estimated total crowd over the five days of the test?

14. A car's GPS estimates a family's trip for Christmas lunch is going to take 45 minutes.
    a. Is it likely to take exactly 45 minutes? Explain your answer.
    b. How would this time be estimated by the GPS?

15. Given the size of the crowds at the Boxing Day test in question **13**, what is the difference between the leading digit estimate and the actual crowd over the five days?

16. Rhonda went to buy a new school uniform. She decided she needed a new blazer, a new winter dress and a new shirt costing $167, $89 and $55 respectively.
    a. Use the leading digit method to estimate how much Rhonda spent, and find the difference in the total she spent and the leading digit estimate.
    b. Was the estimate more or less than the actual price? Give a reason for why this was the case.

1.4 Decimal significance

A decimal number consists of both a whole number part and a fractional part. These parts are separated by a decimal point.

A place-value table can be extended to include decimal place values and can include as many decimal place columns as required.

Whole number part — **73.064** — Fractional part — Decimal point

Thousands	Hundreds	Tens	Ones	Tenths	Hundredths	Thousandths	Ten Thousandths
1000	100	10	1	$\frac{1}{10}$	$\frac{1}{100}$	$\frac{1}{1000}$	$\frac{1}{10\,000}$

The decimal number 73.064 represents 7 tens, 3 ones, 0 tenths, 6 hundredths and 4 thousandths. In expanded form, 73.064 is written as:

$$(7 \times 10) + (3 \times 1) + \left(0 \times \frac{1}{10}\right) + \left(6 \times \frac{1}{100}\right) + \left(4 \times \frac{1}{1000}\right)$$

When reading decimals, the whole number part is read normally and each digit to the right of the decimal point is read separately. For example, 73.**064** is read as 'seventy-three point zero six four'.

The number of decimal places in a decimal number is the number of digits after the decimal point. The number 73.**064** has 3 decimal places. The zero (0) in 73.064 means that there are no tenths. The zero must be written to hold the place value; otherwise the number would be written as 73.64, which does not have the same value.

WORKED EXAMPLE 7

Write the value of the 3 in each of the following decimal numbers.
a. **59.378**　　　　　　　b. **1.0003**　　　　　　　c. **79.737**

THINK		WRITE
a. 1. Identify how many places after the decimal point the 3 lies.		a. 59.**3**78
2 The 3 is the first number after the decimal point, which represents tenths.		$\frac{3}{10}$

---

b. 1. Identify how many places after the decimal point the 3 lies.	b. 1.000**3**
2. The 3 is the fourth number after the decimal point, which represents tens of thousandths.	$\dfrac{3}{10\,000}$
c. 1. Identify how many places after the decimal point the 3 lies.	c. 79.7**3**7
2. The 3 is the second number after the decimal point, which represents hundredths.	$\dfrac{3}{100}$

1.4.1 Rounding decimals

When rounding to a required number of decimal places, look at the first digit after the required number of decimal places.

If this digit is 0, 1, 2, 3 or 4 (less than 5), **do not** change the digit in the last required decimal place.

If this digit is 5, 6, 7, 8 or 9 (greater or equal to 5), **increase** the digit in the last required decimal place by 1.

WORKED EXAMPLE 8

Round each of the following to 3 decimal places.
a. 21.569 42   **b. 35.731 60**

THINK

a. 1. As we are rounding to 3 decimal places, look at the fourth decimal place (in this case 4).

    2. Since this number is less than 5, drop off the numbers after the third decimal place.

b. 1. As we are rounding to 3 decimal places, look at the fourth decimal place (in this case 6).

    2. Since this number is greater or equal to 5, add one to the third decimal number.

WRITE

a. 21.569**4**2

    21.569

b. 35.731**6**0

    35.732

1.4.2 Fractions as decimals

A fraction can be expressed as a decimal by dividing the numerator by the denominator.

$$\frac{3}{8} = 3 \div 8 = 0.375$$

If a decimal does not terminate and has decimal numbers repeating in a pattern, it is called a **recurring decimal**.

A recurring decimal with only one repeating digit can be written by placing a dot above the digit. When there is more than one repeating digit, the recurring decimal can be written by placing a dot over the first and last digits of the repeating part, or by placing a bar above the entire repeating pattern.

$$\frac{1}{3} = 1 \div 3 = 0.333\,333\,33\ldots = 0.\dot{3}$$

$$\frac{3}{7} = 3 \div 7 = 0.428\,574\,285\,7\ldots = 0.\dot{4}285\dot{7} \text{ or } 0.\overline{42857}$$

Write the following fractions as decimals.

a. $\dfrac{5}{8}$

b. $\dfrac{3}{9}$

THINK	WRITE
a. 1. Input the division into the calculator.	a. $5 \div 8$
2. Write the decimal answer.	$= 0.625$
b. 1. Input the division into the calculator.	b. $3 \div 9 = 0.333\,333\,33\ldots$
2. Write the answer as a recurring decimal.	$= 0.\dot{3}$

---

on Resources

 eLesson: Place value (eles-004)

 Interactivity: Place value (int-3921)

 Interactivity: Decimal parts (int-3975)

 Digital document: SkillSHEET Place value (doc-6409)

Interactivity: Conversion of fractions to decimals (int-3979)

Exercise 1.4 Decimal significance

1. **WE7** Write the value of the 7 in each of the following decimals.
   a. 45.871
   b. 81.710
   c. 33.007

2. Write the value of the 9 in each of the following decimals.
   a. 21.090
   b. 0.009
   c. 47.1059

3. Write the value of 7 in the following decimals.
   a. 34.075
   b. 1.7459
   c. 0.1007
   d. 1945.27
   e. 450.07
   f. 2.7
   g. 0.007
   h. 2.170

4. State the number of decimal places on each of the following numbers.
   a. 2.195
   b. 394.7
   c. 104.25
   d. 0.0003
   e. 1997.4
   f. 125.333
   g. 69.15
   h. 2.6936

5. State whether the following statements are TRUE or FALSE.
   a. $0.125 = 0.521$
   b. $0.025 = 0.0250$
   c. $45.0005 = 45.005$
   d. $3.333 = 3.333\,000$
   e. $6.666 = 6.6666$
   f. $20.123 = 2.123$

6. **WE8** Round each of the following to 3 decimal places.
   a. 59.123 45
   b. 72.216 81

7. Round each of the following to 2 decimal places.
   a. 2.069 14
   b. 0.7962

---

8. Round each of the following to 3 decimal places.
   a. 39.184 23
   b. 3.232 323 23
   c. 99.4358
   d. 125.294 61
   e. 100.0035
   f. 0.0479

9. Round the following to 2 decimal places.
   a. 58.123
   b. 23.345
   c. 71.097
   d. 0.8686
   e. 30.9999
   f. 125.8929

10. WE9 Write the following fractions as decimals.
    a. $\dfrac{3}{4}$
    b. $\dfrac{2}{3}$

11. Write the following as decimals to 3 decimal places.
    a. $\dfrac{7}{8}$
    b. $\dfrac{1}{7}$

12. Convert the following fractions to decimals.
    a. $\dfrac{3}{10}$
    b. $\dfrac{3}{5}$
    c. $\dfrac{3}{12}$
    d. $\dfrac{7}{25}$
    e. $\dfrac{33}{100}$
    f. $\dfrac{45}{50}$

13. Convert the following fractions to recurring decimals.
    a. $\dfrac{1}{3}$
    b. $\dfrac{5}{9}$
    c. $\dfrac{1}{12}$
    d. $\dfrac{4}{7}$
    e. $\dfrac{3}{13}$
    f. $\dfrac{1}{21}$

14. Which packet in the photograph below contains more mince?

15. a. Using the table below, list the heights of the five American basketball team players in ascending order.
    b. Add the height of the five players and round to the first digit.

Player	Height
Kobe Bryant	1.98 m
Kevin Durant	2.06 m
Russell Westbrook	1.91 m
Carmelo Anthony	2.02 m
Lebron James	2.03 m

16. Narelle is environmentally conscious and has fitted water tanks to her house to water the garden and flush the toilets. She checks the amount of rainfall each week to see how much water the tanks have collected. In the month of March, the rainfall readings were 12.48 mm, 8.82 mm, 27.51 mm and 44.73 mm. Round the total rainfall for March to 1 decimal place.

17. The Nguyen family decided to install solar panels on the roof to reduce their electricity bill. They compared the prices per kWh of three different suppliers to see how much they would save by using their panels. The prices quoted by companies A, B and C per kWh were 26.78 cents, $25\frac{3}{4}$ cents and $25\frac{7}{9}$ cents respectively.
    a. List the companies from cheapest to dearest.
    b. Write the fractions as decimal numbers rounded to 1 decimal place.

18. The following table shows items that were purchased at the supermarket and their price.

Item	Price
Salmon	$4.99
Chicken fillets	$7.35
Shampoo	$8.95
Ham	$4.42
Milk	$4.29
Bread	$3.60

    a. Find the total cost of the items purchased.
    b. Find the total cost of the items purchased and round to the nearest dollar.
    c. Round the price of each item to the nearest 5 cents and then find the total cost.

1.5 Calculator skills

Your calculator has the ability to complete many different mathematical operations that would be difficult to calculate in your head or with pen and paper. This includes dealing with decimals and fractions.

Even though your calculator will always give the correct answer, it relies on you inputting the information correctly. To make sure you get the correct answer, it's always good to estimate your answer to check for reasonableness and accuracy.

You will notice that your calculator automatically follows the correct order of operations (BIDMAS).

Note: The keys used may vary between different calculators.

Calculate the value of the following expressions using your calculator.

a. $140 + 23 \times 8$ b. $7(28 + 43) - 18$ c. $283 - 13.78 \times 11.35$

THINK	WRITE
a. 1. Estimate your answer using leading digit approximation.	a. $140 + 23 \times 8$ $\approx 100 + 20 \times 10$ ≈ 300
2. Enter the exact equation into your calculator. Compare with your estimate to check for reasonableness.	$140 + 23 \times 8 = 324$
b. 1. Estimate your answer using leading digit approximation.	b. $7(28 + 43) - 18$ $\approx 10(30 + 40) - 20$ $\approx 10(70) - 20$ $\approx 700 - 20$ ≈ 680
2. Enter the exact expression into your calculator. Compare with your estimate to check for reasonableness.	$7(28 + 43) - 18 = 479$
c. 1. Estimate your answer using leading digit approximation.	c. $283 - 13.78 \times 11.35$ $\approx 300 - 10 \times 10$ $\approx 300 - 100$ ≈ 200
2. Enter the exact expression into your calculator. Compare with your estimate to check for reasonableness.	$283 - 13.78 \times 11.35 = 126.597$

When using your calculator, first estimate the result to compare with the answer from the calculator. Always check that what you type into the calculator matches the question.

Evaluate the following expressions using your calculator, rounding your answer to 4 decimal places.

a. $(2.56 + 3.83)^2 + 45.93$ b. $\sqrt{4.39 - 2.51} + (3.96)^2$ c. $\dfrac{4}{7} + \dfrac{5}{9} + 5$

THINK	WRITE
a. 1. Estimate your answer using leading digit approximation.	a. $(2.56 + 3.83)^2 + 45.93$ $\approx (3 + 4)^2 + 50$ $\approx 7^2 + 50$ $\approx 49 + 50$ ≈ 99
2. Enter into the calculator as shown. Use the x^2 key to raise the brackets to the power of two. Compare with your estimate to check for reasonableness.	$(2.56 + 3.83)^2 + 45.93$ $= 86.7621$

▶

b. 1. Estimate your answer using leading digit approximation.

b. $\sqrt{4.39 - 2.51} + (3.96)^2$

$\approx \sqrt{4 - 3} + 4^2$

$\approx \sqrt{1} + 16$

$\approx 1 + 16$

≈ 17

2. Enter into the calculator as shown.
Use the $\sqrt{}$ and x^2 keys. Make sure the square root covers both numbers.
Compare with your estimate to check for reasonableness.

$\sqrt{4.39 - 2.51} + (3.96)^2$

$= 17.0527$

c. 1. Estimate your answer using leading digit approximation.

c. $\dfrac{4}{7} + \dfrac{5}{9} + 5$

$\approx 1 + 1 + 5$

≈ 7

2. Enter into the calculator as shown.
Use the fraction key to enter each fraction.
Compare with your estimate to check for reasonableness.

$\dfrac{4}{7} + \dfrac{5}{9} + 5$

$= 6.1270$

1.5.1 Scientific notation

Scientists, economists, statisticians and mathematicians often need to work with very large or very small numbers.

Scientific notation (sometimes called standard form) is a way of writing a very large or a very small number using less space. In scientific notation, the number is written as a multiple of a power of 10. For example, $4500 = 4.5 \times 10^3$. The 10^3 has a **positive power of 3**, which moves the decimal point **3 places to the right**.

An example of a very small number written using scientific notation is $0.0000067 = 6.7 \times 10^{-6}$. The 10^{-6} has a **negative power of 6**, which moves the decimal point **6 places to the left**.

Scientific notation is displayed on some calculators and spreadsheets using an 'E'. These show a number between 1 and 10, followed by an E and the power to which 10 is raised. For example:

3.45×10^{18} may appear as 3.45E18 on the calculator display 5.9E − 03 represents the number 5.9×10^{-3}.

$4500 = 4.5\overset{\times 10^3}{\curvearrowright}0\,0$
$= 4.5 \times 10^3$

$0.0000067 = 0.00000\overset{\times 10^{-6}}{\curvearrowleft}6.7$
$= 6.7 \times 10^{-6}$

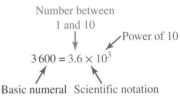

Number between 1 and 10 ↓ ↗Power of 10
$3\,600 = 3.6 \times 10^3$
↗ ↖
Basic numeral   Scientific notation

WORKED EXAMPLE 12

Calculate the value of the following expressions using scientific notation.
a. $7.3 \times 10^4 + 5.8 \times 10^5$ **b.** $3.5 \times 10^8 \times 5.9 \times 10^6$

THINK

a. 1. Enter the expression using scientific notation.

2 Put the calculator in scientific notation mode. (This is usually SCI mode.)
To raise to the power of 10 use the ˆ key.

WRITE

a. $7.3 \times 10^4 + 5.8 \times 10^5$

$= 6.53 \times 10^5$

b. 1. Enter the expression into the calculator.

2. Put the calculator in scientific notation mode. (This is usually SCI mode.) To raise to the power of 10 use the ^ key.

b. $3.5 \times 10^8 \times 5.9 \times 10^6$
$= 2.065 \times 10^{15}$

 Resources

 Interactivity: Scientific notation (int-6031)

Digital document: SkillSHEET Multiplying and dividing by powers of 10 (doc-101815)

Exercise 1.5 Calculator skills

1. **WE10** Calculate the value of the following expressions using your calculator.
   a. $269 + 12 \times 16$
   b. $9(78 + 61) - 45$
   c. $506.34 - 39.23 \times 17.04$

2. Calculate the value of the following expressions using your calculator.
   a. $497 - 13 \times 24 \div 6$
   b. $13(194 - 62) + 4 \times 8$
   c. $459.38 \div 18.5 \times 17.04 - (34.96 + 45.03)$

3. Evaluate the following expressions using your calculator.
   a. $3 + 6 \times 7$
   b. $47 + 8 \times 6$
   c. $285 + 21 \times 16$
   d. $2859 + 178 \times 79$

4. Evaluate the following expressions using your calculator.
   a. $12 - 5 \times 2$
   b. $68 - 4 \times 9$
   c. $385 - 16 \times 9$
   d. $1743 - 29 \times 45$

5. Evaluate the following expressions using your calculator.
   a. $4(6 + 24) - 58$
   b. $5(23.5 - 18.3) + 23$
   c. $2.56(89.43 - 45.23) - 92.45$
   d. $6(45.89 - 32.78) - 3(65.89 - 59.32)$

6. **WE11** Evaluate the following expressions using your calculator, rounding your answer to 4 decimal places.
   a. $(5.89 + 2.16)^2 + 67.99$
   b. $\sqrt{8.77 - 3.81} + (5.23)^2$
   c. $\frac{3}{5} + \frac{7}{8} + 10$

7. Evaluate the following expressions using your calculator, rounding your answer to 4 decimal places.
   a. $\left(\frac{7}{12}\right)^2 - \frac{3}{4}$
   b. $\sqrt{3.56 + 8.28} - \sqrt{5.29 - 3.14}$
   c. $\left(\frac{4}{5}\right)^2 + \sqrt{\frac{7}{12}}$

8. Evaluate the following expressions using your calculator, rounding your answers to 4 decimal places.
   a. $16.9 + 5.2^2 \div 4.3$
   b. $9.3^2 \div 4.5$
   c. $(3.7 + 5.9)^2 - 15.5$
   d. $\frac{(7.2)^2}{4.2}$

9. Evaluate the following expressions using your calculator, rounding your answers to 4 decimal places.
   a. $\sqrt{5.67 + 8.34}$
   b. $2.5 \times \sqrt{8.64} - 2.5$
   c. $12.8 + \sqrt{3.5 \times 5.8} \times 1.2$
   d. $\frac{\sqrt{4.7 - 3.6}}{5}$

10. Evaluate the following expressions to 2 decimal places using your calculator.
    a. $\frac{1}{2} + \frac{1}{4}$
    b. $\frac{3}{7} - \frac{1}{9}$
    c. $\frac{18}{13} - \frac{2}{5} \times \frac{2}{7}$
    d. $15 - \frac{7}{5} \times \frac{5}{3} + 7$

11. **WE12** Calculate the value of the following expressions in scientific notation.
    a. $7.9 \times 10^4 + 4.9 \times 10^4$
    b. $3.7 \times 10^6 \times 8.3 \times 10^9$

12. Calculate the value of the following expressions in scientific notation.
    a. $5.3 \times 10^6 - 8.1 \times 10^5$
    b. $2.7 \times 10^5 \times 5.6 \times 10^3$

---

13. Evaluate the following expressions using your calculator, leaving your answer in scientific notation.
    a. $1.45 \times 10^3 + 5.82 \times 10^3$
    b. $6.89 \times 10^6 - 4.71 \times 10^5$
    c. $3.1 \times 10^5 \times 8.49 \times 10^3$
    d. $4.21 \times 10^4 \div (8.32 \times 10^2)$
    e. $7.38 \times 10^5 \div (2.62 \times 10^8)$
    f. $3.82 \times 10^3 + (9.27 \times 10^5 \times 4.5 \times 10^2)$

14. Cathy is 7 years older than Marie, but Marie is twice the age of her younger brother Fergus. How old is Cathy if Fergus is 3 years old?

15. After a football party night some empty soft drink cans were left and had to be collected for recycling. Of the cans found, 7 were half full, 6 cans were one-quarter full and the remaining 8 cans were completely empty. Use your calculator to find out how many equivalent full cans were wasted.

16. NASA landed the Curiosity Rover on Mars in 2012. The average distance from the Sun to Earth is 1.5×10^8 km and the average distance from the Sun to Mars is 2.25×10^8 km. What is the average distance between Earth and Mars when Earth is between Mars and the Sun?

17. Carbon is an atom that has a radius of 7.0×10^{-11} m. The circumference of a carbon atom can be calculated by evaluating $2 \times 3.14 \times 7.0 \times 10^{-11}$ m (that is $2\pi r$). Calculate the circumference of a carbon atom in scientific form.

18. Jupiter is the largest planet in our solar system with a radius of 7.1492×10^7 m. The volume ($v = \frac{4}{3}\pi r^3$) of Jupiter can be calculated by evaluating $\frac{4}{3} \times 3.14 \times (7.1492 \times 10^7)^3 \text{m}^3$. Calculate the volume of Jupiter in scientific form.

1.6 Approximation

Approximating or estimating an answer is useful when an accurate answer is not necessary.

You can approximate the size of a crowd at a sporting event based on an estimate of the fraction of seats filled. This approach can be applied to a variety of situations, e.g. approximating the amount of water in a water tank.

When approximating size, it is important to use an appropriate unit, e.g. for the distance you drive you would use 24 km instead of 2 400 000 cm.

Approximations can also be made by using past history to indicate what might happen in the future.

WORKED EXAMPLE 13

Approximate the size of the crowd at the first day of the cricket test at the Sydney Cricket Ground, if it is estimated to be at $\dfrac{8}{10}$ of its capacity of 46 000.

THINK	WRITE
1. Write down the information given in the question.	Estimate of crowd $= \dfrac{8}{10}$ of capacity
2. Calculate the fraction of the total capacity.	$\dfrac{8}{10}$ of $46\,000 = \dfrac{8}{10} \times 46\,000$
	$= 0.8 \times 46\,000$
	$= 36\,800$
3. Write the answer.	The approximate crowd is 36 800.

1.6.1 Approximation from graphs

When approximating from graphs:
- if the data is shown, read from the graph, noting the scale on the axes
- if the data is not shown, follow the trend (behaviour of the graph) to make an estimate of future or past values.

WORKED EXAMPLE 14

The following graph shows the farmgate price of milk in cents/litre. Approximate the price in 2019/2020.

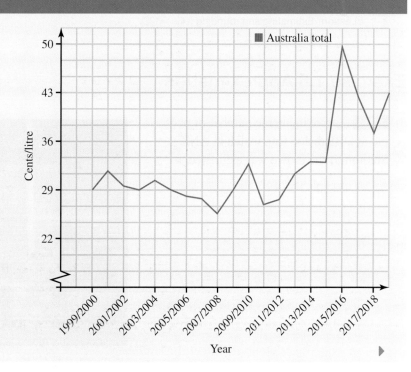

THINK	WRITE
1. Check what each axis represents.	Vertical axis: cents/litre Horizontal axis: years
2. Look for the trend that the graph has followed in previous years. Draw a line to represent this trend and extend it to the required period of 2019/2020.	The graph has been on an upward trend for the past 8 years.

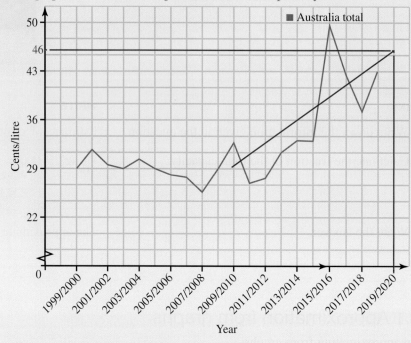

3. Read the answer from the graph.	Approximately 46 cents/litre.

 Resources

eLesson: Estimating and rounding (eles-0822)

Exercise 1.6 Approximation

1. **WE13** Approximate the number of people at a Bledisloe Cup match at ANZ Stadium if the crowd was estimated at $\frac{7}{10}$ of its capacity of 83 500.

---

2. When full, a pool holds 51 000 litres of water. Approximate the number of litres in the pool when it is estimated to be 90% full.

3. Choose the most appropriate unit from the options below for each of the following measurements.
centimetres (cm), metres (m), kilometres (km), millilitres (mL), litres (L), grams (g), kilograms (kg)
   a. The length of a football field
   b. The weight of a packet of cheese
   c. The volume of coke in a can
   d. The distance you travel to go on a holiday
   e. The size of an *LED* TV
   f. The weight of a rugby league player

4. **MC** What is the best approximation of the volume of milk in the container on the right?
   A. 500 mL
   B. 750 mL
   C. 1L
   D. 2L

5. **MC** What is the best approximation of the weight of the tub of butter?
   A. 500 kg   B. 125 g
   C. 500 g   D. 750 g

6. **WE14** From the graph of the cost of tuition fees for one school, predict the approximate tuition fees in 2019.

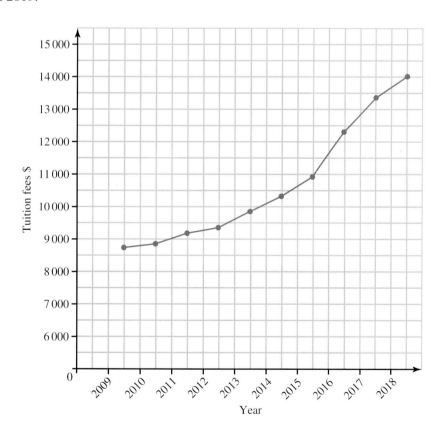

7. Use the following graph to approximate the volume of domestic air travel emissions in 2021.

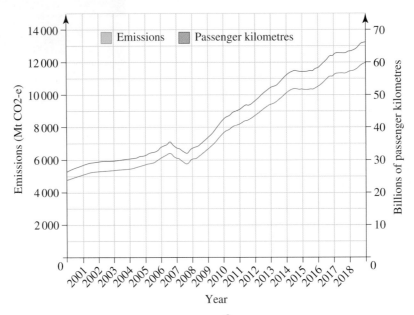

8. If the stadium at the Sydney Olympics swimming was $\frac{9}{10}$ full and had a capacity of 17 000, what was the approximate crowd on this night?

9. Given the capacity at the Staple Centre, where the Los Angeles Lakers play, is 18 118, what is the approximate crowd at the game shown in the picture?

10. The capacity of Rod Laver Arena is 14 820. Approximate the crowd at this match from the Australian Open.

11. The picture shown is of Tiger Woods at the US Masters. From the picture, why is it difficult to estimate the size of the Masters crowd on this day?

12. Explain in your own words the difference between an estimate and a guess.
13. Describe a way to approximate how many students are at your school. Investigate the actual number and compare this to your estimate. How could you improve your initial estimate?
14. From the graph of the population of Australia over the past 51 years, approximate how many people will live in Australia in 2020.

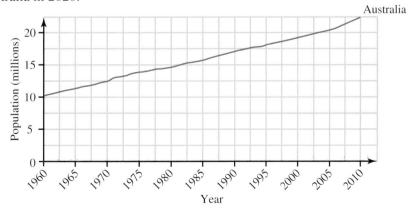

15. From the graph of the value of the Australian dollar against the American dollar, approximate its value in January 2019.

AUDUSD — Australian dollar exchange rate

16. From the graph of Aboriginal and Torres Strait Islander population, predict the approximate population in:
    a. 2025
    b. 2030

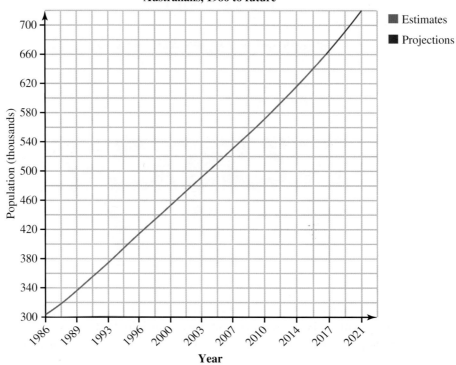

Historical population of Aboriginal and Torres Strait Islander Australians, 1986 to future

1.7 Review: exam practice

1.7.1 Calculations: summary

Number operations

- When adding or subtracting positive and negative numbers remember:

Adding a positive	$+\,+\,=\,+$
Adding a negative	$+\,-\,=\,-$
Subtracting a positive	$-\,+\,=\,-$
Subtracting a negative	$-\,-\,=\,+$

- When multiplying or dividing positive and negative numbers remember:

When multiplying or dividing two numbers with the same signs, the answer is positive.	$+\times+=+$ or $+\div+=+$ $-\times-=+$ or $-\div-=+$
When multiplying or dividing two numbers with different signs, the answer is negative.	$+\times-=-$ or $+\div-=-$ $-\times+=-$ or $-\div+=-$

- When completing arithmetic calculations it is important to follow the correct order of operations, as described by **BIDMAS**.

Brackets
Indices (powers and roots)
Division and **M**ultiplication (left to right)
Addition and **S**ubtraction (left to right)

- When calculating mathematical answers it is always important to understand the question so you have an idea of what a **reasonable** answer should be.

Estimation

- To round to the nearest 10, think about which multiple of 10 the number is closest to.

   For example, if 34 is rounded to the nearest 10, the result is 30 because 34 is closer to 30 than it is to 40.

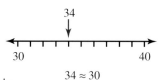

$34 \approx 30$

- To round to the nearest 100, think about which multiple of 100 the number is closest to.
- When rounding to the leading (first) digit the second digit needs to be looked at:
  - If the second digit is 0, 1, 2, 3 or 4 the first digit stays the same and all the following digits are replaced with zeros.
  - If the second digit is 5, 6, 7, 8 or 9 the first digit is increased by 1 (rounded up) and all the following digits are replaced with zeros.

Decimal significance

- A place-value table can be extended to included decimal place values, and can include as many decimal place columns as required.

Thousands	Hundreds	Tens	Ones		Tenths	Hundredths	Thousandths	Ten Thousandths
1000	100	10	1	$\cdot$	$\dfrac{1}{10}$	$\dfrac{1}{100}$	$\dfrac{1}{1000}$	$\dfrac{1}{10000}$

- When rounding to a required number of decimal places, look at the first digit after the number of decimal places required.
  - If this digit is 0, 1, 2, 3 or 4 (less than 5) **do not** change the digit in the last required decimal place.
  - If this digit is 5, 6, 7, 8 or 9 (greater or equal to 5) **increase** the digit in the last required decimal place by 1.
- A fraction can be expressed as a decimal by dividing the numerator by the denominator.
- If a decimal does not terminate and has decimal numbers repeating in a pattern, it is called a recurring decimal.

Calculator skills

- When using your calculator, first estimate the result so you can compare your estimate to the result from the calculator. Always check that what you have typed in the calculator matches what the question states.

Approximation

- You can approximate the size of a crowd at a sporting event based on an estimate of the fraction of seats filled.
- When approximating size it is important to use an appropriate unit.
- Approximations can be made by using past history to give an indication of what might happen in the future.

Exercise 1.7 Review: exam practice

Simple familiar

1. MC The value of the expression $17 + 3 \times 7$ is:
   - A. 27
   - B. 140
   - C. 38
   - D. 28

2. MC The value of the expression $-3(5 - -10) + 50$ is:
   - A. 95
   - B. 5
   - C. 65
   - D. -5

3. MC The number 47 321 rounded using leading digit is:
   - A. 47 000
   - B. 4000
   - C. 5000
   - D. 50 000

4. MC If three purchases were made with the value $7.34, $18.05 and $2.69, using leading digit estimation on the three purchases, the total purchase will approximately cost:
   - A. $30
   - B. $28.08
   - C. $28
   - D. $28.10

5. MC Using your calculator, the value of $\sqrt{6.78 + 19.83} + 3.91$ to two decimal places is closest to:
   - A. 5.16
   - B. 9.07
   - C. 5.52
   - D. 1.25

6. **MC** The volume of Earth can be calculated using the volume of a sphere formula. Since Earth has a radius of 6.38×10^6 its volume can be calculated by evaluating $\frac{4}{3} \times 3.14 \times (6.38 \times 10^6)^3$ m³.

The volume of Earth to two decimal places and in scientific notation is closest to:

A. 1.70×10^{14} m³
B. 3.26×10^{21} m³
C. 1.09×10^{14} m³
D. 1.09×10^{21} m³

7. **MC** The number 37.395 851 rounded to four decimal places is:

A. 37.396
B. 38.396
C. 37.3958
D. 37.3959

8. **MC** The fraction $\frac{7}{9}$ is most accurately written as:

A. 0.7
B. 0.9
C. $0.\dot{7}$
D. 0.8

9. **MC** The approximate volume of Coca-Cola in the can shown is closest to:

A. 500 mL
B. 375 mL
C. 1L
D. 250 mL

10. The number of Indigenous qualifications in Queensland has grown since 2005. From the graph shown the approximate Cert 3+ awards in 2013–14 is closest to:

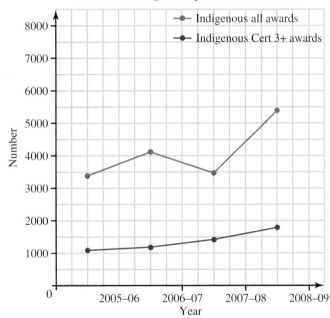

Number of Indigenous qualifications awarded

A. 3000
B. 5000
C. 6000
D. 7000

Complex familiar

11. Calculate the following:
    a. $-18 - -25$ b. -8×6
    c. $36 \div 9 \times 8 - 12$ d. $-4(-4 + -6) - 33$

12. Use leading digit approximation to calculate the following:
    a. 49×821 b. $1396 + 183$ c. $\dfrac{7563}{676}$ d. $17 \times 873 + 47$

13. Use you calculator to evaluate the following to two decimal places:
    a. $\sqrt{3(5.94 - 1.48)}$ b. $(5.74)^2 - \dfrac{5}{8}$
    c. $\dfrac{6}{9} + \dfrac{2}{5} \times \dfrac{3}{7}$ d. $\sqrt{(3.25)^2 + 1.5}$

14. Round the following fractions to two decimal places:
    a. $\dfrac{1}{8}$ b. $\dfrac{37}{50}$ c. $\dfrac{6}{25}$ d. $\dfrac{2}{3}$

15. Round the following to three decimal places:
    a. 0.5555 b. 1.03639 c. 33.1047 d. 10.9999

16. Approximate each of the following:
    a. a ground estimated to be $\dfrac{3}{4}$ full with a capacity of 50 000 people.
    b. a water tank estimated to be 80% full and has a capacity of 7500 litres.

Complex unfamiliar

17. Yi Rong is saving for her end of year holiday with her school friends. Her parents said if she did work around the house they would help contribute to the holiday expenses. They agreed to pay her \$25 a week for doing the dishes and \$15 a week for doing the washing, however for extra incentive they said they would double her pay if she did each of the jobs for the next 4 weeks without missing a day. Assume Yi Rong didn't miss a day, how much money did she earn in the 4 weeks?

18. Bob the builder intends to build a new decking out the back of his house so he can entertain friends. He wants to get a rough idea of how much it is going to cost him in materials so he can see if he can afford to go ahead with the job. The hardware store gave him prices on the materials he needs as shown.

Material	Price
Timber	$789
Ready-mix bags	$32.50
Nails	$67.25
Decking stain	$77.95
General equipment	$65.45

    a. What is the rough price of materials if rounding each price to the leading digit?
    b. What is the rough price if rounding each price to the nearest dollar?
    c. Which of the two rounded prices is greater? Explain why this is so.

19. The surface area of a sphere is calculated by multiplying 12.56 by the radius squared.

    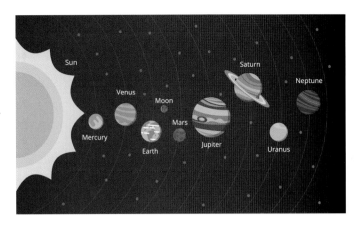

    a. Calculate the surface area of the following planets with the given radii, correct to 2 decimal places.

Radius of Mercury = 2.44×10^6 m
Radius of Mars = 3.40×10^6 m
Radius of Earth = 6.38×10^6 m
Radius of Jupiter = 7.15×10^7 m
Radius of Neptune = 2.48×10^7 m

    b. Which planet has the greatest surface area?
    c. Which planet has the smallest surface area?

20. A roller coaster uses its stored potential energy and coverts it to kinetic energy as it picks up speed on its descent.

    Calculate the following:

    a. The roller coaster's potential energy, PE, measured in Joules, at its highest point, given $PE = 750 \times 9.8 \times 15$.

    b. The roller coaster's kinetic energy, KE, as it is just about to go into a loop, given $KE = \dfrac{1}{2} \times 750 \times 15^2$.

    c. The velocity of the roller coaster, v m/s given
    $$v = \sqrt{\dfrac{120\,000 \times 2}{750}}$$

    d. The height, h the roller coaster is when it has a potential energy of 58 800 Joules of energy, given $h = \dfrac{58\,800}{750 \times 9.8}$.
    What unit do you think the height is measured in?

Answers

Chapter 1 Calculations

Exercise 1.2 Number operations

1. a. 12   b. 8   c. 23
   d. 4   e. 28   f. 30
   g. 38   h. 300

2. a. -18 b. -10 c. -42
   d. -2 e. -108 f. 20
   g. -72 h. -36

3. a. -20 b. -72 c. 49
   d. 60   e. 150   f. -72
   g. -2000 h. 132

4. a. -8 b. -11 c. 20
   d. -3 e. -6 f. -25
   g. 9   h. -10

5. a. -7 b. 0   c. 45
   d. -6 e. -56 f. -58
   g. -7 h. -12

6. a. 10   b. -5 c. -2
   d. 44   e. 14   f. -3
   g. 0   h. -14

7. a. -23 b. -29 c. 40
   d. 9   e. -12 f. 6
   g. 10   h. -324

8. $290

9. $41
   Answer reasonable since roughly chips $\approx$ $10,
   Coke $\approx$ $20, biscuits $\approx$ $5 and dips $\approx$ $5.
   This totals $40, which reduces to $38 after the discount.
   This value is close to $41, thus it is reasonable.

10. $62
    Answer reasonable since roughly timber $\approx$ $60,
    nails $\approx$ $10 and liquid nails $\approx$ $10.
    This totals $40, which reduces to $25 after the discount.
    This value is close to $26, thus it is reasonable.

11. a. 85km   b. 15km north

12. a. -8
    b. $4 + 8 \div -(2)^2 - 7 \times 2 = -12$
       $(4 + 8) \div -(2)^2 - 7 \times 2 = -17$

13. 2

14. -13

Exercise 1.3 Estimation

1. a. 930   b. 13 000   c. 90 000

2. a. 350   b. 86 600   c. 70 000

3. a. 50   b. 80   c. 130
   d. 160   e. 250   f. 2460
   g. 4840   h. 10

4. a. 0   b. 100   c. 100
   d. 200   e. 3300   f. 56 300
   g. 86 600   h. 214 000

5. a. 1000   b. 3000   c. 1000
   d. 6000   e. 8000   f. 13 000
   g. 436 000   h. 728 000

6. a. 10   b. 20   c. 50
   d. 200   e. 1000   f. 20 000
   g. 20 000   h. 500 000

7. 35 000 000

8. 225

9. a. 1400   b. 900   c. 1100
   d. 2000   e. 45 000   f. 180 000
   g. 70   h. $\dfrac{40}{9}$

10. Since it is leading digit approximation, you look at the second digit. If it is 5 or greater then you increase the leading digit by 1 and replace the rest of the digits with zeros. If the second digit is less than 5, then replace it and all of the other digits with zeros.

11. Adding the cost of groceries when shopping.
    Calculating the cost of petrol needed for a trip.
    Determining the size of a crowd.

12. $35

13. 300 000

14. a. It could, but the GPS cannot predict traffic or if you will get red lights on the way, so it is only an estimate.
    b. By calculating the distance of the trip and the speed limits on the roads to come up with an estimate for time.

15. Actual crowd $= 294\,495$
    Estimated crowd $\approx 300\,000$
    Difference $= 5505$

16. a. Estimated $= 350
       Actual $= 311
       Difference $= 39
    b. The estimate was more than the actual price since the items purchased were rounded up.

Exercise 1.4 Decimal significance

1. a. $\dfrac{7}{100}$ b. $\dfrac{7}{10}$ c. $\dfrac{7}{1000}$

2. a. $\dfrac{9}{100}$ b. $\dfrac{9}{1000}$ c. $\dfrac{9}{10\,000}$

3. a. $\dfrac{7}{100}$ b. $\dfrac{7}{10}$ c. $\dfrac{7}{10\,000}$
   d. $\dfrac{7}{100}$ e. $\dfrac{7}{100}$ f. $\dfrac{7}{10}$
   g. $\dfrac{7}{1000}$ h. $\dfrac{7}{100}$

4. a. $2.195 \rightarrow 3$ decimal places
   b. $394.7 \rightarrow 1$ decimal place
   c. $104.25 \rightarrow 2$ decimal places
   d. $0.0003 \rightarrow 4$ decimal places
   e. $1997.4 \rightarrow 1$ decimal place
   f. $125.333 \rightarrow 3$ decimal places
   g. $69.15 \rightarrow 2$ decimal places
   h. $2.6936 \rightarrow 4$ decimal places

5. a. False   b. True   c. False
   d. True   e. False   f. False

6. a. 59.123   b. 72.217

7. a. 2.07   b. 0.80

8. a. 39.184   b. 3.232   c. 99.436
   d. 125.295   e. 100.004   f. 0.048

9. a. 58.12   b. 23.35   c. 71.10
   d. 0.87   e. 31.00   f. 125.89

10. a. 0.75   b. $0.\dot{6}$

11. a. 0.875   b. 0.143

12. a. 0.3   b. 0.6   c. 0.25
   d. 0.28   e. 0.33   f. 0.9

13. a. $0.\dot{3}$ b. $0.\dot{5}$ c. $0.08\dot{3}$
   d. $0.\overline{571428}$ e. $0.\overline{230769}$ f. $0.\overline{047619}$

14. The one on the right since 0.630 g > 0.584 g

15. a.
   1. Russell Westbrook (1.91 m)
   2. Kobe Bryant (1.98 m)
   3. Carmelo Anthony (2.02 m)
   4. Lebron James (2.03 m)
   5. Kevin Durant (2.06 m)
   b. 10 m

16. 93.5 mm

17. a. Company B: $25\frac{3}{4}$ (25.75), Company C: $25\frac{7}{9}$ ($25.\dot{7}$),
   Company A: 26.78

   b. $25\frac{7}{9} = 25.\dot{7}$

     ≈ 25.8

     $25\frac{3}{4} = 25.75$

     ≈ 25.8

18. a. $33.60   b. $34.00   c. $33.60

Exercise 1.5 Calculator skills

1. a. 461   b. 1206   c. −162.1392

2. a. 445   b. 1748   c. 343.136227

3. a. 45   b. 95   c. 621
   d. 16 921

4. a. 2   b. 32   c. 241
   d. 438

5. a. 62   b. 49   c. 20.702
   d. 58.95

6. a. 132.7925   b. 29.5800   c. 11.4750

7. a. −0.4097   b. 1.9746   c. 1.4038

8. a. 23.1884   b. 19.22   c. 76.66
   d. 12.3429

9. a. 3.7430   b. 4.8485   c. 18.2067
   d. 0.2098

10. a. 0.75   b. 0.32   c. 1.27
   d. 19.67

11. a. 1.28×10^5 b. 3.071×10^{16}

12. a. 4.49×10^6 b. 1.512×10^9

13. a. 7.27×10^3 b. 6.419×10^6
   c. 2.6319×10^9 d. 5.06×10^1
   e. 2.8168×10^{-3} f. 4.1715×10^8

14. Cathy is 13 years old

15. 5 cans

16. 7.5×10^7 km

17. 4.396×10^{-10} m

18. 1.53×10^{24} m^3

Exercise 1.6 Approximation

1. 58 450

2. 45 900 L

3. a. m   b. g   c. mL
   d. km   e. cm   f. kg

4. D

5. C

6. $14 500

7. 14 000 Mt CO_2-e

8. 15 300

9. Around 16 000

10. Around 12 000

11. Even though the crowd behind Tiger Woods looks large, we don't know the capacity of the golf course and even if we did the picture doesn't show us the crowd at the other holes. There might not be many people watching the other groups because they are not as popular as Tiger Woods.

12. A guess doesn't really have any thought or logic behind it, however an estimate does.

13. Estimate the size of each year level and add them up. This could be improved by finding how many homeroom or tutor groups there are at each year level, estimating the number of students in each, and adding them all up.

14. Estimate: 25 000 000

15. Estimate: $1.04

16. a. Estimate: 666 000   b. Estimate: 820 000

1.7 Review: exam practice

1. C

2. B

3. D

4. A

5. B

6. D

7. D

8. C

9. B

10. A

11. a. 7   b. −48   c. 20
   d. 7

12. a. 40 000   b. 1200   c. $\frac{80}{7}$
   d. 18 050

13. a. 3.66   b. 32.32   c. 0.84
   d. 4.75

14. a. 0.13   b. 0.74   c. 0.24
   d. 0.67

15. a. 0.556   b. 1.036   c. 33.105
   d. 11.000

16. a. 37 500 people
    b. Between 5000 and 7000 litres (6000 L)
17. $320
18. a. Around $1050
    b. $1032
    c. The price in part a is higher since four of the five values were rounded up using first digit rounding.
19. a. Mercury: 7.48×10^{13} m^2
       Mars: 1.45×10^{14} m^2
       Earth: 5.11×10^{14} m^2
       Jupiter: 6.42×10^{16} m^2
       Neptune: 7.72×10^{15} m^2
    b. Jupiter has the greatest surface area of 6.42×10^{16} m^2
    c. Mercury has the smallest surface area of 7.48×10^{13} m^2
20. a. 110 250 J   b. 84 375 J   c. 17.89 m/s   d. 8 m

CHAPTER 2
Ratios

2.1 Overview

LEARNING SEQUENCE

CONTENT

In this chapter, students will learn to:
- demonstrate an understanding of the fundamental ideas and notation of ratio
- understand the relationship between fractions and ratio
- express a ratio in simplest form using whole numbers
- find the ratio of two quantities in its simplest form
- divide a quantity in a given ratio [complex]
- use ratio to describesimple scales [complex].

Fully worked solutions for this chapter are available in the Resources section of your eBookPLUS at www.jacplus.com.au.

2.2 Introduction to ratios

Ratios are used in our daily lives both at home and at work. The use of ratios is required in areas such as building and construction, cooking, engineering, hair dressing, landscaping, carpentry and pharmaceuticals, just to name a few.

2.2.1 What is a ratio?

A **ratio** is a way of comparing two or more numbers or quantities. The numbers are separated by a colon and they are always whole numbers.

The picture shown represents two scoops of chocolate ice-cream and one scoop of vanilla ice-cream. The ratio of chocolate ice-cream to vanilla ice-cream is 2 scoops to 1 scoop or $2:1$ (two to one).

The basic notation for ratio is

$$a:b$$

where a and b are two whole numbers, and is read as 'a is to b'.

NOTE: When writing a ratio, the order in which it is written is very important.

The ratio chocolate ice-cream : vanilla ice-cream $= 2:1$, while the ratio vanilla ice-cream : chocolate ice-cream $= 1:2$

There are also ratios with more than two values. For example, the ratio flour to sugar to water in a recipe could be flour : sugar : water $= 3:1:4$. In other words, for every 3 parts of flour the recipe requires 1 part of sugar and 4 parts of water.

WORKED EXAMPLE 1

What is the ratio of the black to the coloured layers of licorice in the picture?

THINK	WRITE/DRAW
1. Determine the order in which to write the ratio.	Black layers : coloured layers
2. Write the ratio in simplest form.	$2:3$

2.2.2 Ratios as fractions

Ratios can be written in fraction form, $a:b = \dfrac{a}{b}$. Fractions should always be written in simplest form.

WORKED EXAMPLE 2

What is the ratio of red-bellied macaw parrots to orange-bellied macaw parrots in the picture? Express the ratio in fraction form.

THINK	WRITE/DRAW
1. Count the number of red-bellied macaws.	2 red-bellied macaw parrots
2. Count the number of orange-bellied macaws.	5 orange-bellied macaw parrots
3. Express the ratio in simplest form and in order.	Red-bellied macaws : orange-bellied macaws $= 2:5$ $= \dfrac{2}{5}$

2.2.3 Equivalent ratios

Equivalent ratios are equal ratios. To find equivalent ratios multiply or divide both sides of the ratio by the same number.

Some equivalent ratios to $2:5$ are:

$$2 \times 2:5 \times 2 = 4:10$$
$$2 \times 3:5 \times 3 = 6:15$$
$$2 \times 10:5 \times 10 = 20:50$$

Some equivalent ratios to $100:40$ are:

$$100 \div 2:40 \div 2 = 50:20$$
$$100 \div 20:40 \div 20 = 5:2$$
$$100 \div 10:40 \div 10 = 10:4$$

Equivalent fractions are calculated in the same way as equivalent ratios.

$$\frac{2}{5} \times \frac{3}{3} = \frac{6}{15} \text{ and } \frac{100}{40} \div \frac{5}{5} = \frac{20}{8}$$

WORKED EXAMPLE 3

Use equivalent ratios to determine the values of the pronumerals in the following.

a. $3:10 = 24:a$ b. $45:81:27 = b:c:3$

THINK	WRITE/DRAW
a. 1. The left-hand sides of both ratios are given. To determine the factor, divide 24 by 3. The result is 8.	$3:10 = 24:a$
Hence, multiply both sides of the first ratio by 8.	$3 \times 8:10 \times 8 = 24:80$
2. Write the answer.	$a = 80$
b. 1. The right-hand sides of both ratios are given. To determine the factor, divide 27 by 3. The result is 9.	$45:81:27 = b:c:3$
Hence, divide all sides of the first ratio by 9.	$45 \div 9:81 \div 9:27 \div 9 = 5:9:3$
2. Write the answer.	$b = 5$ and $c = 9$

2.2.4 Simplifying ratios

There are times when the ratio is not written in its simplest form. In the picture below, there are 16 chocolates: 4 white chocolates and 12 milk chocolates.

The ratio between the white and the milk chocolates is 4 to 12.

$$\text{White chocolates : milk chocolates} = 4:12$$

Notice that this ratio can be simplified by dividing both values by 4 because 4 is the highest common factor of both 4 and 12. This ratio becomes:

$$\text{white chocolates : milk chocolates} = 1:3.$$

In other words, this means that for every white chocolate there are 4 milk chocolates.

We can say that $4:12 = 1:3$.

Expressed in fraction form the ratio becomes:

$$\frac{4}{12} = \frac{1}{3}$$

Unless otherwise stated, ratios should always be written in their simplest form.

WORKED EXAMPLE 4

a. **Examine the image and express the ratio of antique chairs to tables in its simplest form.**
b. **Express the ratio in fraction form.**

THINK	WRITE/DRAW
a. 1. Count the number of the first part of the ratio (chairs).	10 chairs
2. Count the number of the second part of the ratio (tables).	4 tables
3. Write the ratio.	Chairs : tables $= 10:4$
4. Both 10 and 4 can be divided exactly by 2. So, the highest common factor of 10 and 4 is 2. Write the ratio in its simplest form by dividing both sides by their highest common factor.	$= \dfrac{10}{2}:\dfrac{4}{2}$ $= 5:2$
b. 1. Write the ratio in fraction form.	$5:2 = \dfrac{5}{2}$

2.2.5 Ratios and decimal numbers

What do we do with ratios written in the form $2:3.5$? We said before that ratios must be expressed as whole numbers. We can convert 3.5 into a whole number by multiplying it by 10:

$$3.5 \times 10 = 35$$

If we multiply the right-hand side of the ratio by 10 we have to multiply the left-hand side of the ratio by 10 also in order to keep the ratio unchanged.

$$2 \times 10:3.5 \times 10 = 20:35$$

Ratios should be expressed in simplest form so divide both sides of the ratio by the highest common factor of 20 and 35, which is 5.

$$20 \div 5:35 \div 5 = 4:7$$

Decimal numbers with two decimal places will need to be multiplied by 10^2 or 100.
Decimal numbers with three decimal places will need to be multiplied by 10^3 or 1000.
The number of decimal places will give the power of ten required.

WORKED EXAMPLE 5

Express the ratio $4.9:0.21$ in its simplest form using whole numbers.

THINK	WRITE/DRAW
1. Multiply both side of the ratio by 100 because the largest number of decimal places is two on the right-hand side of the ratio.	$4.9 \times 100:0.21 \times 100 = 490:21$
2. Divide both sides of the ratio by the highest common factor of the two numbers. The highest common factor of both 490 and 21 is 7.	$490 \div 7:21 \div 7 = 70:3$
3. Write the ratio in its simplest form.	$4.9:0.21 = 70:3$

Exercise 2.2 Introduction to ratios

1. **WE1** Determine the ratio of white flowers (camomile) to purple flowers (bluebell) in the picture shown.

2. The abacus shown has white, green, yellow, red and blue beads. State the following as ratios:
   a. blue beads to yellow beads.
   b. red beads to green beads.
   c. white beads to red beads.

3. **WE2** Convert the following ratios into fractions.
   a. $3:5$ b. $4:28$
   c. $6:11$ d. $5:2$

4. Using the letters in the word MATHEMATICS, express the following as ratios.
   a. vowels to consonants
   b. consonants to vowels
   c. letters E to letters A
   d. letters M to letters T

5. In the picture at right there are eleven students. Express the following as ratios:
   a. girls to boys.
   b. boys to girls.

6. There are 12 shapes shown below.

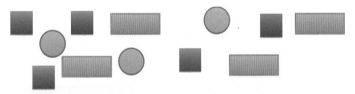

   Express the following as ratios:
   a. circles to quadrilaterals.
   b. squares to rectangles.
   c. rectangles to circles.
   d. squares to circles.

7. Express the following ratios in fraction form.
   a. $2:3$ b. $9:8$ c. $15:49$ d. $20:17$

8. **WE3** Determine the value of the pronumerals in the following equivalent ratios.
   a. $121:66 = a:6$ b. $3:2 = b:14$

9. Determine the value of the pronumerals in the following equivalent ratios.
   a. $2:7:8 = 10:m:n$ b. $81:a:54 = 9:5:b$

10. Determine the value of the pronumerals in the following equivalent ratios.
    a. $7:4 = x:16$ b. $a:9:b = 1:3:7$ c. $5:b = 15:30$ d. $18:12 = 3:m$

11. **MC** Which of the following ratios is not an equivalent ratio to $12:8$?

 A. $6:4$ **B.** $24:16$ **C.** $60:40$ **D.** $11:7$

12. **MC** Which of the following fractions is an equivalent fraction to $\dfrac{39}{65}$?

 A. $\dfrac{78}{120}$ **B.** $\dfrac{13}{21}$ **C.** $\dfrac{6}{10}$ **D.** $\dfrac{13}{195}$

13. **WE4** What is the ratio of orange balls to black and white balls in the picture shown? Express your answer in its simplest fraction form.

14. Convert the ratio $56:42$ into a fraction in its simplest form.

15. Simplify the following ratios.

 a. $\dfrac{120}{36}$ b. $1000:200$

 c. $\dfrac{33}{2200}$ d. $58:116$

 e. $36:24:60$ f. $315:180:360$

16. **WE5** Express the given ratios in simplest form using whole numbers.

 a. $2.5:1.5$ b. $3.6:4.2$ c. $0.11:1.1$

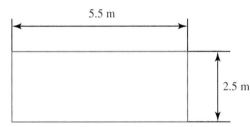

17. Express the following ratios in simplest form using whole numbers.

 a. $1.8:0.32$ b. $1.6:0.24$ c. $1.15:13.8$

18. Express the following decimal ratio in simplest form using whole numbers.

 a. $0.3:1.2$ b. $7.5:0.25$ c. $0.64:0.256$ d. $1.2:3:0.42$

19. In a class there are 28 students. If the number of girls in this class is 16, determine

 a. the number of boys in the class.

 b. the ratio of boys to girls in its simplest form.

20. a. State the ratio length to width for the rectangle below.

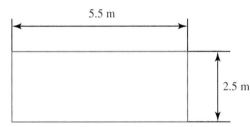

 b. State the ratio length to width to height for the rectangular prism below.

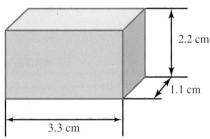

21. Using a calculator, a spreadsheet or otherwise express the following ratios as fractions in simplest form.

 a. $18:66$ b. $20:25$ c. $30:21$ d. $56:63$

 e. $10:100$ f. $728:176$ g. $550:660$ h. $3200:7300$

22. Using a calculator, a spreadsheet or otherwise determine ten equivalent fractions to the following fractions.

 a. $\dfrac{3}{5}$ b. $\dfrac{1080}{840}$

2.3 Ratios of two quantities

Ratios are used to compare the quantities of ingredients when making a mixture. For example, the mortar necessary for building a brick wall is usually prepared by the ratio of cement to sand as $1:3$. Pharmacists are required to use ratios in the preparation of medicines. Chefs use ratios in every meal they cook.

2.3.1 Weight ratios

Food is a very important part of everyone's life. A recipe involves ratios. If a recipe says 3 cups of flour and 1 cup of sugar, it means that the recipe requires 3 cups of flour for every 1 cup of sugar, so the ratio flour : sugar $= 3:1$.

What if the recipe requires 300 grams of flour and 100 grams of sugar? We can calculate the ratio flour to sugar by simplifying the quantities given.

flour : sugar $= 300\,\text{g} : 100\,\text{g}$

The highest common factor of 300 and 100 is 100. To simplify the ratio, divide both sides of the ratio by 100.

flour : sugar $= 3:1$

NOTE:

1. Usually, units are not written in ratios.
2. To find ratios of quantities always write all quantities in the same unit.

Metric conversions of weight

Metric units of weight are converted in the following way:

$$1\,\text{kg} = 1000\,\text{g}$$
$$1\,\text{tonne} = 1000\,\text{kg}$$

WORKED EXAMPLE 6

Peas and corn is a common combination of frozen vegetables. Determine the ratio of 500 g peas to 400 g corn.

THINK	WRITE/DRAW
1. Ensure that the two quantities are written in the same units (both are in grams).	500 g peas 400 g corn
2. Write the quantities as a ratio without units.	Peas : corn $= 500:400$
3. Simplify the ratio by dividing both sides by their highest common factor. As 100 is the highest common factor of 500 and 400, divide both sides of the ratio by 100.	$= 500 \div 100 : 400 \div 100$ $= 5:4$
4. Write the ratio in its simplest form.	Peas : corn $= 5:4$

2.3.2 Length ratios

Leonardo da Vinci was a famous painter, architect, sculptor, mathematician, musician, engineer, inventor, anatomist and writer. He painted the famous Mona Lisa whose face has the 'golden ratio'. The golden ratio distributes height and width according to the ratio $34 : 21$. For the Mona Lisa this is

Height of face : width of face $= 34 : 21$

Ratios of lengths are calculated in the same way as the ratios for weights.

Metric conversions of length

Recall the relationship between the metric units of length. The following diagram can be used for quick reference.

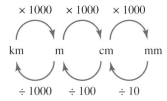

WORKED EXAMPLE 7

The window in the picture has a width of 1000 mm and a height of 1.75 m. Calculate the ratio of width to height.

THINK	WRITE/DRAW
1. Convert both lengths into cm. The most appropriate unit is cm because converting 1.75 m into cm will change the decimal number into a whole number.	$1000 \text{ mm} = 100 \text{ cm}$ $1.75 \text{ m} = 175 \text{ cm}$
2. Write the lengths as a ratio without units.	width : height $= 100 : 175$
3. Simplify the ratio by dividing both sides by their highest common factor. As 25 is the highest common factor of 100 and 175, divide both sides of the ratio by 25.	$= \dfrac{100}{25} : \dfrac{175}{25}$ $= 4 : 7$
4. Write the ratio in its simplest form.	width : height $= 4 : 7$

2.3.3 Area ratios

Areas are enclosed surfaces. Many municipalities have strict building and landscaping guidelines. Landscape architects use ratios to calculate area of land to area of building and area of pavement to area of grass.

The diagram at right is a sketch of a paved backyard (blue) with an area of 50 m^2 and a garden (green) with an area of 16 m^2.

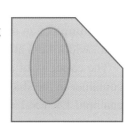

The ratio of the paved area to the garden area is

$$\text{paved area} : \text{garden area} = 50\,\text{m}^2 : 16\,\text{m}^2$$
$$= 50 : 16$$
$$= 25 : 8$$

Metric conversions of area

Recall the relationship between the metric units of area. The diagram shown can be used for quick reference.

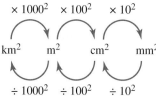

WORKED EXAMPLE 8

Kazem has a bedroom with an area of $200\,000\,\text{cm}^2$. The whole house has an area of $160\,\text{m}^2$.

    Determine the ratio of the bedroom's area to the area of the whole house.

THINK	WRITE/DRAW
1. Convert both areas into m^2 because converting $200\,000\,\text{cm}^2$ into m^2 will simplify this value.	Area of bedroom $= 200\,000\,\text{cm}^2 = 20\,\text{m}^2$ Area of house $= 16\,\text{m}^2$
2. Write the areas as a ratio without units.	Area of bedroom : area of house $= 20 : 160$
3. Simplify the ratio by dividing both sides by their highest common factor of 20.	$= \dfrac{20}{20} : \dfrac{160}{20}$
4. Write the ratio in its simplest form.	Area of bedroom : area of house $= 1 : 8$

2.3.4 Volume and capacity ratios

The **volume** of an object is the space taken up by the object. A cube of length 1 cm, width 1 cm and height 1 cm, takes up a space of $1 \times 1 \times 1 = 1\,\text{cm}^3$.

    The **capacity** of an object is the maximum amount a container can hold. A container in the shape of a cube of length 1 cm, width 1 cm and height 1 cm, can hold $1 \times 1 \times 1 = 1\,\text{cm}^3 = 1\,\text{mL}$ of liquid.

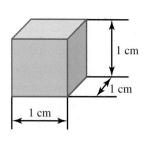

Metric conversions of volume and capacity

Recall the relationship between the metric units of volume.
The diagram shown can be used for quick reference.

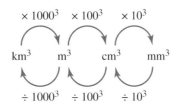

Volume units are converted to capacity units using the following conversions.

$$1\,\text{cm}^3 = 1\,\text{mL}$$

$$1000\,\text{cm}^3 = \frac{1}{1000}\,\text{m}^3 = 1\,\text{L}$$

$$1\,\text{L} = 1000\,\text{mL}$$

WORKED EXAMPLE 9

Salila and Devi have just installed two rain-water tanks; one for household use and one for watering the garden.

   The household tank has a volume of 10 000 000 cm³, while the tank for watering the garden has a volume of 45 m³.

a. Determine the ratio of the volume of the household water tank to the volume of the tank for watering the garden.

b. Determine the ratio of the capacities of the two water tanks.

THINK	WRITE/DRAW
a. 1. Convert both quantities to the same unit.	Volume of household tank $= 10\,000\,000\,\text{cm}^3 = 10\,\text{m}^3$.
The most appropriate unit is m³ because converting 10 000 000 cm³ into m³ will simplify this value.	Volume of garden tank $= 45\,\text{m}^3$
2. Write the volumes as a ratio without units.	Household tank : garden tank $= 10:45$
3. Simplify the ratio by dividing both sides by their highest common factor.	$= \dfrac{10}{5} : \dfrac{45}{5}$
As 5 is the highest common factor of 10 and 45, divide both sides of the ratio by 5.	$= 2:9$
4. Write the ratio in its simplest form.	Household tank : garden tank $= 2:9$
b. 1. Convert both quantities to the same unit.	Capacity of household tanks $= 10\,000\,000\,\text{cm}^3 = 10\,000\,\text{L}$ Capacity of garden tank $= 45\,\text{m}^3 = 45\,000\,\text{L}$

▶

The most appropriate unit is L because converting $10\,000\,000\,\text{cm}^3$ into L will simplify this value.

2. Write the volumes as a ratio without units.

3. Simplify the ratio by dividing both sides by their highest common factor.

   As 5000 is the highest common factor of 10 000 and 45 000, divide both sides of the ratio by 5000.

4. Write the ratio in its simplest form.

Capacity household tank : capacity of garden tank
$$= 10\,000 : 45\,000$$
$$= \frac{10\,000}{5000} : \frac{45\,000}{5000}$$
$$= 2 : 9$$

Capacity household tank : capacity of garden tank $= 2 : 9$

2.3.5 Time ratios

Time is a physical quantity we use for many of our daily activities. Have you ever considered the ratio of the time you spend on homework compared to the time you spend on social media?

If you spend 30 minutes completing homework and 10 minutes on social media, the ratio time for

$$\text{homework} : \text{time on social media} = 30 : 10$$
$$= 3 : 1$$

This means that for every three minutes you spend completing homework, you spend one minute using social media.

Conversions of time

Recall the relationship between the metric units of time. The diagram shown can be used for quick reference.

$$\times 60 \qquad \times 60$$

hour   minutes   seconds

$$\div 60 \qquad \div 60$$

1 hour $= 60$ minutes
1 minute $= 60$ seconds
1 hour $= 60$ minutes $(3600$ seconds$)$

WORKED EXAMPLE 10

Emori is baking lemon cupcakes and chocolate cupcakes. It takes him $\frac{3}{4}$ of an hour to make the lemon cupcakes and 36 minutes to make the chocolate cupcakes. What is the ratio of the time taken to make the chocolate cupcakes to the time taken to make the lemon cupcakes?

THINK	WRITE/DRAW
1. Convert both quantities to the same unit. The most appropriate unit is minutes because converting $\frac{3}{4}$ of an hour into minutes will change the fraction into a whole number.	$\frac{3}{4}$ of an hour $= \frac{3}{4} \times 60$ minutes $= 45$ minutes
2. Write the times as a ratio without units.	Time for chocolate cupcakes : time for lemon cupcakes $= 36 : 45$ $= \frac{36}{9} : \frac{45}{9}$
3. Simplify the ratio by dividing both sides by their highest common factor. As 9 is the highest common factor of 36 and 45, divide both sides of the ratio by 9.	$= 4 : 5$
4. Write the answer.	Time for chocolate cupcakes : time for lemon cupcakes $= 4 : 5$

2.3.6 Money ratios

A budget planner is a way of planning where money earned is going to be spent. Budgets planners are used by people and businesses, as well as by governments.

A family budgets $50 per week for electricity. If the family's weekly wage is $2100, what is the ratio of money spent on electricity per week to the total weekly wage?

$$\begin{aligned} \text{Money on electricity : wage} &= 50 : 2100 \\ &= 1 : 42 \\ &= \frac{1}{42} \end{aligned}$$

Conversions of money

Recall the relationship between dollars ($) and cents.

$$\$1 = 100 \text{ cents}$$

$$\begin{aligned} 1 \text{ cent} &= \$0.01 \\ &= \$\frac{1}{100} \end{aligned}$$

WORKED EXAMPLE 11

Jackie earns $240 per month from her part-time job. She decides to spend $40 per month to buy her lunch from the school canteen. If her monthly phone bill is $20, find the ratio of the lunch money to her phone bill and to her income per month.

THINK	WRITE/DRAW
1. Ensure that all the amounts are written in the same units.	Monthly wage = $240 Monthly lunch money = $40 Monthly phone bill = $20
2. Write the amounts as a ratio without units.	lunch money : phone bill : income $= 40:20:240$
3. Simplify the ratio by dividing all 3 parts by their highest common factor. As 20 is the highest common factor of 20, 40 and 240, divide all 3 parts of the ratio by 20.	$= \dfrac{40}{20}:\dfrac{20}{20}:\dfrac{240}{20}$ $= 2:1:12$
4. Write the answer.	Lunch money : phone bill : income $= 2:1:12$

Exercise 2.3 Ratios of two quantities

1. **WE6** Determine the ratio of 240 000 g sand and 80 kg cement.

2. A recipe for a carrot cake has the ratio 0.5 kg flour, 100 g sugar and 250 g grated carrots. Determine the ratio sugar : flour : grated carrots in its simplest form.

3. A puff pastry recipe requires 225 g flour, 30 g lard and 0.210 kg butter. What is the ratio of
   a. flour to lard?
   b. butter to lard?
   c. flour to butter?
   d. flour to lard to butter?

4. Shane makes a dry ready mix concrete using 2.750 kg of stone and gravel, 2500 g of sand, 1 kg of cement and water.
   Calculate the ratios:
   a. cement and water to stone and gravel.
   b. sand to cement and water.
   c. cement and water to sand to stone and gravel.
   d. cement and water to the total quantity of materials.

5. **WE7** A wheelchair ramp has a length of 21 m and a height of 150 cm. What is the ratio of height: length?

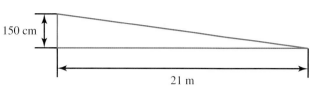

150 cm

21 m

6. A rectangular swimming pool has length 50 m, width 2500 cm and a minimum depth of 1000 mm. Calculate the ratio length: width: depth in its simplest form.

7. Find the ratio length to width of a football field 162 m long and 144 m wide.

8. **WE8** The area of the small triangle, ABC, is 80 000 cm^2 and the area of the large triangle, DEF, is 40 m^2. Write the ratio of the area of the small triangle to the area of the large triangle in its simplest form.

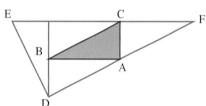

9. Earth's surface area is approximately 510 100 000 km^2. Water covers 361 100 000 km^2. Calculate the ratio of the surface area covered with water to the total surface area of Earth.

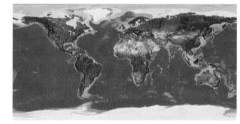

10. The landscape in the diagram at right represents the design of a garden with two ponds of areas 48 m^2 and 640 000 cm^2 respectively. The sand area is 320 000 cm^2 and the area of the table is 4 m^2.
    For the areas given find the following ratios:
    a. the area of the large pond to the area of the small pond.
    b. the area of the table to the area of sand
    c. the area of sand to the area of the large pond.
    d. the area of the small pond to the area of the table.

11. **MC** Which of the conversions below is incorrect?
    A. 3 m = 0.003 km
    B. 8.7 m^2 = 870 000 cm^2
    C. 1.5 km = 1 500 000 mm
    D. 19.6 L = 19 600 cm^3

12. **WE9** The capacity of the glass in the picture is 250 mL and the capacity of the carafe is 1.5 L. What is the ratio of capacity of the carafe to the capacity of the glass?

13. The two jugs shown hold 0.300 L of oil and 375 mL of water. What is the ratio volume of water to volume of oil required for this recipe?

14. The volumes of the orange juice boxes in the figure shown are 1.250 L and 750 mL respectively. Write the ratio of the large volume to the smaller volume in simplest fraction form.

15. Write the following ratios in simplest fraction form.
    a. $70 \, cm^3 : 20 \, cm^3$
    b. $560\,000 \, cm^3 : 0.6 \, m^3$
    c. $15 \, cm^3 : 45\,000 \, mm^3$
    d. $308\,000\,000 \, cm^3 : 385 \, m^3$

16. **WE10** The time it takes Izzy to get to school by bus is 22 minutes while the time it takes her by train is 990 seconds. What is the ratio of the time taken by bus to the time taken by train?

17. If a person spends 8.5 hours sleeping and 15.5 hours being awake, what is the ratio of sleeping time to the time being awake?

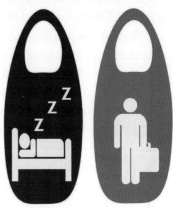

18. **MC** The ratio of 1 h 25 min to 55 min in its simplest form is:

   A. $\dfrac{11}{17}$ **B.** $\dfrac{85}{55}$ **C.** $\dfrac{5}{11}$ **D.** $\dfrac{17}{11}$

2.4.3 Dividing areas into ratios

Surveyors divide areas into ratios. Architects use these ratios to design accurate plans.

If a property is $800\,m^2$ and must be divided into the ratio house: backyard $= 5:3$, then we can calculate the area for the house and the area for the backyard.

The total number of parts is $5 + 3 = 8$.

The number of m^2 per part is $\dfrac{\text{total area}}{\text{total number of parts}} = \dfrac{800}{8}$

$$= 80\,m^2 \text{ per part}$$

The corresponding areas are $5 \times 100 = 500\,m^2$ for the house, and $3 \times 100 = 300\,m^2$ for the backyard.

WORKED EXAMPLE 15

The playground area shown must be divided into a ratio of grass to sandpit of $7:2$. If the area of the playground is $270\,m^2$, what is the area required for the sandpit.

THINK	WRITE/DRAW
1. Calculate the total number of parts of the ratio.	$7 + 2 = 9$
2. Determine the area per part by dividing the total area by the total number of parts.	The total area is $270\,m^2$. $\dfrac{\text{area of playground}}{\text{total number of parts}} = \dfrac{270}{9}$ $= 30\,m^2 \text{ per part}$
3. Multiply each number in the ratio by the area per part.	Grass : sandpit $= 7:2 = 7 \times 30:2 \times 30$ $= 210:60$
4. Check the answer by adding up all areas in the ratio. This sum should be equal to the total area.	$210 + 60 = 270\,m^2$
5. Write the answer	The area of the sandpit is $60\,m^2$.

2.4.4 Dividing volumes and capacities into ratios

Recipes use mass for solid ingredients and volume or capacity for liquid ingredients.

Volume and capacity are measured in different units.

Consider a 2 L cordial drink containing 500 mL cordial and 1.5 L water in the ratio cordial : water $= 1:3$. The cordial drink can be made in a $2000\,cm^3$ jug where the volume of water fills $1500\,cm^3$ of the jug and the cordial fills $500\,cm^3$ of the jug.

The ratio for the volumes filled by cordial and water in the jug are $1:3$ also.

The two containers shown have a total volume of 343 750 cm³. The volumes of the two containers are in a ratio of 4 : 7.
What are the volumes of the two containers?

THINK	WRITE/DRAW
1. Calculate the total number of parts of the ratio.	$4 + 7 = 11$
2. Determine the volume per part by dividing the total volume by the total number of parts.	The total area is $343\,750\text{ cm}^3$. $\dfrac{\text{total volume}}{\text{total number of parts}} = \dfrac{343\,750}{11}$ $= 31\,250\text{ cm}^3 \text{ per part}$
3. Multiply each number in the ratio by the volume per part.	Small container: Large container $= 4 : 7$ $= 4 \times 31\,250 : 7 \times 31\,250$ $= 125\,000 : 218\,750$
4. Check the answer by adding up the volumes in the ratio. This sum should be equal to the total volume.	$125\,000 + 218\,750 = 343\,750\text{ cm}^3$
5. Write the answer.	The volume of the smaller box is $125\,000\text{ cm}^3$. The volume of the larger box is $218\,750\text{ cm}^3$.

2.4.5 Dividing time into ratios

There are times when we travel by various modes of transport.

Suppose that you catch a bus and then a train to visit a friend. The whole trip takes 40 minutes. If the ratio of the time travelling by bus to the time travelling by train is 3 : 1, what will be the travel time on the bus?

The total number of parts of the ratio is $3 + 1 = 4$.

The number of minutes per part is $\dfrac{40}{4} = 10$ minutes per part.

The time travelled on the bus: bus time $= 3 \times 10 : 1 \times 10$
$= 30 : 10$

This means that travel time on the bus is 30 minutes and the travel time on the train is 10 minutes.

There are three stages involved in the manufacture of shoes: cutting out all parts, stitching them together and then attaching the sole. The average time taken to produce a basic pair of shoes is 1 hour. The times for the three stages, cutting, stitching and attaching, are in the ratio $4:5:1$.

How long would it take to cut out all the pieces for one pair of shoes?

THINK	WRITE/DRAW
1. Calculate the total number of parts of the ratio.	$4 + 5 + 1 = 10$
2. Determine the time per part by dividing the total time by the total number of parts. As we are going to divide 1 hour by 10 parts, it is easier to work with minutes than hours.	The total time is 1 hour $= 60$ minutes $$\frac{\text{total time}}{\text{total number of parts}} = \frac{60}{10}$$ $= 6$ minutes per part
3. Multiply each number in the ratio by the number of minutes per part.	$4:5:1 = 4 \times 6 : 5 \times 6 : 1 \times 6$ $= 24 : 30 : 6$
4. Check the answer by adding up the times in the ratio. This sum should be equal to the total time.	$24 + 30 + 6 = 60$ minutes $= 1$ hour
5. Write the answer.	The time taken to cut out all the pieces for a pair of shoes is 24 minutes.

Exercise 2.4 Divide into ratios

1. **WE12** The bouquet in the picture is made of 30 tulips. Flora wants to make two smaller bouquets in a ratio $2:3$. How many tulips are there in each smaller bouquet?

2. The school canteen has 250 mandarins. The mandarins must be divided in the ratio $5:3:2$. How many mandarins are there in each group?

3. Determine the number of children in a group of 1320 people if the ratio children to adults is:
   a. $3:1$ b. $5:1$
   c. $3:2$ d. $7:4$

4. Sophia bought 5.2 kg of fruits and vegetables in the ratio fruits to vegetables of $3:2$.
   a. How many kilograms of vegetables did Sophia buy?
   b. How many kilograms of fruit did she buy?

5. Carbon steel contains amounts of manganese, silicon and copper in a ratio $8:3:2$.
   a. Calculate the quantity of manganese required to produce carbon steel if the total amount of manganese, silicon and copper is 936 kg.
   b. Calculate the number of kilograms of silicon in this quantity.

6. **WE13** Shaun is a chef and has a recipe for a homemade soup that requires 500 g of tomatoes and capsicum in the ratio $3:2$. What quantity of tomatoes and capsicum should Shaun buy to make this soup?

7. The ratio sand to cement is $3:1$ to make mortar. A bricklayer wants to make 17 kg of mortar. Calculate the quantities of sand and cement required.

8. **WE14** Monika has 4.5 m of fabric to make a top and a skirt. She wants to divide the fabric in a ratio $2:3$. Determine the lengths of the two pieces of fabric.

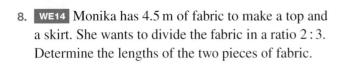

9. A 1.4 m steel beam is cut into three lengths in the ratio 1 : 2 : 4. How long are the shorter beams? Give the answer in mm.

10. The picture frame shown has a perimeter of 123 cm. Calculate the size of the width of the picture frame if the ratio width to length is 1 : 3.

11. **MC** The ratio length to width to height of the container shown is 4 : 2 : 3. If the sum of the three dimensions is 6.75 m, the height of the container is

A. 2.25 m.
B. 150 cm.
C. 1.50 m.
D. 2250 cm.

12. **WE15** The area of the window in the figure shown is 1.5 m^2. The ratio of the areas of the glass used is 1 : 2 : 3. Calculate the area of each part of glass.

13. The total area of a tennis court is approximately 260 m^2 for doubles matches. The ratio of the court area for singles matches and the extension area is approximately 13 : 3. Calculate the area of the court used for singles matches.

14. **MC** Three areas add up to 26000 m^2. If they are in a ratio of 3 : 2 : 8, the smallest area is:
A. 4 km^2
B. 16 000 m^2
C. 6 km^2
D. 4000 m^2

15. The total surface area of Victoria and Western Australia is approximately 2 760 000 km^2. If the ratio of Victoria's surface area and Western Australia's surface area is 1 : 11, what is the approximate surface area of Victoria?

16. **WE16** The ratio the volume of the container with the green lid to the volume of the container with the red lid is 5 : 6. If the total volume of the two containers is 506 cm^3, find the individual volumes of the two containers.

17. Hot air balloons are fitted with propane fuel tanks. There are three common types of fuel tanks with a total volume of 171 L. Their individual volumes are in a ratio 2 : 3 : 4. What are the individual volumes of the fuel tanks?

18. If the total capacity of three rainwater tanks is 42 500 L, find the capacity of each tank if their capacities are in a ratio of 2 : 5 : 10.

19. **WE17** Claire has 25 minutes to tidy up her bedroom and have breakfast before she goes to school. If the ratio of the tidy up time and the time for breakfast is 3 : 2, how many minutes will she have to tidy up her room?

20. Rich has 1 hour and 45 minutes to complete his Science, English and Maths homework. If the time he needs to spend on the three subjects is in a ratio 1 : 2 : 4, how long does he have to spend on his Maths homework?

21. Bayside College has a ratio of class time to lunch to recess 10 : 2 : 1.
    If the total time for a school day at Bayside College is 6 hours and 30 minutes, calculate
    a. the class time per day.   b. the time for lunch.   c. recess time.

22. Find how much a family will spend on mortgage, car loan and health insurance if they allow a budget of $1225 per month and the ratio of mortgage to car loan to health insurance is $10 : 3 : 1$.

23. A company spends $3 875 000 on wages and company cars in a ratio of $30 : 1$. How much money does the company spend on the company cars?

24. Three friends won a prize of $1520. How much money did each of them win if they contributed to the price of ticket in the ratio $2 : 5 : 9$ to buy the winning ticket?

25. Using a calculator, a spreadsheet or otherwise divide the quantities below in the ratios given.
    a. $420 in a ratio $7 : 3$.
    b. 6 hours in a ratio $1 : 5 : 6$.
    c. 18 293 m in a ratio $5 : 2 : 4$.
    d. $15 000 \, \text{m}^2$ in a ratio $10 : 3 : 2$.
    e. 960 L in a ratio $7 : 13$.
    f. 7752 in a ratio $6 : 2 : 9$.
    g. $25 519 in a ratio $3 : 2 : 8$.
    h. $540 \, \text{m}^3$ in a ratio $1 : 2 : 3 : 6$.

26. Using a calculator, a spreadsheet or otherwise find the quantities obtained when $5040 is divided in the following ratios:
    a. $\frac{1}{2}$.
    b. $\frac{2}{3}$.
    c. $\frac{1}{5}$.
    d. $\frac{2}{7}$.
    e. $\frac{3}{4}$.
    f. $2 : 3 : 4$.
    g. $1 : 2 : 5$.
    h. $1 : 2 : 3 : 4$.

2.5 Scales, diagrams and maps

2.5.1 Introduction to scales

Scales are used to draw maps and represent objects, for example cars and boats, where it is not feasible to construct full-size models.

A **scale** is a ratio of the length on a drawing or map to the actual distance or length of an object. It is the length on drawing : actual length.

Scales are usually written with no units. If a scale is given in two different units, the larger unit is usually converted into the smaller unit.

A **scale factor** is the ratio of two corresponding lengths in two similar shapes. The scale factor of $\frac{1}{2}$ or a scale (ratio) of $1 : 2$ means that 1 unit on the drawing represents 2 units in actual size. The unit can be mm, cm, m or km.

WORKED EXAMPLE 18

Calculate the scale of a drawing where 2 cm on the diagram represents 1 km in reality.

THINK	WRITE/DRAW
1. Convert the longer unit into the smaller unit.	$1 \text{ km} = 100\,000 \text{ cm}$
2. Write down the scale of the drawing with the new units.	$2 : 100\,000$
3. Simplify the scale by dividing both sides of the scale by the highest common factor.	$\dfrac{2}{2} : \dfrac{100\,000}{2}$
We divide both sides of the scale by 2 as 2 is the highest common factor of both 2 and 100 000.	$= 1 : 50\,000$
4. Write the answer.	$1 : 50\,000$

2.5.2 Calculating dimensions

The scale factor is always required when a diagram is drawn. It is used to calculate the dimensions needed.

To calculate the actual dimensions, measure the dimensions on the diagram and then multiply these by the scale factor.

WORKED EXAMPLE 19

Calculate the length and the width of the bed shown if the scale of the drawing is 1 : 100.

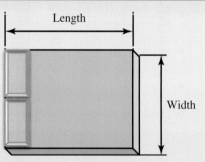

THINK	WRITE/DRAW
1. Measure the length and the width of the diagram.	Length = 2 cm Width = 1.5 cm
2. Write down the scale of the drawing.	$1 : 100$ This means that every 1 cm on the drawing represents 100 cm of the real bed.
3. Multiply both dimensions by the scale factor.	Length of the bed $= 2 \times 100$ $= 200 \text{ cm}$ Width of the bed $= 1.5 \times 100$ $= 150 \text{ cm}$
4. Write the answer.	The length is 200 cm and the width is 150 cm.

2.5.3 Scale drawing

To be able to draw to scale we need to know the dimensions on the diagram.

To calculate these dimensions, we divide the actual dimension by the scale factor.

WORKED EXAMPLE 20

A drawing of a car uses a scale of 1 : 50. Calculate the length of the diameter of the car wheels on the diagram if their real diameter is 60 cm.

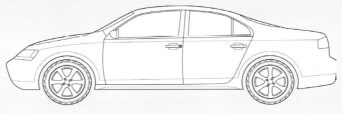

THINK	WRITE/DRAW
1. Calculate the length on the diagram.	Length on the diagram $= \dfrac{\text{real measurement}}{\text{scale factor}}$
	$= \dfrac{60}{50}$
	$= 1.2\,\text{cm}$
2. Write the answer.	A 60 cm actual diameter will be 1.2 cm on a diagram of scale 1 : 50.

2.5.4 Maps and scales

Maps are always drawn at a smaller scale. All maps have the scale written or drawn on the map.

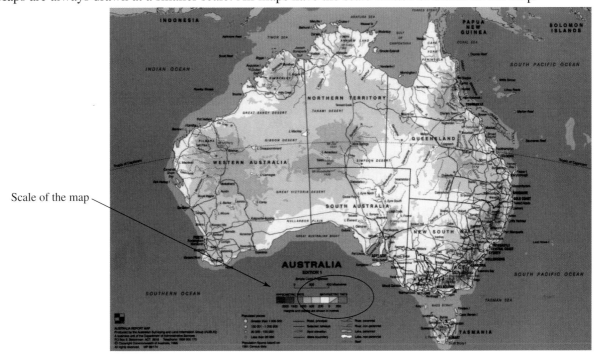

Scale of the map

Source: http://australia.gov.au/topics/science-and-technology/mapping

Determine the scale of the map as a ratio using the information in the diagram.

0   200   400 km

THINK	**WRITE/DRAW**
1. Write the scale as a ratio.	$1\,cm : 200\,km$
2. Convert the larger unit into the smaller unit.	The larger unit is km. This has to be converted into cm.
	$1\,cm : 200\,km = 1\,cm : 20\,000\,000\,cm$
3. Write the answer.	The scale is $1 : 20\,000\,000$.

2.5.5 Maps and distances

Both actual distances and distances on the map can be calculated if the scale of the map is known.
- To calculate actual distances, measure the lengths on the map and then multiply these by the scale factor.
- To calculate dimensions on the map, divide the actual dimension by the scale factor.

The scale of an Australian map is $1 : 40\,000\,000$.
a. What is the actual distance if the distance on the map is 3 cm?
b. What is the distance on the map if the actual distance is 2500 km?

THINK	**WRITE/DRAW**
a. 1. Write the scale.	Distance on the map : actual distance $1 : 40\,000\,000$
2. Set up the ratios for map ratio and actual ratio.	Map ratio $1 : 40\,000\,000$
	Actual ratio $3 : x$
3. Construct equivalent fractions.	$\dfrac{x}{3} = \dfrac{40\,000\,000}{1}$
	$x = 3 \times 40\,000\,000$
	$\quad = 120\,000\,000$
	The distance on the map is $120\,000\,000\,cm$.
4. Convert the answer into km.	$120\,000\,000\,cm = 1200\,km$
5. Write the answer.	The actual distance represented by 3 cm on the map is $1200\,km$.
b. 1. Write the scale.	Distance on the map : actual distance $= 1 : 40\,000\,000$
2. The actual distance 2500 km. Convert this to cm.	$2500\,km = 250\,000\,000\,cm$
3. Construct equivalent fractions.	$1 : 40\,000\,000$
	$x : 250\,000\,000$
	$\dfrac{x}{250\,000\,000} = \dfrac{1}{40\,000\,000}$
	$x = \dfrac{250\,000\,000}{40\,000\,000}$
	$\quad = 6.25$

4. Write the answer.   The distance on the map of 6.25 cm represents an actual
                                            distance of 2500 km.

Exercise 2.5 Scales, diagrams and maps

1. **WE18** Calculate the scale of a drawing where 100 mm on the diagram represents 2 m in reality.
2. If the scale of a diagram is 5 cm : 100 km, write the scale using the same units.
3. Calculate the scales of a drawing where
    a. 4 cm on the diagram represents 5 km in reality.
    b. 20 mm on the diagram represents 100 m in reality.
    c. 5 cm on the diagram represents 10 km in reality.
    d. 3 cm on the diagram represents 600 m in reality.
4. **WE19** Estimate the diameter of the table top shown if the scale of the diagram is 1 : 50.

5. If the scale of the diagram shown is 1 : 50, calculate the dimensions, in metres, of the Australian
   Aboriginal flag-style design shown.

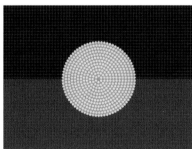

6. Calculate the actual dimensions of the nut and the bolt shown if the scale of the drawing is 1 : 2 and the
   dimensions on the diagram are:

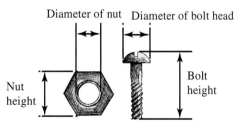

   • Diameter of the nut is 5 mm
   • Diameter of bolt head is 7 mm
   • Nut height is 10 mm
   • Bolt height is 14 mm
7. **WE20** The floor plan of a kitchen is drawn to a scale of 1 : 100. What is the width of the kitchen on the
   plan if its real width is 5 m?

8. The real height of a building is 147 m. Calculate the height of the building on a diagram drawn at a scale of 1 : 3000.

9. The floor plan shown is drawn at a scale of 1 : 200. The actual dimensions of the house are shown on the diagram. The floor plan is not drawn to scale. Calculate the lengths of the dimensions shown if the floor plan was drawn to scale.

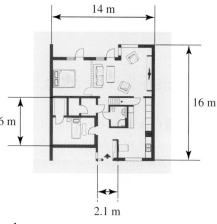

10. Calculate the actual lengths for the following lengths measured on a diagram with the given scale.
    a. 12 mm, scale 1 : 1000
    b. 3.8 cm, scale 1 : 150
    c. 27 mm, scale 1 : 200
    d. 11.6 cm, scale 1 : 7000

11. **WE21** Find the scale of the map as a ratio using the information in the diagram.

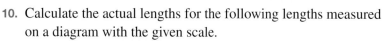

12. Find the scale of the map as a ratio using the information in the diagram.

13. **MC** The scale of the map shown is:

    A. 1 : 2 500 000   **B.** 1 : 100 000   **C.** 1 : 500 000   **D.** 1 : 5 000 000

14. For each of the map scales shown, state the scale and find the actual distance for the map distance given.
    a. 5.1 cm

    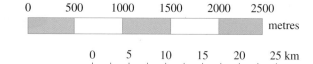

    b. 27 mm

    c. 38 mm

    d. 9.6 cm

15. **WE22** The scale of a street map has a scale of 1 : 500 000.
    a. What is the actual distance if the distance on the map is 2 cm?
    b. How long is the distance on the map if the actual distance is 10 km?

16. The scale of a map has a scale of 1 : 1 000 000.
    a. What is the actual distance if the distance on the map is 1.2 cm?
    b. How long is the distance on the map if the actual distance is 160 km?

17. **MC** The road from Perth to Adelaide is 2693 km. If this distance was drawn on a scale of 1 cm : 250 km, the distance on the map would be:
    A. 10 772 mm.   **B.** 10 772 cm.   **C.** 0.10 772 cm.   **D.** 10.772 cm.

18. If the scale of a map is 1 : 20 000, what is the distance on the map that represents
    a. an actual distance of 1 km?   b. an actual distance of 15 km?

19. A map of Australia has a scale of 10 cm : 5000 km.
    a. What is this scale written in the same units?
    b. Calculate the distances, in cm, correct to one decimal place, on the map between:
        i. Canberra and Sydney with an actual distance of 290 km.
        ii. Sydney and Brisbane with an actual distance of 925 km.
        iii. Brisbane and Darwin with an actual distance of 3423 km.
        iv. Darwin and Perth with an actual distance of 4042 km.

     v. Perth and Adelaide with an actual distance of 2693 km.

    vi. Adelaide and Melbourne with an actual distance of 727 km.

   vii. Melbourne and Canberra with an actual distance of 663 km.

20. A street map is drawn at a scale of 1 : 150 000. How long are the actual distances given by the following lengths on the map?

   a. 2 cm   **b.** 1.6 cm   **c.** 54 mm   **d.** 37 mm

21. Using a calculator, a spreadsheet or otherwise, calculate the actual dimensions of the following dimensions measured on a diagram. The corresponding scales are given.

   a. 1 cm, scale 1 : 1000   **b.** 5 cm, scale 1 : 8000   **c.** 18 mm, scale 1 : 1 000 000

   d. 9.7 cm, scale 1 : 30 000   **e.** 45 mm, scale 1 : 120 000   **f.** 58 mm, scale 1 : 50 000

   g. 19 cm, scale 1 : 200   **h.** 6.8 cm, scale 1 : 24 000

22. Using a calculator, a spreadsheet or otherwise, calculate the diagram lengths for the following actual lengths. The corresponding scales are given.

   a. 100 m, scale 1 : 5000   **b.** 27 km, scale 1 : 1 500 000

   c. 260 km, scale 1 : 4 000 000   **d.** 600 m, scale 1 : 200 000

   e. 3600 km, scale 1 cm : 3000 km   **f.** 55 km, scale 10 mm : 10 000 m

   g. 8400 m, scale 10 cm : 60 000 m   **h.** 6.8 m, scale 10 cm : 10 m

2.6 Review: exam practice

2.6.1 Ratios: summary

Introduction to ratios

- A ratio is a way of comparing two or more values, numbers, quantities etc.
  - The numbers are separated by a colon and they are always whole numbers.
  - The basic notation for ratio is $a : b$ read 'a to b', where a and b are two whole numbers.
  - Ratios can be written in fraction form, $a : b = \dfrac{a}{b}$.
- Equivalent ratios are nothing else but equal ratios. To find equivalent ratios, multiply or divide all sides of the ratio by the same number.
- Ratios are always written in their simplest form.
- Ratios with decimal numbers have to be multiplied first by a power of 10 . The number of decimals is used to determine the power of 10 required.
  - Decimal numbers with two decimal places will have to be multiplied by 10^2 or 100.
  - Decimal numbers with three decimal places will have to be multiplied by 10^3 or 1000.

Ratios of two quantities

- Ratios are written without units.
  - In order to find ratios of quantities always write all quantities in the same unit.
  - Metric units of weight are tonnes, kilograms and grams.
  - 1 tonne $= 1000$ kg, 1 kg $= 1000$ g
  - Conversions of metric units

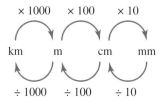 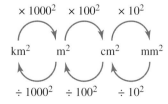 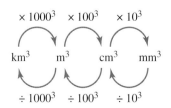

- $1\,\text{cm}^3 = 1\,\text{mL}$, $1000\,\text{cm}^3 = \dfrac{1}{1000}\,\text{m}^3 = 1\text{L}$, $1\text{L} = 1000\,\text{mL}$
- The volume of an object is the space taken up by the object.
- The capacity of an object is the maximum amount a container can hold.
- Conversions of time

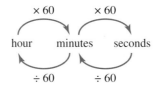

- Conversions of money
- $\$1 = 100\,\text{cents}$, $1\,\text{cent} = \$0.01 = \$\dfrac{1}{100}$

Divide into ratios

- The steps to follow when dividing a quantity into ratios are:
  - Step 1. Add up all parts in the ratios
  - Step 2. Find the quantity per part using the formula $\dfrac{\text{total quantity}}{\text{total number of parts}}$
  - Step 3. Multiply the parts of the ratios by the quantity per part.
  - Step 4. Check the answer by adding app all the quantities from Step 3.

Scales, diagrams and maps

- A scale is a ratio of the length on a drawing to the actual length.
  - Scale = Length of drawing : Actual length
  - Scales are usually written with no units.
  - If a scale is given in two different units, the larger unit has to be converted into the smaller unit.
- A scale factor of $\dfrac{1}{2}$ or a scale (ratio) of $1:2$ means that 1 unit on the drawing represents 2 units in actual size. The unit can be mm, cm, m or km.
- To calculate actual dimensions, we measure the dimensions on the diagram and then multiply these by the scale factor.
- To calculate dimensions on the diagram, we divide the actual dimension by the scale factor.

Exercise 2.6 Review: exam practice

Simple familiar

1. The ratio blue pawns to red pawns, in simplest form, in the picture shown is:
   - **A.** $6:9$.
   - **B.** $3:2$.
   - **C.** $2:3$.
   - **D.** $9:6$.
2. The ratio $385:154$ in simplest fraction form is:
   - **A.** $\dfrac{5}{2}$.
   - **B.** $\dfrac{25}{14}$.
   - **C.** $7:5$.
   - **D.** $\dfrac{2}{5}$.
3. The ratio of $364\,\text{m}^2$ to $455\,\text{m}^2$ to $819\,\text{m}^2$, in simplest form, is:
   - **A.** $28:35:63$.
   - **B.** $4:5:9$.
   - **C.** $63:35:28$.
   - **D.** $52:65:117$.

4. The ratio between the volume of the sphere and the volume of the cube in the diagram shown, in simplest fractional form, is:

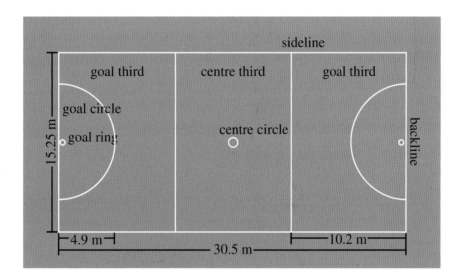

$V_{sphere} = 0.450$ m^3

$V_{cube} = 1$ m^3

   A. $\dfrac{45}{100}$.
   B. $9:20$.
   C. $45:100$.
   D. $\dfrac{9}{20}$.

5. If an area of $2730\,\text{cm}^2$ is divided into a ratio $5:3:7$, then the corresponding areas are:
   A. $546\,\text{cm}^2$, $390\,\text{cm}^2$ and $910\,\text{cm}^2$.
   B. $455\,\text{cm}^2$, $273\,\text{cm}^2$ and $637\,\text{cm}^2$.
   C. $546\,\text{cm}^2$: $1274\,\text{cm}^2$ and $910\,\text{cm}^2$.
   D. $910\,\text{cm}^2$, $546\,\text{cm}^2$ and $1274\,\text{cm}^2$.

6. If a road map is drawn to a scale of $1:900\,000$, then 3.5 cm on the map represents a road length of:
   A. 31.5 cm.
   B. 3.9 km.
   C. 3.5 km.
   D. 31.5 km.

7. A netball court is 30.5 m long and 15.25 m wide. If the dimensions of the court in the diagram shown are 6.1 cm and 3.05 cm, then the scale factor of the diagram is:

sideline		
goal third	centre third	goal third
goal circle	centre circle	
goal ring		backline

15.25 m

4.9 m — 30.5 m — 10.2 m

   A. $\dfrac{1}{5}$.
   B. $\dfrac{15.25}{3.05}$.
   C. $\dfrac{1}{500}$.
   D. $\dfrac{6.1}{30.5}$.

8. The ratio $4\dfrac{1}{3}:1\dfrac{2}{3}$ in simplest form is:
   A. $13:5$.
   B. $3:13$.
   C. $4:1$.
   D. $5:13$.

9. Three areas add up to $56\,000\,\text{m}^2$. If they are in a ratio of $5:3:8$, the smallest area is:
   A. $4\,\text{km}^2$.
   B. $10\,500\,\text{m}^2$.
   C. $3500\,\text{km}^2$.
   D. $14\,000\,\text{m}^2$.

10. The ratio length to width to height of a storage container is $3:1:5$.
    If the sum of the three dimensions is 6.75 m, the length of the container is
    A. 2.25 m.
    B. 150 cm.
    C. 1.50 m.
    D. 2250 cm.

11. Two bus stops, A and B, are 45 km apart. Two new bus stops, C and D, are going to be placed in a ratio $2:3:4$ between A and B. How far apart are bus stops C and D?

A   C   D   B

12. While camping, the Blake family use powdered milk. They mix the powder with water in the ratio of $1:24$. How much of each ingredient would they need to make up:
    a. 600 mL   b. 1.2 L?

13. Yoke made 3 L of a refreshing drink using orange juice, apple juice and carrot juice in a ratio $5:8:2$.
    a. How much orange juice did she use?
    b. How much apple juice did she use?
    c. How much carrot juice did she use?
    d. What is the ratio of carrot juice to apple juice?

14. The ratio of oxygen atoms to hydrogen atoms in a molecule of water is $1:2$. If there are 3840 atoms altogether, how many of these are hydrogen atoms?

15. Calculate the following unknowns:
    a. $160 : x = 5:7$ b. $258 : a = 3:10$ c. $b : 39.6 = 2:9$
    d. $5:4 = 21.5 : x$ e. $8:1 = 134.9 : a$

16. State the following ratios in simplest fraction form.
    a. $945 \, \text{m}^2 : 1365 \, \text{m}^2$ b. $1386 \, \text{m} : 2.079 \, \text{km}$ c. $10 \, \text{h} \; 12 \, \text{min} : 61 \, \text{h} \; 12 \, \text{min}$
    d. $2.592 \, \text{m}^3 : 3.240 \, \text{m}^3$ e. $13.68 \, \text{L} : 12.24 \, \text{L}$

17. Three swimming pools have the dimensions shown in the table.

Pool	Length (m)	Width (m)	Depth (m)
Large	12.6	3.50	1.20
Medium	8.4	3.50	1.20
Small	4.2	3.50	1.20

    a. What is the ratio between the lengths of the three pools, small, medium and large?
    b. Calculate the volumes of the three swimming pools.
    c. What is the ratio between the volumes of the three swimming pools, small, medium and large?
    d. Convert the volume measurements $\left(\text{m}^3\right)$ into capacity measurements (L).
    e. If water is poured into the swimming pools at a rate of 5 L/min, how long would it take, to the nearest hour, to fill in the three swimming pools?

18. For a main course at a local restaurant, guests can select from a chicken, fish or vegetarian dish. On Friday night the kitchen served 72 chicken plates, 56 fish plates and 48 vegetarian plates.
    a. Express the number of dishes served as a ratio in the simplest form.
    b. On a Saturday night the restaurant can cater for 250 people. If the restaurant was full, how many people would be expected to order a non-vegetarian dish?
    c. The Elmir family of five and the Cann family of three dine together. The total bill for the table was $268.
       i. Calculate the cost of dinner per head.
       ii. If the bill is split according to family size, what proportion of the bill will the Elmir family pay?

19. The floor plan shown is drawn to a scale of 1 : 200. The dimensions written represent the actual dimensions in millimetres (i.e. 4000 = 4 m).

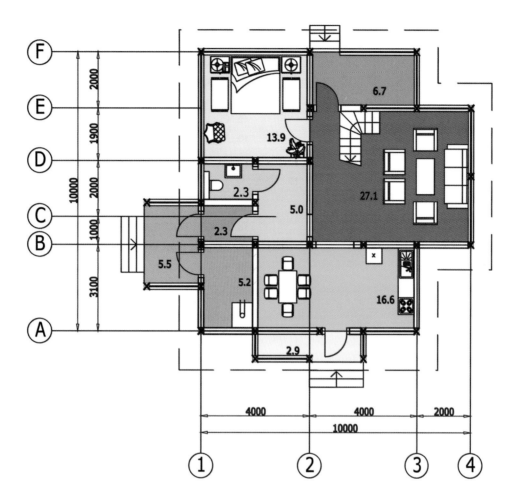

a. What are the actual dimensions, length and width, of the bedroom?
b. Calculate the area of the bedroom.
c. How long are the dimensions from part A on the diagram?
d. What are the actual dimensions, length and width, of the lounge room?
e. Calculate the area of the lounge room.
f. How long are the dimensions from part D on the diagram?
g. Calculate the same dimensions from parts A and D if the scale of the diagram was 1 : 150.
h. What is the ratio between the area of the bedroom to the area of the lounge room?

20. A recipe for shortbread cookies requires 500 g flour, 250 g unsalted butter, 125 g of sugar, 150 g walnuts and 200 g chocolate chips.
a. What is the total amount of ingredients?
b. In what ratio are flour and sugar?
c. What is the ratio of chocolate chips to flour?
d. In what ratios are all the ingredients in the total mixture?
e. Find the amount of each ingredient if the total quantity of ingredients is 2.205 kg.
f. What is the total cost for the original recipe, correct to two decimal places, if the costs of the ingredients are: 1 kg flour $1.50, 250 g unsalted butter $2.10, 1 kg sugar $1.25, 1 kg walnuts $9.50 and 200 g chocolate chips $3.25?

Answers

Chapter 2 Ratios

Exercise 2.2 Introduction to ratios

1. $5:4$
2. a. $8:5$ b. $5:9$ c. $2:1$
3. a. $\dfrac{3}{5}$ b. $\dfrac{1}{7}$ c. $\dfrac{6}{11}$ d. $\dfrac{5}{2}$
4. a. $4:7$ b. $7:4$ c. $1:2$ d. $1:1$
5. a. $4:7$ b. $7:4$
6. a. $1:3$ b. $5:4$ c. $4:3$ d. $5:3$
7. a. $\dfrac{2}{3}$ b. $\dfrac{9}{8}$ c. $\dfrac{15}{49}$ d. $\dfrac{20}{17}$
8. a. $a=11$ b. $b=21$
9. a. $m=35, n=40$ b. $a=45, b=6$
10. a. $x=28$ b. $a=3, b=21$
    c. $b=10$ d. $m=2$
11. D
12. C
13. Orange balls : black and white balls $= \dfrac{1}{2}$
14. $\dfrac{4}{3}$
15. a. $\dfrac{10}{3}$ b. $5:1$ c. $\dfrac{3}{200}$
    d. $1:2$ e. $3:2:5$ f. $7:4:8$
16. a. $5:3$ b. $6:7$ c. $1:10$
17. a. $45:8$ b. $20:3$ c. $1:12$
18. a. $1:4$ b. $30:1$ c. $5:2$ d. $20:50:7$
19. a. 12 boys b. $3:4$
20. a. $11:5$ b. $3:1:2$
21. a. $\dfrac{3}{11}$ b. $\dfrac{4}{5}$ c. $\dfrac{10}{7}$ d. $\dfrac{8}{9}$
    e. $\dfrac{1}{10}$ f. $\dfrac{91}{22}$ g. $\dfrac{5}{6}$ h. $\dfrac{32}{73}$
22. a. Some equivalent fractions could be $\dfrac{6}{10}, \dfrac{9}{15}, \dfrac{12}{20}, \dfrac{15}{25},$
    $\dfrac{18}{30}, \dfrac{27}{45}, \dfrac{30}{50}, \dfrac{33}{55}, \dfrac{300}{500}, \dfrac{3000}{5000}.$
    b. Some equivalent fractions could be $\dfrac{540}{420}, \dfrac{270}{210}, \dfrac{180}{140},$
    $\dfrac{135}{105}, \dfrac{108}{84}, \dfrac{90}{70}, \dfrac{54}{42}, \dfrac{45}{35}, \dfrac{18}{14}, \dfrac{9}{7}.$

Exercise 2.3 Ratios of two quantities

1. $3:1$
2. $2:10:5$
3. a. Flour : lard $= 15:2$
   b. Butter : lard $= 7:1$
   c. Flour : butter $= 15:14$
   d. Flour : lard : butter $= 15:2:14$
4. a. $4:11$ b. $5:2$ c. $4:10:11$ d. $4:25$
5. $1:14$

6. $50:25:1$
7. $9:8$
8. AreaΔABC : AreaΔDEF$= 1:5$
9. Area$_{water}$: Area$_{total}= 3611:5101$
10. a. $4:3$ b. $1:8$ c. $1:2$ d. $12:1$
11. B
12. $C_{carafe} : C_{glass} = 6:1$
13. $V_{water} : V_{oil} = 5:4$
14. $\dfrac{5}{3}$
15. a. $\dfrac{7}{2}$ b. $\dfrac{14}{15}$ c. $\dfrac{1}{3}$ d. $\dfrac{4}{5}$
16. Time by bus : Time by train $= 4:3$
17. Sleeping time : Awake time $= 17:31$
18. D
19. a. $3:4$ b. $8:7$ c. $7:6$ d. $6:7:8$
20. a. $7:9:3$ b. $7:9$ c. $3:1$ d. $7:3$
21. a. $5:1$ b. $3:2$ c. $2:15$ d. $15:3:2$
22. a. $1:40$ b. $5:2$
23. a. $\dfrac{4}{5}$ b. $\dfrac{7}{3}$ c. $\dfrac{9}{16}$ d. $\dfrac{2}{11}$ e. $\dfrac{8}{3}$
24. a. $38:14, 57:21, 76:28, 95:35, 114:42, 133:49, 152:56,$
    $171:63, 190:70.$ Other answers are possible.
    b. $3:2, 6:4, 12:8, 72:48, 720:480, 1872:1248, 5040:3360,$
    $18\,018:12\,012, 32\,760:21\,840, 55\,440:36\,960.$
    Other answers are possible

Exercise 2.4 Divide into ratios

1. 12 tulips and 18 tulips
2. 125 mandarins, 75 mandarins and 50 mandarins
3. a. 990 children and 330 adults
   b. 1100 children and 220 adults
   c. 792 children and 528 adults
   d. 840 children and 480 adults
4. a. 2.08 kg b. 3.12 kg
5. a. 576 kg b. 216 kg
6. 300 g tomatoes, 200 g capsicum
7. 12.75 kg sand, 4.25 kg cement
8. 1.8 m and 2.7 m
9. 200 mm, 400 mm, 800 mm
10. 15.375 cm
11. A
12. $0.25\,m^2, 0.5\,m^2, 0.75\,m^2$
13. $211.25\,m^2$
14. D
15. $230\,000\,km^2$
16. $230\,cm^3, 276\,cm^3$
17. 38 L, 57 L, 76 L
18. 5000 L, 12\,500 L, 25\,000 L
19. 15 minutes

20. 60 minutes

21. a. 5 hours   b. 1 hour   c. 30 minutes

22. $875, $262.50, $87.50

23. $125 000

24. $190, $475, $855

25. a. $294, $126
    b. 30 minutes, 2.5 hours, 3 hours
    c. 8315 m, 3326 m, 6652 m
    d. 10 000 m^2, 3000 m^2, 2000 m^2
    e. 336 L, 624 L
    f. 2736, 912, 4104
    g. $5889, $3926, $15 704
    h. 45 m^3, 90 m^3, 135 m^3, 270 m^3

26. a. $1680, $3360
    b. $2016, $3024
    c. $840, $4200
    d. $1120, $3920
    e. $2160, $2880
    f. $1120, $1680, $2240
    g. $630, $1260, $3150
    h. $504, $1008, $1512, $2016

Exercise 2.5 Scales, diagrams and maps

1. $1:20$

2. $1:2\,000\,000$

3. a. $1:125\,000$ b. $1:5\,000$
   c. $1:200\,000$ d. $1:20\,000$

4. 75 cm

5. Length = 1.8 m, width = 0.9 m

6. Nut height = 2 cm, diameter of the nut = 1 cm, bolt height = 2.8 cm, diameter of bolt head = 1.4 cm.

7. 5 cm

8. 4.9 cm

9. 7 cm, 8 cm, 3 cm, 1.05 cm

10. a. 12 m   b. 5.7 m   c. 5.4 m   d. 812 m

11. $1:100\,000$

12. $1:50\,000$

13. C

14. a. Scale $3:100\,000\,000$, 1700 km
    b. Scale $1:500\,000$, 13 500 m
    c. Scale $1:20\,000$, 760 m
    d. Scale $1:500\,000$, 48 km

15. a. 10 km   b. 2 cm

16. a. 12 km   b. 16 cm

17. D

18. a. 5 cm   b. 75 cm

19. a. $1:50\,000\,000$
    b. i. 0.6 cm   ii. 1.9 cm   iii. 6.8 cm
       iv. 8.1 cm   v. 5.4 cm   vi. 1.5 cm
       vii. 1.3 cm

20. a. 3 km   b. 2.4 km   c. 8.1 km   d. 5.55 km

21. a. 10 m   b. 400 m   c. 18 km

---

     d. 2910 m   e. 5.4 km   f. 2.9 km
     g. 38 m   h. 1632 m

22. a. 2 cm   b. 18 mm   c. 65 mm
    d. 3 mm   e. 1.2 cm   f. 55 mm
    g. 1.4 cm   h. 6.8 cm

2.6 Review: exam practice

1. C

2. A

3. B

4. D

5. D

6. D

7. C

8. A

9. B

10. A

11. 15 km.

12. a. 24 mL of powder and 576 mL of water
    b. 48 mL of powder and 1152 mL of water

13. a. 1 L   b. 1.6 L   c. 400 mL   d. $1:4$

14. 2560 hydrogen atoms

15. a. 224   b. 860   c. 8.8
    d. 17.2   e. 16.86

16. a. $\dfrac{9}{13}$ b. $\dfrac{2}{3}$ c. $\dfrac{1}{6}$ d. $\dfrac{4}{5}$ e. $\dfrac{19}{17}$

17. a. $1:2:3$
    b. 52.92 m^3, 35.28 m^3 and 17.64 m^3.
    c. $1:2:3$
    d. 52 920 L, 35 280 L and 17 640 L
    e. 7 days 8 hours, 4 days 22 hours and 2 days 11 hours.

18. a. $9:7:6$
    b. 182
    c. i. $33.50   ii. 62.5%

19. a. Length = 4 m and width = 3.9 m
    b. 15.6 m^2
    c. Length = 20 mm and width = 19.5 mm
    d. Length = 6 m and width = 3.9 m
    e. 23.4 m^2
    f. Length = 30 mm and width = 19.5 mm
    g. Length of bedroom = 26.7 mm and width of bedroom = 26 mm
       Length of lounge room = 40 mm and width of loungeroom = 26 mm
    h. $2:3$

20. a. 1.225 kg
    b. $4:1$
    c. $2:5$
    d. $20:10:5:6:8$
    e. 900 g flour, 450 g unsalted butter, 225 g sugar, 270 g walnuts and 360 g chocolate chips
    f. $7.68

CHAPTER 3
Rates

3.1 Overview

LEARNING SEQUENCE

3.1 Overview
3.2 Identifying rates
3.3 Conversion of rates
3.4 Comparison of rates
3.5 Rates and costs
3.6 Review: exam practice

CONTENT

In this chapter, students will learn to:
- review identifying common usage of rates, including km/h
- convert between units for rates
- complete calculations with rates, including solving problems involving direct proportion in terms of rate [complex]
- use rates to make comparisons
- use rates to determine costs.

Fully worked solutions for this chapter are available in the Resources section of your eBookPLUS at www.jacplus.com.au.

3.2 Identifying rates

A **rate** is a form of a fraction that compares the size of one quantity in relation to another quantity. Unlike ratios, rates have units. Examples of rates are cost per kilogram or kilometres per hour.

In a rate, the two quantities being compared may have different units.

The unit for a rate contains the word **per**, which is written as '/'. For example, $60\,km/h$ is a rate that compares distance travelled to the time taken to travel that distance. The distance unit (km) and the time unit (h) are combined to form the unit of the rate as (km/h).

$$\textbf{Rate} = \frac{\textbf{one quantity}}{\textbf{another quantity}}$$

To calculate a rate, divide the first quantity by the second quantity.

Common rates involving time:

$$\textbf{Speed} = \frac{\textbf{distance}}{\textbf{time}}$$

$$\textbf{Flow rate} = \frac{\textbf{litres}}{\textbf{minute}}$$

$$\textbf{Heart rate} = \frac{\textbf{number of heartbeats}}{\textbf{minute}}$$

on Resources

Interactivity: Rates (int-3738)

WORKED EXAMPLE 1

A car travels 80 kilometres in 2 hours. Calculate the rate at which the car is travelling using units of:

a. kilometres per hour

b. metres per second (to one decimal place).

THINK	WRITE
a. 1. Identify the units in the rate. The rate 'kilometres per hour' indicates that the number of kilometres should be divided by the number of hours taken.	a. $\text{Rate} = \dfrac{\text{Distance (km)}}{\text{Time (h)}}$
2. Write the rate using the correct units.	$\text{Rate} = \dfrac{80\,\text{km}}{2\,\text{h}}$
3. Simplify the rate by dividing the numbers.	$\text{Rate} = \dfrac{40\,\text{km}}{1\,\text{h}}$
4. Write the answer.	80 kilometres in 2 hours is equivalent to a rate of $\text{Rate} = 40\,\text{km/h}$.

▶

b. 1. Identify the units in the rate. The rate 'metres per second' indicates that the number of metres should be divided by the number of seconds taken.

2. Convert the distance to metres by multiplying by 1000.
Convert time to seconds by multiplying by 60 to change to minutes and 60 again to change to seconds.

3. Write the rate using the correct units.

4. Simplify the rate.

5. Write the answer.

b. Rate $= \dfrac{\text{Distance (m)}}{\text{Time (s)}}$

$80\,\text{km} = 80 \times 1000 = 80\,000\,\text{m}$
$2\,\text{h} = 2 \times 60 \times 60 = 7200\,\text{s}$

Rate $= \dfrac{80\,000\,\text{m}}{7200\,\text{s}}$

Rate $= 11.1\ \text{m/s}$

80 kilometres in 2 hours is equivalent to a rate of 11.1 m/s.

When comparing the difference between two quantities you need to subtract the smaller quantity from the larger quantity.

WORKED EXAMPLE 2

According to the growth chart shown, at what rate was the child growing between the ages of 10 and 13? Give your answer in centimetres per year.

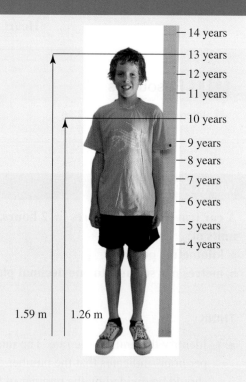

1.59 m   1.26 m

THINK

1. Amount of growth = final height − initial height. To find how much he has grown, calculate the difference in heights.

2. The final rate is in centimetres per year, so convert from metres to centimetres by multiplying by 100.

WRITE

Amount of growth $= 1.59\,\text{m} - 1.26\,\text{m} = 0.33\,\text{m}$

$= 33\,\text{cm}$

---

3. He has grown 33 cm between the ages of 10 and 13, i.e. in 3 years.	$\text{Rate of growth} = \dfrac{\text{amount of growth (cm)}}{\text{time (years)}}$ $= \dfrac{33}{3}$ $= 11$
4. Answer the question.	The child is growing at a rate of 11 cm per year or 11 cm/year.

Some common rates that don't involve time include:

Gradient, which is a measure of steepness. This can be calculated by:

$$\textbf{Gradient} = \dfrac{\textbf{Rise}}{\textbf{Run}}$$

Concentration, which is a measure of how much material is dissolved in a liquid. This can be calculated by:

$$\textbf{Concentration} = \dfrac{\textbf{Amount of solution}}{\textbf{Amount of solvent}}$$

Density, which is how much mass the material has for a certain volume. This can be calculated by:

$$\textbf{Density} = \dfrac{\textbf{Mass}}{\textbf{Volume}}$$

WORKED EXAMPLE 3

A material has a mass of 4.5 kilograms and a volume of 0.75 cubic metres. Calculate the density of the material in grams per cubic metres.

THINK	WRITE
1. Density is the mass (g) divided by the volume (m^3).	$\text{Density} = \dfrac{\text{Mass (g)}}{\text{Volume (m}^3)}$
2. Convert the mass to grams by multiplying by 1000.	$4.5 \text{ kg} = 4.5 \times 1000 \text{ g} = 4500 \text{ g}$
3. Write the rate using the correct units.	$\text{Density} = \dfrac{4500 \text{ g}}{0.75 \text{ m}^3}$
4. Write the answer.	$\text{Density} = 6000 \text{ g/m}^3$

Exercise 3.2 Identifying rates

1. **WE1** A cyclist travels 120 km in 3 hours. Calculate the rate at which the cyclist is travelling using units of:
   a. kilometres per hour
   b. metres per second.

2. An athlete's heart beats 250 beats in 5 minutes. Calculate the athlete's heart rate in beats per minute.

3. Rewrite the fractions below as rates.
   a. $\dfrac{\text{Mass}}{\text{Volume}}$
   b. $\dfrac{\text{Distance}}{\text{Time}}$
   c. $\dfrac{\text{Rise}}{\text{Run}}$
   d. $\dfrac{\text{Amount of solution}}{\text{Amount of solvent}}$

4. Express each of the following as a rate using the units given.
   a. A truck travels 560 km in 7 hours (km/h)
   b. A 5.4 m length of timber costing $21.06 ($/m)
   c. A 9 litre bucket taking 45 seconds to fill (L/s)
   d. A hiker walking 9.6 km in 70 minutes (km/min)

5. A student walks 1.6 km to school in 22 minutes. Calculate the speed she walks to school in m/s.

6. In the 2000 Sydney Olympics, Cathy Freeman won gold in the 400 m race. Her time was 49.11 seconds. Calculate her average speed in km/h.

7. The Bathurst 1000 is a 1000 km car race. In 2010, it was won by Craig Lowndes and Mark Skaife in 6 hours, 12 minutes and 51.4153 seconds. Calculate their average speed in km/h.

8. **WE2** Noah has grown from 1.12 m at 5 years old to 1.41 m at 7 years old. Calculate his growth rate in centimetres per year.

9. If Sabrina has grown 0.32 m in the past 4 years, what is her growth rate in centimetres per year?

10. A child's height at 5 years of age is 85 cm. On reaching 12 years, her height is 128 cm. Calculate her average rate of growth (in cm/year) over the 7 years.

11. A school had 300 students in 2010 and 450 students in 2012. What was the average rate of growth in students per year?

12. **WE3** A material has a mass of 12.5 kilograms and a volume of 0.25 cubic metres. Calculate the density of the material in grams per cubic metres.

13. A 2 litre drink container was filled with a mix of 200 millilitres of cordial and 1.8 litres of water. Calculate the concentration of the cordial mix in millilitres per litre.

14. Three sugar cubes, each with a mass of 1 gram, are placed into a 300 millilitre mug of coffee. Calculate the concentration of sugar in grams per millilitre (g/mL).

15. Beaches are sometimes unfit for swimming if heavy rain washes pollution into the water. A beach is declared unsafe for swimming if the concentration of bacteria is more than 5000 organisms per litre. A sample of 20 millilitres was tested and found to contain 55 organisms. Calculate the concentration in the sample (in organisms/litre) and state whether or not the beach should be closed.

16. Uluru is in the middle of Australia and is 348 m high and 3.6 km long. Calculate the average gradient (to one decimal place) from the bottom of Uluru to the top, assuming the top is in the middle.

17. A kettle is filled with tap water at 16 °C and then turned on. It boils (at 100 °C) after 2.75 minutes. Calculate the average rate of increase in temperature in °C/s.

18. Simon weighs 88 kg and has a volume of $0.11 \, \text{m}^3$, whereas Ollie weighs 82 kg and has a volume of $0.08 \, \text{m}^3$.
    a. Calculate and simplify Simon's density.
    b. Calculate and simplify Ollie's density.
    c. Who is more dense and by how much?

3.3 Conversion of rates

A rate is a measure of the change in one quantity with respect to another.

$$\text{The rate of change of Y with respect to X} = \frac{\text{Change in Y}}{\text{Change in X}}$$

WORKED EXAMPLE 4

A car travels 320 km in 4 hours. Express this rate in km/h.

THINK	WRITE
1. Write out the rate formula in terms of the two units.	$\text{Rate} = \dfrac{\text{Change in Y}}{\text{Change in X}}$ $= \dfrac{\text{Change in km}}{\text{Change in hours}}$

▶

| 2. Substitute the known quantities into the formula. | $\text{Rate} = \dfrac{320\,\text{km}}{4\,\text{h}}$ |
| 3. Simplify the rate. | $\text{Rate} = 80\,\text{km/h}$ |

3.3.1 Equivalent rates

The units of rates can be converted to satisfy particular problems, e.g. for speed we can convert km/h to m/s.

To change the unit of a rate:

- Convert to the new unit for the numerator of the rate.
- Convert to the new unit for the denominator of the rate.
- Divide the new numerator by the new denominator.

WORKED EXAMPLE 5

Covert the following rates as shown to two decimal places.
a. 60 km/h to m/s.   b. 25 L/h to mL/min.

THINK

WRITE

a. 1. Convert the numerator from km to m by multiplying by 1000.

a. $60\,\text{km} = 60 \times 1000 = 60\,000\,\text{m}$

2. Convert the denominator from hours to seconds by multiplying first by 60 to change to minutes and by 60 again to change to seconds.

$1\,\text{h} = 1 \times 60 \times 60 = 3600\,\text{sec}$

3. Divide the numerator by the denominator.

$\text{Rate} = \dfrac{60\,000}{3600}$

$= 16.6667\,\text{m/s}$

4. Write the new rate to two decimal places.

$60\,\text{km/h} = 16.67\,\text{m/s}.$

b. 1. Convert the numerator from L to mL by multiplying by 1000.

b. $25\,\text{L} = 25 \times 1000 = 25\,000\,\text{mL}$

2. Convert the denominator from hours to minutes by multiplying by 60.

$1\,\text{h} = 1 \times 60 = 60\,\text{min}$

3. Divide the numerator by the denominator.

$\text{Rate} = \dfrac{25\,000}{60}$

$= 416.6667\,\text{mL/min}$

4. Write the new rate to two decimal places.

$25\,\text{L/h} = 416.67\,\text{mL/min}$

3.3.2 Using rates

Rates are often used to calculate quantities such as costs and distances.

For example, if you are charged at $15/ min then for 10 minutes you would be charged $150, since you need to multiply the rate by the number of minutes.

**An athlete has an average heartbeat of 54 beats/minute.
How many times will their heart beat in a week?**

THINK	WRITE
1. Write the rate in terms of the quantities: heartbeats and time.	$\text{Rate} = \dfrac{\text{Heartbeats}}{\text{Time}}$
2. Substitute in the known quantities in the formula. Adjust the time for 1 week into minutes; that is, 7 days multiplied by 24 hours multiplied by 60 minutes.	$54 = \dfrac{\text{Heartbeats}}{7 \times 24 \times 60}$ $54 = \dfrac{\text{Heartbeats}}{10\,080}$
3. Solve the equation by multiplying both sides by the value of the denominator.	$\text{Heartbeats} = 54 \times 10\,080$ $\text{Heartbeats} = 544\,320$
4. Write your answer.	The athlete's heart beats 544 320 beats in a week.

Exercise 3.3 Conversion of rates

1. **WE4** A cyclist travels 102 km in 3 hours. Express this rate in km/h.
2. A person purchases a 2.5 kg bag of chocolates for $37.50. Express this rate in $/kg.
3. Given the following information, express the rate in the units shown in the brackets.
   a. A car travels 240 km in 4 hours (km/h).
   b. A hiker covers 1400 m in 7 minutes (m/min).
   c. 9 litres of water runs out of a tap in 30 seconds (L/s).
   d. Grass has grown 4 mm in 2 days (mm/day).
4. Given the following information, express the rate in the units shown in the brackets.
   a. A runner travels 800 m in 4 minutes (km/min).
   b. Water runs into a water tank at 800 mL in 2 minutes (L/min).
   c. A phone call cost 64 cents for 2 minutes ($/min).
   d. Apples cost $13.50 for 3 kg (cents/gram).
5. **WE5** Convert the following rates as shown to two decimal places.
   a. 100 km/h to m/s.   b. 40 L/h to mL/min.
6. Convert the following rates as shown to two decimal places.
   a. 5.50 $/m to cents/cm.   b. 10 m/s to km/h.
7. Convert the following rates as shown.
   a. 70 km/h to m/h   b. 40 kg/min to g/min   c. $120/min to $/h   d. 27 g/mL to g/L

8. Convert the following rates as shown, giving your answer to two decimal places.
   a. 80 km/h to m/s.
   b. 25 m/s to km/h.
   c. 50 g/ml to kg/L.
   d. 360 mL/min to L/h.
9. **WE6** If a person has an average heartbeat of 72 beats/minute, how many times would their heart beat in a week?
10. If a phone call to America costs 10c/min, how much would a 1.5 hour call to America cost?
11. If an electricity company charges 26.5 cents/kWh, how much would they charge if you used 750 kWh?
12. Quentin knows his average heart rate is 63 beats/minute. Calculate how many times his heart would beat in one year (assume 365 days in a year).

13. Christi wants to hire a jukebox for her party and received a quote of $45 per hour. If Christi is having a party that goes from 8.30 pm to 1 am, how much will she have to pay for the jukebox?

14. John was driving and, on average, it took him 45 minutes to travel 60 km.
    a. What was John's average speed in km/h?
    b. If John travelled from 11.30 am to 3.45 pm continuously, how far did he travel?
15. A gum tree grows at a rate of 80 cm/year. How long will it take to grow from the ground to a height of 10 m?

16. Tiger cubs put on approximately 100 g of weight per day. If Simba the tiger cub was born at a weight of 950 g, how long would it take him to weigh 2 kg? Give your answer in days and hours.

3.4 Comparison of rates

Rates are commonly used to make comparisons. For example, useful comparisons for consumers could be comparing the price per 100 grams of different sizes of packaged cheese in the supermarket, or the fuel economy of a different models of cars in litres per 100 kilometres.

When you are comparing two rates it is important to have both rates in the same units.

WORKED EXAMPLE 7

Johan wants to purchase a new car and fuel economy is important when considering which car to buy. The first car he looked at used 34.56 litres for 320 kilometres and the second car used 29.97 litres for 270 kilometres. Calculate the fuel consumption rate for each car in L/100 km and determine which car has the better fuel economy.

THINK	WRITE
1. The first car used 34.56 L/320 km.	First car rate $= 34.56 \, \text{L}/320 \, \text{km}$
2. Convert the rate to L/100km by dividing the numerator and denominator by 3.2.	First car rate $= \dfrac{34.56 \div 3.2}{320 \div 3.2}$ $= \dfrac{10.8}{100}$ $= 10.8 \, \text{L}/100 \, \text{km}$
3. The second car used 29.97 L/270 km.	Second car rate $= 29.97 \, \text{L}/270 \, \text{km}$
4. Convert the rate to L/100 km by dividing the numerator and denominator by 2.7.	Second car rate $= \dfrac{29.97 \div 2.7}{270 \div 2.7}$ $= \dfrac{11.1}{100}$ $= 11.1 \, \text{L}/100 \, \text{km}$
5. Now the rates have the same units, a direct comparison can be made.	The first car has better fuel economy than the second car.

Exercise 3.4 Comparison of rates

1. Two cars used fuel as shown.
   - Car A used 52.89 litres to travel 430 kilometres.
   - Car B used 40.2 litres to travel 335 kilometres.

   Determine which car has the better fuel economy by comparing the rates of L/100 km.
2. **WE7** Car A uses 43 L of petrol travelling 525 km. Car B uses 37 L of petrol travelling 420 km. Show mathematically which car is more economical.

3. Josh wants to be an iron man, so he eats iron man food for breakfast. Nutri-Grain comes in a 290 g pack for $4.87 or a 805 g pack for $8.67. Compare the price per 100 g of cereal for each pack, and hence calculate how much will be saved per 100 g if he purchases the 805 g pack.

4. A solution has a concentration of 45 g/L. The amount of solvent is doubled to dilute the solution. Calculate the concentration of the diluted solution. Explain your reasoning.

5. Tea bags in a supermarket can be bought for $1.45 per pack of 10 or for $3.85 per pack of 25. Which is the cheaper way of buying tea bags?

6. Coffee can be bought in 250 g jars for $9.50 or in 100 g jars for $4.10. Which is the cheaper way of buying the coffee and how much is the saving?

7. Eating a balanced diet is important for your overall health. One area that should be looked at is minimising the amount of total fats you eat. From the two labels below, which product is the healthier according to the amount of total fats?

Product A:

Nutrition Facts
Serving Size 1 Cake (43g)
Servings Per Container 5

Amount Per Serving

Calories 200 Calories from Fat 90

	% Daily Value*
Total Fat 10g	15%
Saturated Fat 5g	25%
Trans Fat 0g	
Cholesterol 0mg	0%
Sodium 100mg	4%
Total Carbohydrate 26g	9%
Dietary Fiber 0g	0%
Sugars 19g	
Protein 1g	

Vitamin A 0%	•	Vitamin C 0%
Calcium 0%	•	Iron 2%

* Percent Daily Values are based on a 2,000 calorie diet. Your daily values may be higher or lower depending on your calorie needs:

	Calories:	2,000	2,500
Total Fat	Less than	65g	80g
Sat. Fat	Less than	20g	25g
Cholesterol	Less than	300mg	300mg
Sodium	Less than	2,400mg	2,400mg
Total Carbohydrate		300g	375g
Dietary Fiber		25g	30g

Product B:

Nutrition Facts
Serving Size 1/4 Cup (30g)
Servings Per Container About 38

Amount Per Serving

Calories 200 Calories from Fat 150

	% Daily Value*
Total Fat 17g	26%
Saturated Fat 2.5g	13%
Trans Fat 0g	
Cholesterol 0mg	0%
Sodium 120mg	5%
Total Carbohydrate 7g	2%
Dietary Fiber 2g	8%
Sugars 1g	
Protein 5g	

Vitamin A 0%	•	Vitamin C 0%
Calcium 4%	•	Iron 8%

*Percent Daily Values are based on a 2,000 calorie diet.

8. In cricket, a batsman's strike rate indicates how fast a batsman is scoring his runs by comparing the number of runs scored per 100 balls. Shane Watson scored 72 runs in 56 balls and David Warner scored 42 runs in 31 balls. Correct to two decimal places,

   a. what is the strike rate of Shane Watson?

   b. what is the strike rate of David Warner?

   c. who has the best strike rate and by how much?

9. Two athletes had their heart rates tested as shown.

   • Athlete A: Heart beats 255 beats in 5 minutes.

   • Athlete B: Heart beats 432 beats in 9 minutes.

      a. Calculate Athlete A's heart rate in beats/minute.

      b. Calculate Athlete B's heart rate in beats/minute.

      c. Who has the lower heart rate and by how many beats/minute?

      d. How many times would each of the athlete's heart beat in one day?

10. a. The Earth is a sphere with a mass of 6.0×10^{24} kg and a radius of 6.4×10^6 m.

      i. Use the formula $V = \dfrac{4}{3}\pi r^3$ to calculate the volume of the Earth.

      ii. Hence calculate the density of the Earth.

   b. Jupiter has a mass of 1.9×10^{27} kg and a radius of 7.2×10^7 m.

      i. Calculate the volume of Jupiter.

      ii. Calculate the density of Jupiter.

   c. Different substances have their own individual density. Do your results suggest that the Earth and Jupiter are made of the same substance? Explain.

11. The Stawell Gift is a 120 m handicap footrace. Runners who start from scratch run the full 120 m; for other runners their handicap is how far in front of scratch they start. Joshua Ross has won the race twice. In 2003, with a handicap of 7 m, his time was 11.92 seconds. In 2005, from scratch, he won in 12.36 seconds. In which race did he run faster?

12. Supermarkets use rates to make it easier for the consumer to compare prices of different brands and different sized packages. To make the comparison they give you the price of each product per 100 grams, known as unit pricing. Two different brands have the following pricing:

   • Brand A comes in a 2.5 kg pack for $3.10.

   • Brand B comes in a 1 kg pack for $1.05.

   Calculate the unit price (the price per 100 g) of each of the brands and determine which brand is best value.

13. Two car hire companies offer two different rates as shown:

   • Company A charges an upfront fee of $45 and then 32 cents/km.

   • Company B charges an upfront fee of $85 and then 26 cents/km.

      a. Calculate the charges for companies A and B if a car travels 120 km.

      b. Calculate the charges for companies A and B if a car travels 265 km.

      c. Which company would be better to choose if you travelled 100 km?

      d. Which company would be the better option if you were only planning to travel a small number of kilometres?

3.5 Rates and costs

As consumers, we all need to be able to compare costs. Rates help us compare costs by ensuring we are comparing the same quantities. If we need work done around the home we always ask for two or three quotes from different tradespeople to compare costs.

3.5.1 Costs for trades

Most tradespeople charge a call out fee and then a rate per hour. These costs may vary from one tradesperson to another.

Some tradespeople have fixed rates for certain jobs. An electrician, for example, may charge a $60 callout fee or fixed cost and then $75 per hour. If the job requires 3 hours to be completed, the final cost will be:

$$\text{Cost} = 60 + 75 \times 3$$
$$= 60 + 225$$
$$= \$285$$

Cost = fixed fee + charge per hour × hours

WORKED EXAMPLE 8

Adam and Jess are both electricians. Adam charges $70 for a callout fee and $80 per hour, while Jess charges a $50 callout fee and $85 per hour. Which electrician will have a cheaper quote for a
a. 2 hour job?
b. 5 hour job?

THINK	WRITE/DRAW
a. 1. Write the formula.	Cost = fixed cost + charge per hour × hours
2. Substitute the known values for each electrician and simplify.	Adam's cost = 70 + 80 × 2 = $230 Jess' cost = 50 + 85 × 2 = $220
3. Compare the two costs	Adam charges an extra $10 for this two-hour job than Jess does, so Jess is cheaper.
b. 1. Write the formula.	Cost = fixed cost + charge per hour × hours
2. Substitute the known values for each electrician and simplify.	Adam's cost = 70 + 80 × 5 = $470 Jess' cost = 50 + 85 × 5 = $475
3. Compare the two costs.	Jess charges an extra $5 for this five-hour job than Adam does, so Adam is cheaper.

3.5.2 Living costs: food

To calculate how much we spend per week on food we simply add up all the money we spend on food products over a week.

However, food products like sugar and cooking oil are not all going to be consumed in that week. The money spent per week on these products could be estimated.

If we use 1 litre of cooking oil that costs $6 over a period of 4 weeks, then the price estimate is $1.50 per week. This is regardless of whether we use more cooking oil in one week than another.

NOTE: Throughout this section we will use an average of 30 days per month.

WORKED EXAMPLE 9

Georgia drinks 1L of milk per day, eats a loaf of bread per week and uses one 500 g container of butter per month. Calculate Georgia's daily, weekly and monthly spending for the three types of food if 1 L of milk costs $1, a loaf of bread costs $4.20 and a 500 g container of butter costs $3.15.

THINK

1. Draw a table with foods listed in the first vertical column and time periods listed on the first horizontal row.

WRITE/DRAW

Type of food	Time period		
	Day ($)	Week ($)	Month ($)
milk			
bread			
butter			
Total			

2. Fill in the table with the information given.

We know that Georgia drinks 1 L of milk per day at a cost of $1 per litre. This means that the milk costs her $1 per day.
We also know that she eats a loaf of bread per week. This means that she spends $4.20 per week on bread. She also spends $3.15 on butter per month.

Type of food	Time period		
	Day ($)	Week ($)	Month ($)
milk	1		
bread		4.20	
butter			3.15
Total			

▶

3. Calculate the missing entries.

Price of milk per week $= \$1 \times 7$ days
$$= \$7$$
Price of milk per month $= \$1 \times 30$ days
$$= \$30$$
Price of bread per day $= \$4.20 \div 7$ days
$$= \$0.60$$
Price of bread per month $= \$0.60 \times 30$ days
$$= \$18$$
Price of butter per day $= \$3.15 \div 30$ days
$$= \$0.105$$
Price of butter per week $= \$0.105 \times 7$ days
$$= \$0.735$$

4. Fill in the rest of the table.

Type of food	Time period		
	Day ($)	Week ($)	Month ($)
milk	1	7	30
bread	0.60	4.20	18
butter	0.105	0.735	3.15
Total	1.705	11.935	51.15

5. Answer the question.

Georgia spends $1.71 on food per day, $11.94 on food per week and $51.15 on food per month.

3.5.3 Living costs: transport

Means of transport to and from school are personal cars, bicycles or public transport. There are people who go to work or school using one or more forms of transport.

The costs for transport when using a personal car include petrol, maintenance and insurance.

NOTE: Throughout this section we will use an average of 30 days per month.

WORKED EXAMPLE 10

Tim uses his personal car as a means of transport. He spends $720 per year on car insurance, $64.40 per week on petrol and $42 per month on maintenance. Tabulate this data and calculate Tim's daily, weekly, monthly and yearly spending for running the car.

THINK

1. Construct a table with the types of service listed in the first vertical column and time periods listed on the first horizontal row.

WRITE/DRAW

Type of service	Time period			
	Day ($)	Week ($)	Month ($)	Year ($)
petrol				
insurance				
maintenance				
Total				

2. Fill in the table with the information given.

We know that Tim spends $720 per year on insurance, $64.40 of petrol per week and $42 for maintenance per month.

Type of service	Time period			
	Day ($)	Week ($)	Month ($)	Year ($)
petrol		64.40		
insurance				720
maintenance			42	
Total				

3. Calculate the missing entries.

$$\text{Price of petrol per day} = \$64.40 \div 7 \text{ days}$$
$$= \$9.20$$
$$\text{Price of petrol per month} = \$9.20 \times 30 \text{ days}$$
$$= \$276$$
$$\text{Price of petrol per year} = \$276 \times 12 \text{ months}$$
$$= \$3312$$
$$\text{Price of insurance per month} = \$720 \div 12 \text{ months}$$
$$= \$60$$
$$\text{Price of insurance per day} = \$60 \div 30 \text{ days}$$
$$= \$2$$
$$\text{Price of insurance per week} = \$2 \times 7 \text{days}$$
$$= \$14$$
$$\text{Price of maintenance per day} = \$42 \div 30 \text{days}$$
$$= \$1.40$$
$$\text{Price of maintenance per week} = \$1.4 \times 7 \text{ days}$$
$$= \$9.80$$
$$\text{Price of maintenance per year} = \$42 \times 12 \text{ months}$$
$$= \$504$$

4. Fill in the rest of the table.

Type of service	Time period			
	Day ($)	Week ($)	Month ($)	Year ($)
petrol	9.20	64.40	276	3312
insurance	2	14	60	720
maintenance	1.40	9.80	42	504
Total	12.60	88.20	378	4536

5. Answer the question.

Tim's car costs him $12.60 per day, $88.20 per week, $378 per month and $4536 per year.

3.5.4 Living costs: clothing

There are people who like to buy clothes on a regular basis to keep up with trends. Other people are not concerned about trends and buy only what they need.

How much do you spend on clothing?

Vince is creating a budget for clothing. He made four categories: small items, trousers, shirts and jumpers. He added up the money he spent on these items and recorded them in the two-way table shown. Complete the table.

Category	Time period		
	Day ($)	Week ($)	Month ($)
small items		12.60	
trousers	5.00		
shirts		35.70	
jumpers			93
Total			

THINK

1. Calculate the missing entries.

WRITE/DRAW

Cost of small items per day = $12.60 ÷ 7days
$$= \$1.80$$
Cost of small items per month = $1.80 × 30days
$$= \$54$$
Cost of trousers per week = $5 × 7days
$$= \$35$$
Cost of trouser per month = $5 × 30days
$$= \$150$$
Cost of shirts per day = $35.70 ÷ 7days
$$= \$5.10$$
Cost of shirts per month = $5.10 × 30days
$$= \$153$$
Cost of jumpers per day = $93 ÷ 30days
$$= \$3.10$$
Cost of jumpers per week = $3.10 × 7days
$$= \$21.70$$

2. Complete the table.

Category	Time period		
	Day ($)	Week ($)	Month ($)
small items	1.80	12.60	54
trousers	5	35	150
shirts	5.10	35.70	153
jumpers	3.10	21.70	93
Total	15	105	450

3.5.5 Living costs on spreadsheets

Excel spreadsheets are a practical and useful tool when calculating budgets. For the purpose of this exercise we are going to consider the same situation as in Worked example 10.

Tim uses his personal car as a means of transport. He spends $720 per year on car insurance, $64.40 per week on petrol and $42 per month on maintenance. Using a spreadsheet tabulate this data and calculate Tim's daily, weekly, monthly and yearly spending for running the car.

THINK

1. Construct a table in a spreadsheet with the type of service in the first vertical column, column A, and time periods on the horizontal row, row 2.

2. Fill in the table with the information given.

3. Set up the calculations required. Remember to start each formula with an = sign. Click in cell B3, type = **C3/7** and Enter.

4. Click in cell D3, type = **B3 ∗ 30** and Enter.

5. Click in cell E3, type = **D3∗12** and Enter.

6. In the bottom right corner of cell E3 you will notice a small circle. Drag it down with your mouse to copy the formula in the cell below.

WRITE/DRAW

	A	B	C	D	E
1		Time period			
2	Type of service	Day	Week	Month	Year
3	Petrol				
4	Maintenance				
5	Insurance				
6					

	A	B	C	D	E
1		Time period			
2	Type of service	Day	Week	Month	Year
3	Petrol		64.4		
4	Maintenance			42	
5	Insurance				720
6					

	A	B	C	D	E
1		Time period			
2	Type of service	Day	Week	Month	Year
3	Petrol	=C3/7	64.4		
4	Maintenance			42	
5	Insurance				720
6					

	A	B	C	D	E
1		Time period			
2	Type of service	Day	Week	Month	Year
3	Petrol	9.2	64.4	=B3*30	
4	Maintenance			42	
5	Insurance				720
6					

	A	B	C	D	E
1		Time period			
2	Type of service	Day	Week	Month	Year
3	Petrol	9.2	64.4	276	=D3*12
4	Maintenance				42
5	Insurance				720
6					

7. Click in cell B4, type = **D4/30** and Enter.

	A	B	C	D	E
1				Time period	
2	Type of service	Day	Week	Month	Year
3	Petrol	9.2	64.4	276	3312
4	Maintenance			42	504
5	Insurance				720
6					

	A	B	C	D	E
1				Time period	
2	Type of service	Day	Week	Month	Year
3	Petrol	9.2	64.4	276	3312
4	Maintenance	=D4/30		42	504
5	Insurance				720
6					

8. Drag down the small circle in the bottom right corner of cell B4 to copy the formula in the cell below.

	A	B	C	D	E
1				Time period	
2	Type of service	Day	Week	Month	Year
3	Petrol	9.2	64.4	276	3312
4	Maintenance	1.4		42	504
5	Insurance	0			720
6					

9. You will notice a '0' in cell B5. This is because at the moment there is no value in cell D5.
Click in cell D5, type = **E5/12** and Enter.

	A	B	C	D	E
1				Time period	
2	Type of service	Day	Week	Month	Year
3	Petrol	9.2	64.4	276	3312
4	Maintenance	1.4		42	504
5	Insurance	0		=E5/12	720

10. Click in cell C4, type = **B4 * 7** and Enter.

	A	B	C	D	E
1				Time period	
2	Type of service	Day	Week	Month	Year
3	Petrol	9.2	64.4	276	3312
4	Maintenance	1.4	=B4*7	42	504
5	Insurance	2		60	720
6					

11. Drag down the small circle in the bottom right corner of cell C4 to copy the formula in the cell below.

	A	B	C	D	E
1				Time period	
2	Type of service	Day	Week	Month	Year
3	Petrol	9.2	64.4	276	3312
4	Maintenance	1.4	9.8	42	504
5	Insurance	2	14	60	720
6					

12. Now, we are going to add a Total row by typing 'Total' in cell A6. Click in cell B6, type = **SUM(B3:B5)** and Enter. This command will add up the values in cells B3, B4 and B5.

	A	B	C	D	E
1		Time period			
2	Type of service	Day	Week	Month	Year
3	Petrol	9.2	64.4	276	3312
4	Maintenance	1.4	9.8	42	504
5	Insurance	2	14	60	720
6	Total	12.6			

13. This time drag the small circle in the bottom right corner of cell B6 to the right to copy the formula in cells C6, D6 and E6.

	A	B	C	D	E
1		Time period			
2	Type of service	Day	Week	Month	Year
3	Petrol	9.2	64.4	276	3312
4	Maintenance	1.4	9.8	42	504
5	Insurance	2	14	60	720
6	Total	12.6	88.2	378	4536

Exercise 3.5 Rates and costs

1. Compare the two quotes given by two computer technicians, Joanna and Dimitri, for a four-hour job. Joanna charges a $65 fixed fee and $80 per hour. Dimitri charges a $20 fixed fee and $100 per hour.

2. Hamish and Hannah are two tilers. Hamish charges $200 for surface preparation and $100 per square metre, while Hannah charges $170 for surface preparation and $120 per square metre. Compare their charges for job that covers an area of 10 m^2.

3. Calculate the following costs for a tradesperson.
   a. A fixed fee of $35 and $50 per hour for 3 hours.
   b. A fixed fee of $45 and $65 per hour for 2 hours.
   c. A fixed fee of $20 and $30 per half an hour for 1.5 hours.
   d. A fixed fee of $80 and $29.50 per 15 minutes for 30 minutes.
   e. A fixed fee of $70 and $37.50 per half an hour for 2.5 hours.
   f. A fixed fee of $100 and $50.25 per 15 minutes for 45 minutes.

4. **WE8** Lynne needs a plumber to fix the drainage of her house. She obtained three different quotes:
   Quote 1: Fixed fee of $80 and $100 per hour.
   Quote 2: Fixed fee of $25 and $120 per hour.
   Quote 3: Fixed fee of $70 and $110 per hour.
   a. Which quote is the best if the job requires 2 hours?
   b. Which quote is the best if the job requires 7 hours?

5. a. Calculate the fuel consumption (in L/100 km) for the
      following three cars:
      i. a sedan using 35 L of petrol for a trip of 252 km
      ii. a wagon using 62 L of petrol for a trip of 428 km
      iii. a 4WD using 41 L of petrol for a trip of 259 km.
   b. Which of the three cars is most economical?

6. A lawyer charges a $250 application fee and $150 per
   hour. How much money would you spend for a case that requires 10 hours?

7. Calculate and compare the fuel consumption rates for a sedan travelling in city conditions for 37 km on
   5 L and a 4WD travelling on the highway for 216 km on 29 L.

8. **WE9** Ant spends $129 a month on fruit, $18.20 a week on vegetables and $0.30 per day on eggs.
   Calculate Ant's daily, weekly and monthly spending for the three types of food and construct a table to
   display these costs.

9. Shirley spends $2.80 a week on potatoes, $45 a month on tomatoes
   and $1.60 per day on cheese. Calculate Shirley's daily, weekly and
   monthly spending for the three types of food and construct a table to
   display these costs.

10. **WE10** Halina spends $4.10 per day on petrol, $18.20 a week on
    insurance and $39 per month on maintenance for her car. Calculate
    Halina's daily, weekly, monthly and yearly spending for the three types
    of services and construct a table to display these costs.

11. Andy spends $17.50 per week on petrol, $72 a month on insurance and $612 per year on maintenance
    for his car. Calculate Andy's daily, weekly, monthly and yearly spending for the three types of services
    and construct a table to display these costs.

12. Elaine spends $15 a day on food, Tatiana spends $126 a week on food and Anya spends $420 per month
    on food.
    a. How much money does Anya spend on food per day?
    b. How much money does Tatiana spend on food per day?
    c. Who, out of the three friends, spends the most on food per month? Explain your answer.

13. **WE11** Carol has noticed that she spends too much money buying tops and wants to cut down her
    spending. She separated the tops into three categories: singlets, casual and formal. She added up the
    money she spent on these items and recorded them in the table shown. Fill in the rest of the table.

	Time period		
Category	Day ($)	Week ($)	Month ($)
singlets	3.42		
casual			198.30
formal		52.50	
Total			

14. Nadhea made a budget for clothing. She made four categories: small items, bottoms (skirts and trousers), dresses and tops. She added up the money she spent on these items and recorded them in the table shown. Fill in the rest of the table.

Category	Time period		
	Day ($)	Week ($)	Month ($)
small items	2.65		
bottoms			325.50
dresses		22.40	
tops	6.20		
Total			

15. Paris uses both her personal car and public transport. She spends $630 per year on car insurance, $9.10 per week on petrol and $21 per month on maintenance. She also spends $5.80 per day on public transport. Tabulate this data and calculate Paris' daily, weekly, monthly and yearly spending for running the car and catching public transport. (Assume Paris catches public transport every day of the year.)

16. If the total cost for running a car for 12 days is $27, what is this cost per day, per week, per month and per year?

17. **WE12** Construct an Excel spreadsheet for 8.

18. Construct an Excel spreadsheet for 14.

19. Using a calculator, a spreadsheet or otherwise calculate the weekly and monthly spending on food, clothes and transport for the following daily costs:
    a. $10 on food, $6.70 on clothes and $5.20 on transport.
    b. $12.15 on food, $9.65 on clothes and $10.80 on transport.
    c. $5.21 on food, $7.26 on clothes and $2.95 on transport.
    d. $8.93 on food, $12.54 on clothes and $3.72 on transport.

20. Using a calculator, a spreadsheet or otherwise calculate the daily and monthly spending on food, clothes and transport for the following yearly costs:
    a. $3720 on food, $2700 on clothes and $2100 on transport.
    b. $5610 on food, $3200 on clothes and $4016 on transport.
    c. $7200 on food, $3410 on clothes and $1800 on transport.
    d. $6960 on food, $4230 on clothes and $3500 on transport.

3.6 Review: exam practice

3.6.1 Rates: summary

Identifying rates

A rate is a type of fraction that compares the size of one quantity in relation to another quantity. Unlike ratios, rates have units.

In a rate, the two quantities being compared may have different units.

$$\text{Rate} = \frac{\text{One quantity}}{\text{Another quantity}}$$

To calculate a rate, divide the first quantity by the second quantity.
Some common rates that involve time include:

$$\text{Speed} = \frac{\text{Distance}}{\text{Time}}$$

$$\text{Flow rate} = \frac{\text{Litres}}{\text{Minute}}$$

$$\text{Heart rate} = \frac{\text{Number of heartbeats}}{\text{Minutes}}$$

Some common rates that don't involve time include:

$$\text{Gradient} = \frac{\text{Rise}}{\text{Run}}$$

$$\text{Concentration} = \frac{\text{Amount of solution}}{\text{Amount of solvent}}$$

$$\text{Density} = \frac{\text{Mass}}{\text{Volume}}$$

Conversion of rates

A rate is a measure of the change in one quantity with respect to another.

$$\text{The rate of change of Y with respect to X} = \frac{\text{Change in Y}}{\text{Change in X}}$$

To change the unit of a rate:
- Convert the numerator to the new rate.
- Convert the denominator to the new rate.
- Divide the new numerator by the new denominator.

Comparison of rates

Rates can be used to make comparisons; however, to make an accurate comparison it is important to have both rates in the same units.

Rates and costs

- Trades costs:
  Cost = fixed fee + charge per hour × hours
- Fuel consumption is the number of litres used to run a vehicle for a distance of 100 km

$$\text{Fuel consumption} = \frac{\text{petrol used}}{\text{km travelled}} \times 100$$

- Fuel consumption varies according to factors such as
  - the type of vehicle, motor size and make
  - road and traffic conditions
  - weather conditions.
- Living costs can be calculated by adding up all the money spent.
- Living costs are estimated per day, per week, per month or per year.
- Excel spreadsheets are a practical and useful tool when calculating budgets.

Exercise 3.6 Review: exam practice

Simple familiar

1. **MC** If a runner travels 3 km in 10 minutes and 20 seconds, the runner's average speed in m/s is:
   A. 5 m/s. **B.** 5.24 m/s. **C.** 4.84 m/s. **D.** 4.64 m/s.

2. **MC** If an escalator rises 3.75 m for a horizontal run of 5.25 m, then the escalator's gradient is:
   A. 1.4. **B.** 1.21. **C.** 0.91. **D.** 0.71.

3. The speed limit around schools is 40 km/h. This rate when converted to m/s is:
   A. 144 m/s. **B.** 120 m/s. **C.** 11.1 m/s. **D.** 12.5 m/s.

4. **MC** On average, Noah kicks 2.25 goals per game of soccer. In a 16-game season the number of goals Noah would kick is:
   A. 32. **B.** 36. **C.** 30. **D.** 38.

5. **MC** If car A travels at 60 km/h and car B travels at 16 m/s, which of the following statements is correct?
   A. They travel at the same speed.
   B. Car A travels 0.67 m/s slower than car A.
   C. Car A travels 44 km/h faster than car B.
   D. Car B travels 0.67 m/s slower than car A.

6. **MC** At the gym Lisa and Simon were doing a skipping exercise. Lisa completed 130 skips in one minute, whereas Simon skipped at a rate of 40 skips in 20 seconds. In 3 minutes of skipping, which statement is correct?
   A. Simon completes 20 more skips.
   B. Lisa completes 20 more skips.
   C. Lisa completes 30 more skips.
   D. Simon completes 30 more skips.

7. The Shinkansen train in Japan travels at a speed of approximately 500 km/h. The distance from Broome to Melbourne is approximately 4950 km. How long, to the nearest minute, would it take the Shinkansen train to cover this distance?
   A. 9 hours 54 minutes **B.** 0.1 hours
   C. 10 hours **D.** 9 hours 56 minutes and 48 seconds

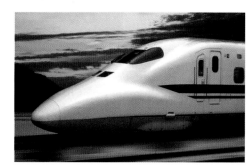

8. If the cost of food per week for a family is $231 and the cost of petrol per month is $159.60, the corresponding costs for food and petrol per day, respectively, are:
   A. $7.7 and $5.32 **B.** $5.32 and $32 **C.** $7.7 and $22.8 **D.** $33 and $5.32

9. **MC** A rate of 86.4 L/h is equivalent to:
   A. 24 mL/sec. **B.** 24 L/sec. **C.** 1.44 mL/min. **D.** 24 mL/min.

10. **MC** A sedan, a wagon and a 4WD are travelling the same distance of 159 km. The sedan uses 20 L for this trip, the wagon uses 22 L and the 4WD uses 25 L. The fuel consumptions for the 4WD, wagon and sedan are respectively:
    A. 13.8 L/km, 15.7 L/100km, and 12.6 L/100 km.
    B. 15.7 L/100km, 12.6 L/km and 13.8 L/100 km.
    C. 15.7 L/100km, 13.8 L/km and 12.6 L/100 km.
    D. 13.8 L/km, 15.7 L/100km and 12.5 L/100 km.

Complex familiar

11. Convert the following rates to the rates shown in the brackets.
    a. 17 m/s (km/h)
    b. 24.2 L/min (L/h)
    c. 0.0125 c/sec ($/min)
    d. 2.5 kg/mL (g/L)

12. A dragster covers a 400 m track in 8.86 sec.
    a. What is its average speed in m/s?
    b. If it could maintain this average speed for 1 km, how long would it take to cover the 1 km?

13. Olivia's hair grows at a rate of 125 mm/month, whereas Emily's grows at 27.25 mm/week. If their hair is initially 5 cm long, how long will their hair be after 1 year (assume 52 weeks in a year)?

14. Igor worked 4 weeks at the rate of $693.75 per week before his wage was increased to $724.80 per week.
    a. How much money was he paid for the 4 weeks?
    b. How much money will he earn at the new rate for the next 4 weeks?
    c. How much more money is Igor receiving per week?
    d. What is the difference between his earnings for the first four weeks compared to the following four weeks?

Complex unfamiliar

15. It takes me 2 hours to mow my lawn. My son takes 2.5 hours to mow the same lawn. If we work together using two lawnmowers, how long will it take to mow the lawn? Give your answer in hours, minutes and seconds.

16. To renovate your house you may require a number of different tradesmen. Given the charges of each tradesman and the hours required to complete their section of the work, calculate the total cost of the renovation.
    - Carpenter at $60 per hour, for 40 hours.
    - Electrician at $80 per hour for 14.5 hours.
    - Tiler at $55 per hour for 9 hours and 15 minutes.
    - Plasterer at $23 per half hour for 8 hours and 20 minutes.
    - Plumber with a call out charge of $85 and $90 per hour for 6.5 hours.

17. Following the Australian tax rates shown, calculate the amount of tax to be paid on the following taxable incomes.

    Tax rates:

Taxable income	Tax on this income
0 – $18 200	Nil
$18, 201–$37 000	19c for each $1 over $18 200
$37 001–$80 000	$3 572 plus 32.5c for each $1 over $37 000
$80 001–$180 000	$17 547 plus 37c for each $1 over $80 000
$180 001 and over	$54 547 plus 45c for each $1 over $180 000

    a. $21 000
    b. $15 600
    c. $50 000
    d. $100 000
    e. $210 000

18. Given the nutrition information in the label shown, calculate the following.

NUTRITION INFORMATION		
SERVINGS PER PACKAGE: 3 SERVING SIZE: 150g		
QUANTITY PER SERVING		QUANTITY PER 100g
Energy	608KJ	405KJ
Protein	4.2g	2.8g
Fat, total	7.4g	4.9g
Saturated	4.5g	3.0g
Carbohydrate	18.6g	12.4g
Sugars	18.6g	12.4g
Sodium	90mg	60mg

    a. If you ate 400 g, how much energy would that include?
    b. If you ate 550 g, how much saturated fat would that include?
    c. If you had one serving, how much protein would this include?
    d. If you had three serves then how much sugar would that include?
    e. If you ate 350 g, what percentage of carbohydrates would you have consumed?
    f. If you ate 250 g, what percentage of total fat would you have consumed?

Answers

Chapter 3 Rates

Exercise 3.2 Identifying rates

1. a. 40 km/h
   b. 11.11 m/s
2. 50 beats/minute
3. a. $\dfrac{\text{Mass}}{\text{Volume}} = \text{Density}$

   b. $\dfrac{\text{Distance}}{\text{Time}} = \text{Speed}$

   c. $\dfrac{\text{Rise}}{\text{Run}} = \text{Gradient}$

   d. $\dfrac{\text{Amount of solution}}{\text{Amount of solvent}} = \text{Concentration}$
4. a. 80 km/h   b. $3.90/m
   c. 0.2 L/s   d. 0.14 km/min
5. 1.21 m/s
6. 29.32 km/h
7. 160.92 km/h
8. 14.5 cm/year
9. 8 cm/year
10. 6.14 cm/year
11. 75 students/year
12. 50 000 g/m^3
13. 111.11 mL//L
14. 0.01 g/mL
15. 2 750 organisms/L; the beach is safe for swimming
16. 0.1
17. 0.51°C/s
18. a. 800 kg/m^3
    b. 1 025 kg/m^3
    c. Ollie is 225 kg/m^3 denser than Simon.

Exercise 3.3 Conversion of rates

1. 34 km/h
2. $15/kg
3. a. 60 km/h   b. 200 m/min
   c. 0.3 L/s   d. 2 mm/day
4. a. 0.2 km/min   b. 0.4 L/min
   c. 0.32 $/min   d. 0.45 cents/g
5. a. 27.78 m/s   b. 666.67 mL/min
6. a. 5.50 cents/cm   b. 36.00 km/h
7. a. 70 000 m/h   b. 40 000 g/min
   c. 7200 $/h   d. 27 000 g/L
8. a. 22.22 m/s   b. 90.00 km/h
   c. 50.00 kg/L   d. 21.60 L/h
9. 725 760 beats
10. $9.00
11. $198.75
12. 33 112 800 beats

13. $202.50
14. a. 80 km/h   b. 340 km
15. 12.5 years
16. 10 days and 12 hours

Exercise 3.4 Comparison of rates

1. Car A rate = 12.3 L/100 km
   Car B rate = 12 L/100 km
   Therefore Car B has the best fuel economy.
2. Car A = 8.19 L/100 km
   Car B = 8.81 L/100 km
   Therefore Car A is more economical.
3. 290 g: rate = $1.68/100 g
   805 g: rate = $1.08/100 g
   Therefore cheaper by: $1.68 − $1.08 = $0.60 per 100 g.
4. 22.5 g/L
   Since the solvent is what the solution is dissolved in, then double the solvent halves the concentration.
5. Pack of 10: $0.145 each
   Pack of 25: $0.154 each
   Technically speaking the pack of 10 is cheaper, but when rounded off to the nearest cent they are the same price for each tea bag.
6. 250 g: 3.80 $/100 g
   100 g: 4.10 $/100 g
   It is cheaper to buy the 250 gram jar by 30 cents per 100 grams.
7. Product A has 15% fat and Product B has 26% fat, therefore Product A is healthier according to total fat.
8. a. 128.57 runs per 100 balls
   b. 135.48 runs per 100 balls
   c. David Warner has the greater strike rate by 6.91 runs per 100 balls.
9. a. 51 beats/min
   b. 48 beats/min
   c. Athlete B has the lower heart rate by 3 beats/minute.
   d. Athlete A: 73 440
      Athlete B: 69 120
10. a. i. 1.10×10^{21} m^3 ii. 5464.15 kg/m^3
    b. i. 1.56×10^{24} m^3 ii. 1215.26 kg/m^3
    c. No because they have different densities
11. Speed 2003: 9.48 m/s
    Speed 2005: 9.71 m/s
    Joshua Ross ran faster in winning the Stawell Gift in 2005.
12. Brand B is better because it is 10.5 cents per grams compared to 12.4 cents per 100 grams for Brand A.
13. a. Company A = $83.40
       Company B = $116.20
    b. Company A = $129.80
       Company B = $153.90
    c. Company A = $77
       Company B = $111
       Therefore Company A is cheaper to travel 100 km.
    d. Company A is cheaper for smaller trips; however Company B would be better for trips greater than 667 km since its rate per kilometre is lower. Both companies have a rate at around 667 kilometres travelled.

Exercise 3.5 Rates and costs

1. Joanna = $385, Dimitri = $420

2. Hanish = $1200, Hannah = $1370,

3. a. $185 b. $175
   c. $110 d. $139
   e. $257.50 f. $250.75

4. a. Quote 2 b. Quote 1

5. a. i. 13.89 L/100 km ii. 14.49 L/100 km
   iii. 15.83 L/100 km
   b. The sedan

6. $1750

7. Sedan − 13.51 L/100 km
   4WD − 13.43 L/100 km
   The fuel consumption rate is better when travelling on a highway compared to city driving.

8.

Type of food	Time period		
	Day ($)	Week ($)	Month ($)
fruit	4.30	30.10	129.00
vegetables	2.60	18.20	78.00
eggs	0.30	2.10	9.00
Total	7.20	50.40	216.00

9.

Type of food	Time period		
	Day ($)	Week ($)	Month ($)
potatoes	0.40	2.80	12.00
tomatoes	1.50	10.50	45.00
cheese	1.60	11.20	48.00
Total	3.50	24.50	105.00

10.

Type of service	Time period			
	Day ($)	Week ($)	Month ($)	Year ($)
petrol	4.10	28.70	123.00	1476.00
insurance	2.60	18.20	78.00	936.00
maintenance	1.30	9.10	39.00	468.00
Total	8.00	56.00	240.00	2880.00

11.

Type of service	Time period			
	Day ($)	Week ($)	Month ($)	Year ($)
petrol	2.50	17.50	75.00	900.00
insurance	2.40	16.80	72.00	864.00
maintenance	1.70	11.90	51.00	612.00
Total	6.60	46.20	198.00	2376.00

12. a. $14 b. $18 c. Tatiana

13.

Category	Time period		
	Day ($)	Week ($)	Month ($)
singlets	3.42	23.94	102.60
casual	6.61	46.27	198.30
formal	7.50	52.50	225.00
Total	17.53	122.71	525.90

14.

Category	Time period		
	Day ($)	Week ($)	Month ($)
small items	2.65	18.55	79.50
bottoms	10.85	75.95	325.50
dresses	3.20	22.40	96.00
tops	6.20	43.40	186.00
Total	22.90	160.30	687.00

15.

Type of service	Time period			
	Day ($)	Week ($)	Month ($)	Year ($)
insurance	1.75	12.25	52.50	630
fuel	1.30	9.10	39	468
maintenance	0.70	4.90	21	252
public transport	5.80	40.60	174	2088
Total	$9.55	$66.85	$286.50	$3438.00

16. Per day: $2.25
    Per week: $15.75
    Per month: $67.50
    Per year: $810

17.

	A	B	C	D
1			Time period	
2	Type of food	Day	Week	Month
3	fruit	4.30	30.10	129.00
4	vegetables	2.60	18.20	78.00
5	eggs	0.30	2.10	9.00
6	Total	7.20	50.40	216.00

18.

	A	B	C	D
1			Time period	
2	Category	Day	Week	Month
3	small items	2.65	18.55	79.50
4	bottoms	10.85	75.95	325.50
5	dresses	3.20	22.40	96.00
6	tops	6.20	43.40	186.00
7	Total	22.90	160.30	687.00

19. a. Weekly: $153.30 b. Weekly: $228.20
       Monthly: $657   Monthly: $978
    c. Weekly: $107.94 d. Weekly: $176.33
       Monthly: $462.60   Monthly: $755.70

20. a. Daily: $23.67
Monthly: $710
c. Daily: $34.47
Monthly: $1034.17
b. Daily: $35.63
Monthly: $1068.83
d. Daily: $40.81
Monthly: $1224.17

3.6 Review: exam practice

1. C   **2.** D   **3.** C   **4.** B   **5.** D
6. C   **7.** A   **8.** D   **9.** A   **10.** C

11. a. 61.2 km/h
b. 1452 L/h
c. 0.0075 $/min
d. 2 500 000 g/L

12. a. 45.15 m/s
b. 22.15 sec

13. Olivia: 155 cm
Emily: 146.7 cm

14. a. $2775
b. $2899.20
c. $31.05
d. $124.20

15. 1 hour 6 minutes 40 seconds

16.

Trade	Calculation	Total
Carpenter	$60 \times 40 = \$2400$	$2400
Electrician	$80 \times 14.5 = \$1160$	$1160
Tiler	$\$55 \times 9\frac{15}{60} = \$508.75$	$508.75
Plasterer	$(\$23 \times 2) \times 8\frac{20}{60} = \$383.33$	$383.33
Plumber	$85 + \$90 \times 6.5 = \$670$	$670
TOTAL		**$5122.08**

17. a. $532
c. $7797
e. $68 047
b. $0
d. 24 947

18. a. 1620 kJ
c. 4.2 g
e. 12.4%
b. 16.5 g
d. 55.8 g
f. 4.9%

CHAPTER 4
Percentages

4.1 Overview

LEARNING SEQUENCE

4.1 Overview
4.2 Calculate percentages
4.3 Express as a percentage
4.4 Percentage decrease and increase
4.5 Simple interest
4.6 Review: exam practice

CONTENT

In this chapter, students will learn to:
- calculate a percentage of a given amount
- determine one amount expressed as a percentage of another for same units
- determine one amount expressed as a percentage of another for different units [complex]
- apply percentage increases and decreases in situations, including mark-ups, discounts and GST [complex]
- determine the overall change in a quantity following repeated percentage changes [complex]
- calculate simple interest for different rates and time periods [complex].

Fully worked solutions for this chapter are available in the Resources section of your eBookPLUS at www.jacplus.com.au.

4.2 Calculate percentages

4.2.1 Converting percentages to fractions

The term **per cent** means 'per hundred' or 'out of 100'. The symbol for per cent is %. For example, 13% means 13 out of 100.

Since all percentages are out of 100, they can be converted to a fraction by dividing the percentage value by 100 and, if required, simplified.

To convert a percentage to a fraction, divide by 100 or multiply by $\frac{1}{100}$ and simplify; for example,

$$35\% = \frac{{}^{7}\cancel{35}}{{}^{20}\cancel{100}} = \frac{7}{20}$$

$$\frac{25}{4}\% = \frac{25}{4 \times 100} = \frac{{}^{1}\cancel{25}}{{}^{16}\cancel{400}} = \frac{1}{16}$$

$$12\frac{4}{5}\% = \frac{64}{5}\% = \frac{64}{5 \times 100} = \frac{{}^{16}\cancel{64}}{{}^{125}\cancel{500}} = \frac{16}{125}$$

$$42.5\% = \frac{42.5}{100} = \frac{42.5 \times 10}{100 \times 10} = \frac{{}^{17}\cancel{425}}{{}^{40}\cancel{1000}} = \frac{17}{40}$$

---

on Resources

Interactivity: Fractions as percentages (int-3994)

Interactivity: Percentages as fractions (int-3992)

WORKED EXAMPLE 1

Write the following percentages as fractions in their simplest form.

a. 35%   b. $\frac{5}{8}\%$ c. $25\frac{1}{4}\%$ d. 0.75%

THINK	WRITE
a. 1. Divide the integer value by 100.	a. $35\% = \frac{35}{100}$
2. Simplify by dividing the numerator and denominator by the highest common factor (in this case 5).	$= \frac{{}^{7}\cancel{35}}{{}^{20}\cancel{100}} = \frac{7}{20}$
b. 1. Convert the fraction percentage to a fraction by multiplying by $\frac{1}{100}$.	b. $\frac{5}{8}\% = \frac{5}{8} \times \frac{1}{100}$
2. Simplify by dividing the numerator and denominator by the highest common factor (in this case 5).	$= \frac{{}^{1}\cancel{5}}{{}^{160}\cancel{800}} = \frac{1}{160}$

c. 1. Convert the mixed number to an improper fraction.

c. $25\dfrac{1}{4}\% = \dfrac{4 \times 25 + 1}{4}\% = \dfrac{101}{4}\%$

2. Convert the improper fraction percentage to a fraction by multiplying by $\dfrac{1}{100}$.

$\dfrac{101}{4}\% = \dfrac{101}{4} \times \dfrac{1}{100}$

3. Simplify if possible, or if not write the fractional answer.

$= \dfrac{101}{400}$

d. 1. Divide the decimal value by 100.

d. $0.75\% = \dfrac{0.75}{100}$

2. Convert the decimal value to an integer by multiplying by an appropriate multiple of 10. Also multiply the denominator by this multiple of 10. (Here 0.75 has 2 decimal places, so to make it an integer multiply by 100.)

$\dfrac{0.75}{100} = \dfrac{0.75 \times 100}{100 \times 100} = \dfrac{75}{10\,000}$

3. Simplify by dividing the numerator and denominator by the highest common factor (in this case 25).

$\dfrac{{}^3\cancel{75}}{{}_{400}\cancel{10\,000}} = \dfrac{3}{400}$

4.2.2 Converting percentages to decimals

In order to convert percentages to decimals divide by 100.

When dividing by 100 the decimal point moves two places to the left.

---

on Resources

Interactivity: Percentages as decimals (int-3993)

Interactivity: Decimals as percentages (int-3995)

WORKED EXAMPLE 2

Write the following percentages as decimals.

a. 67%   **b. 34.8%**

THINK

a. 1. Convert the percentage to a fraction by dividing by 100.

2. Divide the numerator by 100 by moving the decimal point two places to the left. Remember to include a zero in front of the decimal point.

b. 1. Convert the percentage to a fraction by dividing by 100.

WRITE

a. $67\% = \dfrac{67}{100}$

$\dfrac{67}{100} = 67 \div 100 = 0.67$

b. $34.8\% = \dfrac{34.8}{100}$

2. Divide the numerator by 100 by moving the decimal point two places to the left. Remember to include a zero in front of the decimal point.

$$\frac{34.8}{100} = 34.8 \div 100 = 0.348$$

4.2.3 Converting a fraction or a decimal to a percentage

To convert a decimal or fraction to a percentage, multiply by 100.

When multiplying by 100, move the decimal point two places to the right.

WORKED EXAMPLE 3

Convert the following to percentages.

a. **0.51**

b. $\dfrac{2}{5}$

THINK

a. 1. To convert to a percentage, multiply by 100.

    2. Move the decimal point two places to the right and write the answer.

b. 1. To convert to a percentage, multiply by 100.

    2. Simplify the fraction by dividing the numerator and denominator by the highest common factor (in this case 5).

WRITE

a. $0.51 \times 100 = 51\%$

b. $\dfrac{2}{5} \times 100 = \dfrac{\overset{40}{\cancel{200}}}{\underset{1}{\cancel{5}}} = \dfrac{40}{1}$

    $= 40\%$

4.2.4 Determining the percentage of a given amount

Quantities are often expressed as a percentage of a given amount. For example, the percentage of left-handed tennis players is around 10%. So, for every 100 tennis players, approximately 10 of them will be left handed.

To determine the percentage of an amount, convert the percentage to a decimal and multiply by the amount. For example,

$$25\% \text{ of } 40\,\text{kg} = \frac{25}{100} \times 40$$

$$= 0.25 \times 40$$

$$= 10\,\text{kg}$$

WORKED EXAMPLE 4

Calculate the following.

a. **35% of 120**

b. **76% of 478 kg**

THINK	WRITE
a. 1. Write the problem.	35% of 120
2. Express the percentage as a fraction and multiply by the amount.	$= \dfrac{35}{100} \times 120$
3. Write the answer.	$= 42$
b. 1. Write the problem as a decimal.	76% of 478 kg
2. Express the percentage as a fraction and multiply by the amount.	$= \dfrac{76}{100} \times 478$
3. Write the answer. Remember the result has the same units.	$= 363.28$ kg

WORKED EXAMPLE 5

If 60% of Year 11 students said Mathematics is their favourite subject and there are 165 students in Year 11, to the nearest person, how many students claimed that Mathematics was their favourite subject?

THINK	WRITE
1. Write the problem as a mathematical expression.	60% of 165
2. Write the percentage as a fraction and multiply by the amount.	$= \dfrac{60}{100} \times 165$
3. Simplify.	$= 0.60 \times 165$
	$= 99$
4. Write the answer.	99 Year 11 students said their favourite subject was Mathematics.

on Resources

Interactivity: Percentage of an amount using decimals (int-3997)

4.2.5 A shortcut for determining 10% of a quantity

To determine 10% of an amount, divide by 10 or move the decimal point one place to the left; for example,

10% of $23 = $2.30

10% of 45.6 kg = 4.56 kg

This shortcut can be adapted to other percentages; for example, to determine:

- 5% first determine 10% of the amount, then halve this amount.
- 20% first calculate 10% of the amount, then double this amount.
- 15% first calculate 10%, then 5% of the amount, then add the totals together.
- 25% first calculate 10% of the amount and double it, then calculate 5%, then add the totals together.

WORKED EXAMPLE 6

Determine the following:

a. 10% of $84   **b. 5% of 160 kg**   **c. 15% of $162**

THINK	WRITE
a. 1. To calculate 10% of 84, move the decimal point one place to the left.	10% of $84 = $8.40
b. 1. First calculate 10% of 160. To do this, move the decimal point one place to the left.	10% of 160 kg = 16 kg
2. To calculate 5%, divide the 10% value by 2.	$\dfrac{16}{2} = 8$
3. Write the answer.	5% of 160 kg = 8 kg
c. 1. First calculate 10% of 162. To do this, move the decimal point one place to the left.	10% of $162 = $16.20
2. To calculate 5%, divide the 10% value by 2.	5% of 162 = $\dfrac{16.20}{2} = 8.10$
3. Determine 15% by adding the 10% and 5% values together.	$16.20 + 8.10 = 24.30$
4. Write the answer.	15% of $162 = $24.30

 Resources

Interactivity: Common percentages and shortcuts (int-3999)

Exercise 4.2 Calculate percentages

1. **WE1** Write the following percentages as fractions in their simplest form.

   a. 48%   **b.** 26.8%   **c.** $12\frac{2}{5}\%$ **d.** $\frac{4}{5}\%$

---

2. Write the following percentages as fractions in their simplest form.
   a. 92%
   b. 74.125%
   c. $66\frac{2}{3}\%$

3. Write the following percentages as fractions.
   a. 88%
   b. 25%
   c. 0.92%
   d. 35.5%
   e. $30\frac{2}{5}\%$
   f. $72\frac{3}{4}\%$

4. **WE2** Write the following percentages as decimals.
   a. 73%
   b. 94.3%

5. Write the following percentages as decimals.
   a. 2.496%
   b. 0.62%

6. Write the following percentages as decimals.
   a. 43%
   b. 39%
   c. 80%
   d. 47.25%
   e. 24.05%
   f. 0.83%

7. **WE3** Write the following as a percentage.
   a. 0.21
   b. $\frac{4}{5}$

8. Write the following as a percentage.
   a. 0.652
   b. $\frac{8}{25}$

9. Write the following as a percentage.
   a. 0.55
   b. 0.83
   c. $\frac{1}{2}$
   d. $\frac{7}{8}$
   e. $2\frac{1}{5}$
   f. 1.65

10. **WE4** Calculate the following:
    a. 45% of 160
    b. 28% of 288 kg.

11. Calculate the following:
    a. 63% of 250
    b. 21.5% of $134

12. Determine the following to the nearest whole number.
    a. 34% of 260
    b. 55% of 594 kg
    c. 12.5% of 250 m
    d. 45% of 482
    e. 60.25% of 1250 g
    f. 37% of 2585

13. **WE5** If a student spent 70% of their part-time weekly wage of $85 on their mobile phone bill, how much were they charged?

14. If a dress was marked down by 20% and it originally cost $120, by how much has the dress been marked down?

15. **WE6** Determine the following:
   a. 10% of $56
   b. 5% of 250 kg

16. Determine the following:
   a. 15% of 370
   b. 20% of 685.

17. If 35% of the 160 students surveyed in Year 11 prefer the PlayStation 4, without using a calculator calculate how many Year 11 students prefer the PlayStation 4.

18. In 2016 Australia had a population of around 23 million people, with approximately 2.8% of them being Aboriginal and Torres Strait Islander Peoples. Using these figures, how many Aboriginal and Torres Strait Islander Peoples are there in Australia?

19. Bella spends 45% of her $1100 weekly wage on her rent. How much is this?

20. After a good 54 mm rainfall overnight, Jacko's 7500 litre water tank was 87% full. How much water did Jacko have in his water following the rainfall?

21. Last year a loaf of bread cost $3.10. Over the last 12 months the price increased by 4.5%.
   a. How much, to the nearest cent, has the price of bread increased over the past 12 months?
   b. What is the current price of a loaf of bread?
   c. If the price of bread is to increase by another 4.5% over the next year, what will be the increase in price over the next 12 months, to the nearest cent?
   d. What would a loaf of bread cost 12 months from now?

22. Rebecca has had her eyes on her dream pair of shoes and to her surprise they were on a 25% off sale. If the full price of the shoes is $175, what price did Rebecca pay for her dream shoes?

23. The Goods and Services Tax (GST) requires that 10% be added to the price of certain goods and services. Calculate the total cost of each of the following items after GST has been added. (Prices are given as pre-tax prices.)

a.

$1250

b.

$145

c.

$370

24. If a surf shop had a 20% discount sale on for the weekend, what price would you pay for each of the following items?
   a. i. Board shorts $65.00
       ii. T-shirt $24.50
       iii. Hoodie $89.90
   b. How much would you have saved by purchasing these three items on sale compared to full price?

25. Catherine was shopping for the best price for a new mini iPad. She found an ad saying that if you purchase the iPad online you can get 15% off the price. The original price was advertised at $585. How much would Catherine save if she made the purchase online?

26. If John spends 35% of his $1200 weekly wage on rent, compared to Jane who spends 32% of her $1050 weekly wage on rent:
   a. who spends the most money per week on rent?
   b. what is the difference in John and Jane's weekly rent?

4.3 Express as a percentage

4.3.1 Expressing one amount as a percentage of another

To express an amount as a percentage of another, write the two numbers as a fraction with the first number on the numerator and the second number on the denominator. Then convert this fraction to a percentage by multiplying by 100.

WORKED EXAMPLE 7

Express the following to the nearest percentage.
a. 37 as a percentage of 50
b. 18 as a percentage of 20
c. 23 as a percentage of 30.

THINK	WRITE
a. 1. Write the amount as a fraction of the total.	$\dfrac{\text{Amount}}{\text{Total}} = \dfrac{37}{50}$
2. Multiply the fraction by 100 and express as a percentage.	$\dfrac{37}{50} \times 100 = 19.5\%$
b. 1. Write the amount as a fraction of the total.	$\dfrac{\text{Amount}}{\text{Total}} = \dfrac{18}{20}$
2. Multiply the fraction by 100 and express as a percentage.	$\dfrac{18}{20} \times 100 = 90\%$
c. 1. Write the amount as a fraction of the total.	$\dfrac{\text{Amount}}{\text{Total}} = \dfrac{23}{30}$
2. Multiply the fraction by 100 and express as a percentage.	$\dfrac{23}{30} \times 100 = 76.666\%$
3. Round to the nearest percentage.	$= 77$

4.3.2 Expressing one amount as a percentage of another for different units

When expressing one amount as a percentage of another, ensure that both amounts are expressed in the same units.

WORKED EXAMPLE 8

Write 92 cents cents as a percentage of $5, to one decimal place.

THINK	WRITE
1. Since the values are in different units, convert the larger unit into the smaller one.	$\$5 = 500\,\text{cents}$
2. Write the first amount as a fraction of the second.	$\dfrac{92}{500}$

---

3. Multiply the fraction by 100.

$$\frac{92}{500} \times 100$$

4. Calculate the percentage and round to one decimal place as required.

$$= 18.4\%$$

---

on Resources

Interactivity: One amount as a percentage of another (int-3998)

WORKED EXAMPLE 9

If Liz Cambage scored 12 baskets from the free throw line out of her 15 shots, what percentage of her free throws did she get in?

THINK	WRITE
1. Write the amount as a fraction of the total.	$\dfrac{\text{Amount}}{\text{Total}} = \dfrac{12}{15}$
2. Multiply the fraction by 100 and express as a percentage.	$\dfrac{12}{15} \times 100 = 80\%$

4.3.3 Percentage discounts

To determine a percentage discount, determine the fraction: $\dfrac{\text{Amount saved}}{\text{Original price}}$ and multiply by 100.

WORKED EXAMPLE 10

A T-shirt was on sale, reduced from \$89.90 to \$75.45. What percentage was taken off the original price? Give your answer to one decimal place.

THINK	WRITE
1. Calculate the amount saved.	Amount saved $= \$89.90 - \$75.45$ $= \$14.45$

▶

2. Divide the amount saved by the original price.	$\dfrac{\text{Amount saved}}{\text{Original amount}} = \dfrac{14.45}{89.90}$
3. Multiply the fraction by 100 to get the percentage.	$\dfrac{14.45}{89.90} \times 100$
4. Calculate the value of the expression.	16.0734%
5. Round to one decimal place.	16.1% discount on the original price

 Resources

 Interactivity: Discount (int-3744)

Exercise 4.3 Express as a percentage

1. **WE7** Express each of the following to the nearest percentage.
   a. 64 as a percentage of 100
   b. 13 as a percentage of 20
   c. 9 as a percentage of 30
2. Express each of the following to the nearest percentage.
   a. 37 as a percentage of 50
   b. 21 as a percentage of 40
   c. 15 as a percentage of 23
3. **WE8** Express the following to the nearest percentage.
   a. 36 as a percentage of 82
   b. 45 as a percentage of 120
   c. 12 as a percentage of 47
   d. 9 as a percentage of 15
   e. 15 as a percentage of 44
   f. 67 as a percentage of 175
4. Express the following as a percentage, giving your answers to one decimal place.
   a. 23.5 of 69
   b. 59.3 of 80
   c. 45.75 of 65
   d. 23.82 of 33
   e. 0.85 of 5
   f. 1.59 of 2.2
5. Write 53 c as a percentage of $3, to one decimal place.
6. Write 1500 m as a percentage of 5 km.
7. Express the following as a percentage, giving your answers to one decimal place.
   a. 68 c of $2
   b. 31 c of $5
   c. 67 g of 2 kg
   d. 0.54 g of 1 kg
   e. 546 m of 2 km
   f. 477 m of 3 km
   g. 230 mm of 400 cm
   h. 36 min of 3 hours
8. **WE9** If Tom Lynch kicks 6 goals from his 13 shots at goal, what percentage of goals did he kick? Give your answer correct to 2 places.

9. If Greg Inglis has 8 of his team's 32 tackle breaks in a game, what percentage of the team's tackle breaks did Greg have?

10. If a student received 48 out of 55 for a mathematics test, what percentage, correct to 2 decimal places, did they get for their test?

11. Corn Flakes have 7.8 g of protein per 0.1 kg. What percentage of Corn Flakes is protein?

12. Adam Scott hit 16 greens in regulation in his 18-hole round of golf. What was his percentage, correct to 1 decimal place, of greens in regulation for this round?

13. **WE10** If a pair of boots were reduced from $155 to $110, what was the percentage discount offered on the boots? Give your answer to one decimal place.

14. Olivia saved $35 on the books she purchased, which were originally priced at $187.
    a. How much did she pay for the books?
    b. What percentage, correct to 1 decimal place, did she save from the original price?

15. A store claims to be taking 40% off the prices of all items, but your friend is not so sure. Calculate the percentage discount on each of the items shown and determine if the store has been completely truthful!

4.4 Percentage decrease and increase

4.4.1 Percentage decrease

A percentage decrease is often used to discount goods that are on sale. A **discount** is a reduction in price from its original marked price.

When the discount is expressed as a percentage, to determine the amount of the discount we need to multiply the original price by the discount percentage expressed as a decimal.

$$\text{Discount} = \frac{\text{Percentage change}}{100} \times \text{Original price}$$

For example, a 10% discount on an item marked at $150 gives a discount of $\frac{10}{100} \times \$150 = 0.1 \times \$150 = \$15$.

Calculating the discounted price

The discounted price can be calculated in two ways.

Method 1
- Calculate the discount by multiplying the original price by the percentage expressed as a decimal.
- Subtract the discount from the original price.

Method 2:
- Discounted price $= \left(1 - \dfrac{\text{Percentage discount}}{100}\right) \times$ original amount.

WORKED EXAMPLE 11

A store has a 15% off everything sale. Molly purchases an item that was originally priced at $160.
a. What price did Molly pay for the item?
b. How much did she save from the original price?

THINK	WRITE
a. 1. **Method 1:** Discount is 15% of $160, so multiply 0.15 (15% as a decimal) by $160.	a. Discount $= 0.15 \times \$160 = \$24$

---

2. Find the discounted price by subtracting the discount from the original price.

Discounted price $= \$160 - \$24 = \$136$

3. State the discounted price.

Molly paid $136.

4. **Method 2:**

Discounted price $= \left(1 - \dfrac{15}{100}\right) \times 160$

$$= (1 - 0.15) \times 160$$
$$= 0.85 \times 160$$

5. Calculate the discounted price.

$$= \$136$$

b. 1. **Method 1:**
Calculate the saving as a percentage of the original price.

b. Saving $= 0.15 \times \$160 = \$24$

2. State the amount saved.

Molly saved $24.

3. **Method 2:**
Calculate the difference between the original price and the discounted price (found in part a).

Saving $= \$160 - \$136 = \$24$

4. State the amount saved.

Molly saved $24.

4.4.2 Percentage increase

Just as some items are on sale or reduced, some items or services increase in price. Examples of increases in price can be the cost of electricity or a new model of car.

We calculate the percentage increase in price in a similar way to the percentage decrease; however, instead of subtracting the discounted value we add the extra percentage value.

WORKED EXAMPLE 12

Ashton paid $450 for his last electricity bill, and since then the charges have increased by 5%.
a. Assuming he used the same amount of electricity, how much extra would he expect to pay on his next bill?
b. How much will his total bill be?

THINK

a. 1. To calculate the increase amount, multiply the percentage increase by the previous cost.

2. State the amount of the extra charge.

b. 1. Calculate the total cost of the bill by adding the extra charge to the original cost.

2. State the total cost of the bill.

WRITE

5% of $450 $= 0.05 \times \$450$
$$= \$22.50$$

The extra charge is $22.50.

Total bill $= \$450 + \$22.50$

Total bill $= \$472.50$

4.4.3 Repeated percentage change

In many situations the percentage change differs over the same time period. For example, the population of a small town increases by 10% in one year and the following year increases by only 4%. If the initial population of the town is known to be 15 500, then percentage multipliers can be used to calculate the population of the town after the 2 years. The multiplying factor is $1 + \dfrac{r}{100}$, where r is the percentage increase and $1 - \dfrac{r}{100}$, where r is the percentage decease.

An increase of 10% means the population has increased by a factor of $1 + \dfrac{10}{100} = 1.10$ and an increase of 4% means the population has increased by a factor of $1 + \dfrac{4}{100} = 1.04$.

Therefore, the population after 2 years is:

$$15\,500 \times 1.10 \times 1.04 = 17\,732$$

Overall percentage change

To calculate the overall percentage change in the situation described previously, multiply the two percentage factors together, $1.10 \times 1.04 = 1.144$. This is 0.144 greater than 1. Convert 0.144 to a percentage by multiplying by 100. That gives a percentage change of $0.144 - 100 = 14.4\%$.

WORKED EXAMPLE 13

Calculate the overall change in the following quantities.

a. A small tree 1.2 metres in height grew by 22% in the first year and 16% in the second year. What was its height at the end of the second year?

b. An investment of \$12 000 grew by 6% in the first year and decreased by 1.5% in the second year. What was the value of the investment at the end of the second year?

THINK	WRITE
a. 1. The multiplying factors are found using the two percentages 22% and 16% and $1 + \dfrac{r}{100}$	First percentage change $= 1 + \dfrac{22}{100} = 1.22$ Second percentage change $= 1 + \dfrac{16}{100} = 1.16$
2. Multiply the initial height of the tree by 1.22 and 1.16.	$1.2 \times 1.22 \times 1.16 = 1.698$
3. Write the answer.	The height of the tree after two years is 1.70 m correct to two decimal places.
b. 1. The multiplying factors are found using 6% and $1 + \dfrac{r}{100}$ for the increase and 1.5% and $1 - \dfrac{r}{100}$ for the decrease.	First percentage change $= 1 + \dfrac{6}{100} = 1.06$ Second percentage change $= 1 - \dfrac{1.5}{100} = 0.985$

| 2. Multiply the initial investment by 1.06 and 0.985. | $12\,000 \times 1.06 \times 0.985 = 12\,529.20$ |
| 3. Write the answer. | The value of the investment after two years is $12\,529.20. |

Exercise 4.4 Percentage decrease and increase

1. **WE11** A store has a 25% off everything sale. Bradley purchases an item that was originally priced at $595.
   a. What price did Bradley pay for the item?
   b. How much did he save from the original price?

2. Nico purchases a new pair of shoes that were marked down by 20%. Their original price was $139.95.
   a. What price did Nico pay for the new shoes?
   b. How much did he save?

3. A new suit is priced at $550. How much would you pay if you received the following percentage discount?
   a. 5%
   b. 10%
   c. 15%
   d. 20%
   e. 30%
   f. 50%

4. A pool holds 55 000 L of water. During a hot summer week the pool lost 5.5% of its water due to evaporation.
   a. How much water was lost during the week?
   b. How much water did the pool hold at the end of the week?

5. **WE12** This year it costs $70 for a ticket to watch the Australian Open Tennis at the Rod Laver Arena. Tickets to the Australian Open Tennis are due to increase by 10% next year.

   a. How much will the tickets increase by?
   b. How much would you expect to pay for a Rod Laver Arena ticket for the tennis next year?

6. A television costs $2500 this week, however next week there will be a price increase of 7.5%. How much would you expect to pay for the television next week?

7. The Australian government introduced a Goods and Services Tax (GST) in 2006. It added 10% to a variety of goods and services. Calculate the price after GST has been added to the following pre-taxed prices.

   a. $189 car service
   b. $1650 dining table setting
   c. $152.50 pair of runners
   d. $167.85 pair of sunglasses

8. When you purchase a new car you also need to pay government duty on the car. You must pay a duty of 3% of the value of the car for cars priced up to $59 133 and 5% of the value for cars priced over $59 133.

   a. Calculate the duty on each of the following car values.

i. $15 500	ii. $8000
iii. $21 750	iv. $45 950
v. $65 000	vi. $78 750

   b. Calculate the cost of each of the cars including the government duty.

9. A sports shop is having an end-of-year sale on all stock. A discount of 20% is applied to clothing and a discount of 30% is applied to all sports equipment. What discount will you receive on the following items?

   a. A sports jumper initially marked at $175.
   b. A basketball ring marked at $299.
   c. A tennis outfit priced at $135.
   d. A table tennis table initially priced at $495.

10. A clothing store has a rack with a 50% off sign on it. As a weekend special, for an hour the shop owner takes a further 50% off all items in the store. Does that mean you can get clothes for free? Explain your answer.

11. The following items are discounted as shown.

$260 with $33\frac{1}{3}\%$ discount

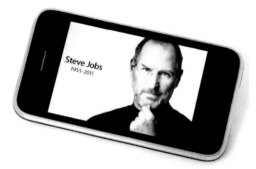

$380 with 25% discount

$450 with 20% discount

$600 with 15% discount

a. Which item has the largest discount?
b. Which of the purchases give the same discount?
c. What is the difference between the largest and smallest discount?
d. What percentage discount would you need if you only have $420 to buy the bike?

12. A new car was purchased for $35 000. It depreciates from its current value each year by the percentage shown in the table.
a. Complete the table.

End year	% depreciation	Depreciation ($)	Value ($)
First	10%	$3500	$31 500
Second	7.5%	$2362.50	
Third	6%		
Fourth	5%		
Fifth	4%		

b. If the car continued to lose 2% each year in the sixth, seventh, eighth and ninth year, how much would the car be worth at the end of the ninth year?

13. **WE13** Calculate the overall change for each of the following situations.
a. A new IT company of 2400 employees grew its workforce by 3% in the first year and 4.5% in the second year. How many employees did the company have at the end of the second year?
b. Roya had invested $250 000 in her superannuation fund. The fund grew at 6% in the first year but had negative growth of 3.5% in the second year. How much was in the superannuation fund at the end of the second year?

14. Chris purchased a new car for $42 000. After 1 year, the value of the car had decreased by 8%. In the following year, the value of the car decreased by a further 5.2%.
    a. Calculate the value of the car at the end of the two years.
    b. What is the overall percentage change of the car at the end of the two years? Give your answer correct to the nearest percent.

15. An amount of $55 000 is invested for 7 years in an account that grows by 6.5% every year. How much is the investment worth at the end of the 7 years?

16. Alex has been saving his money for a new tennis racket. He has been monitoring the website that sells top brands and notices a Yonex Ezone racket for $329.95. The next week he notices there is a 20% off sale on all rackets on this website. After talking to some of his friends he decides to wait until the EOFY sale in case it is marked down further. To his delight, another 15% is taken off the sale price and he then purchases the racket. How much did Alex pay for the racket?

17. Priscilla takes her savings of $17 000 and invests it in a fixed term deposit that pays 5% each year on the money that is in the account. At the end of the fixed term, her savings are worth $20 663.60. Using a trial and error method, determine the length in years of the fixed term.

18. Heidi decided to purchase a bottle of perfume for her mother on Mother's Day. She was able to get her mother's favourite perfume at 20% off the original price of $115.50. What price did Heidi pay for the perfume?

19. Store A had to increase their prices by 10% to cover expenses, whereas store B was having a 15% off sale over the entire store. Sarah wanted to purchase a Blu-ray player and assumed it would be better to get it from store B since it was having a sale. The original price of the Blu-ray player at store A was $185, whereas at store B it was originally priced at $240.

    a. Was Sarah's assumption correct? Prove this mathematically.
    b. What is the difference in the prices of the Blu-ray player?

20. A classmate was completing a discount problem where she needed to calculate a 25% discount on $79. She misread the question and calculated a 20% discount to get $63.20. She then realised her mistake and took a further 5% from $63.20. Is this the same as taking 25% off $79? Use calculations to support your answer.

4.5 Simple interest

4.5.1 Principal, interest and value

When you put money in a financial institution such as a bank or credit union, the amount of money you start with is called the **principal**.

People who place money in a bank or financial institution (**investors**) receive a payment called **interest** from the financial institution in return for leaving their money in the financial institution.

The amount of interest is determined by the **interest rate**.

An interest rate is quoted as a percentage for a given time period, usually a year. For example, a bank might offer 5.8% per year interest on its savings accounts. This is also written as 5.8% per annum, or 5.8% p.a.

4.5.2 Simple interest calculations

Simple interest is the interest paid on the principal amount of an investment.

The amount of interest paid each time period is based on the principal, so the amount of interest is constant. For example, $500 placed in an account that earns 10% simple interest per year earns $50 each year, as shown in the following table.

Time period (years)	Amount of money at the start of the year	Amount of interest after one year
1	$500	$50
2	$550	$50
3	$600	$50
4	$650	$50

The total interest earned is $4 \times 50 = \$200$, so the value of the investment after 4 years is $500 + 200 = \$700$. Simple interest is calculated using the formula below.

$$I = \frac{P \times R \times T}{100}$$

where I = **interest**
P = **principal**
R = **interest rate (as a percentage)**
T = **the number of time periods for which the money is invested.**

If R is given as a percentage per year, then the time, T, must be given in years. If R is given as a percentage per month, then T must be given in months.

The total amount of money is known as the value of the investment.

Value = principal (P) + interest (I)

WORKED EXAMPLE 14

A real estate developer offers investors a chance to invest in their latest development. If $10 000 is invested, the developer will pay 11.5% simple interest per year for 5 years . Calculate the value of the investment.

THINK	WRITE
1. Identify the known quantities.	$P = 10\,000$ $R = 11.5\%$ $T = 5$
2. Calculate the amount of interest (I) using the formula $I = \dfrac{P \times R \times T}{100}$.	$\text{Interest} = \dfrac{10\,000 \times 11.5 \times 5}{100}$ $= \$5750$
3. The value of the investment after 5 years is the sum of the interest ($5750) and the principal ($10 000).	$\text{Value} = 10\,000 + 5750$ $= 15\,750$
4. Answer the question.	The value of the investment after 5 years is $15 750.

In some cases, the time period for which the money is invested is not a multiple of the time quantity that the interest rate is quoted as; for example, if the interest rate is given as a per year rate and the money is invested for 3 months. In these instances it is necessary to calculate the equivalent interest rate for the time units for which the money is invested.

WORKED EXAMPLE 15

A bank offers interest on its savings account at 4% per year. If a Year 11 boy opens an account with $600 and leaves his money there for 4 months, how much interest does he earn?

THINK	WRITE
1. The interest rate period is per year and the amount of time the money is invested for is in months. Calculate an equivalent monthly interest rate. Since there are 12 months in a year, divide the annual interest rate by 12 to get a monthly interest rate.	$\dfrac{4}{12} = \dfrac{1}{3}$ 4% per year is $\dfrac{1}{3}$% per month.
2. Identify the known quantities. Note that we are now dealing in months, rather than years.	$P = 600$ $R = \dfrac{1}{3}$% per month $T = 4$ months
3. Calculate the amount of interest (I) using the formula $I = \dfrac{P \times R \times T}{100}$.	$\text{Interest} = \dfrac{600 \times \frac{1}{3} \times 4}{100}$ $= 8$
4. Answer the question.	The total interest earned for 4 months is $8.

4.5.3 Calculating the interest rate

If the interest (I), principal (P) and time period (T) are known, it is possible to calculate the interest rate using the simple interest formula.

WORKED EXAMPLE 16

A Year 11 girl is paid $65.40 in interest for an original investment of $800 for 2 years . What is the annual interest rate?

THINK	WRITE
1. Identify the known quantities. • interest (I) • principal (P) • time period (T)	$I = \$65.40$ $P = \$800$ $T = 2$ years
2. Substitute the values into the formula $I = \dfrac{P \times R \times T}{100}$.	$65.40 = \dfrac{800 \times R \times 2}{100}$
3. Solve the equation for R to find the annual interest rate by: • multiplying both sides by 100 • dividing both sides by 1600.	$65.40 = \dfrac{1600 \times R}{100}$ $65.40 \times 100 = 1600 \times R$ $\dfrac{6540}{1600} = R$ $R = 4.09$
4. Answer the question.	The annual interest rate is 4.09%.

Similarly, the formula could be used to find P if I, R and T are known:

$$I = \frac{P \times R \times T}{100}$$

$$P = \frac{100I}{R \times T}$$

and it could be used to find T if I, P and R are known:

$$I = \frac{P \times R \times T}{100}$$

$$T = \frac{100I}{P \times R}$$

on Resources

Interactivity: Simple interest (int-6074)

Exercise 4.5 Simple interest

1. In your own words, explain the difference in meaning in the following pairs of terms:
   a. principal and value of the investment
   b. amount of interest and interest rate.

2. **WE14** A film producer offers investors the chance to invest in his latest movie. If $20 000 is invested, the producer will pay 22.3% simple interest per year for 2 years. Determine the value of the investment at the end of the 2 years.

3. **MC** If an investment of $400 pays 8% simple interest per year, then the value of the investment at the end of 3 years is:
   A. $32. **B.** $96. **C.** $432. **D.** $496.

4. Calculate the simple interest paid on the following investments.
   a. $500 at 6.7% per year for 2 years
   b. $500 at 6.7% per year for 4 years
   c. $1000 at 6.7% per year for 4 years

5. Calculate the amount of interest paid on a $1000 investment at 5% for 5 years.

6. **WE15** A bank offers interest on its savings account of 6% p.a. If a Year 11 girl opens an account with $750 and leaves her money there for 5 months, how much interest does she earn?

7. **MC** If the annual interest rate is 8%, then the monthly interest rate is closest to:
   A. 0.8%. **B.** 0.77%.
   C. 0.67%. **D.** 0.6%.

8. Calculate the interest paid on the following investments.
   a. $500 invested at 8% per annum for 1 month
   b. $500 invested at 8% per annum for 3 months
   c. $500 invested at 8% per annum for 6 months

9. Calculate the value of the following investments.
   a. $1000 invested at 10% p.a. for 10 years
   b. $1000 invested at 12% p.a. for 10 months
   c. $1000 invested at 6% p.a. for 3 years

10. A bank offers investors an annual interest rate of 9% if they buy a term deposit. If a customer has $5600 and leaves the money in the term deposit for 2.5 years, what is the value of the investment at the end of the 2.5?

11. **WE16** A Year 11 girl is paid $79.50 in interest for an original investment of $500 for 3 years. What is the annual interest rate?

12. Bank A offers an interest rate of 7.8% on investments, while Bank B offers an interest rate of 7.4% in the first year and 7.9% in subsequent years. If a customer has $20 000 to invest for 3 years, which is the better investment?

13. **MC** If the total interest earned on a $6000 investment is $600 after 4 years, then the annual interest rate is:
    A. 10%. **B.** 7.5%. **C.** 5%. **D.** 2.5%.

14. Determine the annual interest rate on the following investments.
   a. Interest = $750, principal = $6000, time period = 4 years
   b. Interest = $924, principal = $5500, time period = 3 years
   c. Interest = $322, principal = $7000, time period = 3 months

15. An online bank has a special offer on term deposits as shown below.

Term deposit

Special offers:

4.50 % p.a.* for 5 months

4.35 % p.a.* for 6 months

4.50 % p.a.* for 24 months

*Interest is calculated daily and paid at maturity.

The fine print at the bottom of the web page contains the following information:

Rates shown above are the nominal interest rates for a term deposit of $10 000. Interest is payable at maturity and if you choose a 24-month deposit you can request interest to be transferred to you every month, once a quarter, at six months or at maturity.

   a. For each of the offers, use technology to determine the total amount of the term deposit at the end of the term.
   b. Calculate the final amount in a 24-month term deposit that was opened with $10 000 if the interest is calculated on the new total when the interest is added to the account:
      i. annually
      ii. at the end of every month
      iii. at the end of every quarter
      iv. at the end of each six-month period.
   c. Which of the interest payment options from part b would you choose? Why?

16. Your parents decide to borrow money to improve their boat but cannot agree which loan is the better value. They would like to borrow $2550. Your mother goes to Bank A and finds that they will lend the money at $11\frac{1}{3}$ % simple interest per year for 3 years. Your father finds that Bank B will lend the $2550 at 1% per month simple interest.

   a. Which bank offers the best rate over the three years?
   b. Provide reasons for your answer to part a.

The statement below relates to questions 17 and 18.
   A loan is an investment in reverse; you *borrow* money from a bank and are *charged* interest. The value of a loan becomes its total cost.

17. A worker wishes to borrow $10 000 from a bank, which charges 11.5% interest per year. If the loan is over 2 years:
   a. calculate the total interest paid.
   b. calculate the total cost of the loan.

18. Most loans require a monthly payment. The monthly payment of a simple interest loan is calculated by dividing the total cost of the loan by the number of payments made during the term of the loan.
      Determine the monthly payment for the loan in question 17 above.

4.6 Review: exam practice

4.6.1 Percentages: summary

Calculate percentages

The term 'per cent' means per hundred or out of 100.
- A percentage can be converted to a fraction or decimal by dividing the percentage by 100.
- To convert a decimal or fraction to a percentage, multiply by 100.
- When multiplying a decimal by 100, move the decimal point two places to the right.
- To find the percentage of an amount, convert the percentage to a decimal and multiply by the amount.
- To find 10% of an amount, divide by 10 or move the decimal point one place to the left.
- To find 5%, halve 10%.
- To find 20%, double 10%.
- To find 15%, add 10% and 5%.
- To find 25%, double 10% and add 5%.

Express as a percentage

To express an amount as a percentage of another, write the amount as a fraction and then convert the fraction to a percentage by multiplying by 100.
- When expressing one amount as a percentage of another, make sure that both amounts have the same units.
- To find a percentage discount, find the fraction $\dfrac{\text{Amount saved}}{\text{Original price}}$ and multiply by 100.

Percentage decrease and increase

The percentage discounted price can be calculated two ways:

Method 1:
- Calculate the discount by multiplying the original price by the percentage expressed as a decimal.
- Subtract the discount from the original price.

Method 2:
- Discounted amount $= \left(1 - \dfrac{\text{percentage discount}}{100}\right) \times \text{Original price}$

    We calculate the percentage increase in price in a similar way to the percentage decrease, however instead of subtracting the discounted value we add the extra percentage value.

    To calculate the percentage discount, write the fraction as the discounted amount divided by the original amount, then multiply by 100.

Simple interest

Simple interest is the interest paid on the principal amount of an investment.
    Simple interest is calculated using the following formula.

$$I = \frac{P \times R \times T}{100}$$

where $I =$ interest
  $P =$ principal
  $R =$ interest rate (as a percentage)
  $T =$ the number of time periods for which the money is invested.

Exercise 4.6 Review: exam practice

Simple familiar

1. **MC** The percentage 45% expressed as a simplified fraction is:

    A. $\frac{1}{4}$.
    B. $\frac{9}{20}$.
    C. $\frac{3}{20}$.
    D. $\frac{9}{25}$.

2. **MC** 64% of 280 is closest to:

    A. 210.
    B. 79.
    C. 179.
    D. 101.

3. **MC** 270 m of 1.5 km as a percentage is closest to:

    A. 18%.
    B. 24%.
    C. 14%.
    D. 27%.

4. **MC** The cost of a $16 000 new car increased by 7.5%. The amount by which the car increased in price is closest to:

    A. $750.
    B. $950.
    C. $1200.
    D. $1000.

5. **MC** The annual salary of a sales assistant increased from $45 000 to $48 600. The percentage salary increase is:

    A. 4%.
    B. 5%.
    C. 6%.
    D. 8%.

6. **MC** Air contains 21% oxygen, 0.9% argon, 0.1% trace gases and the remainder is nitrogen. The percentage of nitrogen in the air is:

    A. 100%.
    B. 50%.
    C. 35%.
    D. 78%.

7. **MC** A tree was planted when it was 2.3 m tall. In the first year after planting it grew by 30% and in the second year by 12%. At the end of the second year after planting, the height of the tree is closest to:

    A. 3.35 m.
    B. 2.66 m.
    C. 4.56 m.
    D. 3.95 m.

8. **MC** The sale price of the car is:

    A. $25 000.
    B. $27 000.
    C. $28 000.
    D. $28 500.

9. **MC** If a house that cost $200 000 to build is sold for $10 000 less, then the percentage discount is:

    A. 95%.
    B. 5%.
    C. 90%.
    D. 10%.

10. **MC** If an investment of $400 pays 8% simple interest per year, then the value of the investment at the end of 3 years is:

    A. $32.
    B. $96.
    C. $432.
    D. $496.

11. **MC** If the annual interest rate is 8%, then the monthly interest rate is:

    A. 0.8%.
    B. 0.77%.
    C. 0.67%.
    D. 0.6%.

12. **MC** If the total interest earned on a $6000 investment is $600 after 4 years, then the annual interest rate is:

    A. 10%.
    B. 7.5%.
    C. 5%.
    D. 2.5%.

Complex familiar

13. The winner of a football tipping competition wins 65% of the prize pool, second place wins 20% and third place receives what is left over. The prize pool is $860. Calculate the amount:

    a. first prize wins.
    b. second prize wins.
    c. third prize wins.

14. Laptops were on a weekend sale, where they were reduced by 30%. For a laptop valued at $1500, calculate:

    a. the amount of money saved.

    b. the sale price of the laptop.

15. Express the following as percentages.

    a. 756 m of 2.7 km

    b. 45 g of 0.85 kg

    c. 15 c of $2.90

    d. 45 sec of 1.5 min

16. A person invests $100 each month, earning simple interest of 1% per month for 4 months.

    a. Calculate the interest on the first investment, which earns interest for 4 months.

    b. Calculate the interest on the second, third and fourth investments, which earn interest for 3 months, 2 months and 1 month respectively.

    c. Calculate the total value of the investment at the end of the 4 months.

Complex unfamiliar

17. An expensive lamp has a regular price of $200. At that price the shop expects to sell 4 lamps per month.

    - If offered at a discount of 10%, the shop will sell 8 lamps per month.
    - If offered at a discount of 20%, the shop will sell 16 lamps per month.
    - If offered at a discount of 30%, the shop will sell 20 lamps per month.
    - These lamps cost the shop $100 each.

    a. Determine the total value of the sales in each of the 3 discount cases above.

    b. Which discount results in the highest total sales?

18. Anastasia is holding a birthday party and she has invited 25 friends and 15 family members.

    a. Express the number of friends as a percentage of the total number of people invited.

    b. Express the number of family member as a percentage of the total number of people invited.

    c. For catering purposes, 25% of people are vegetarians and 75% of people like desserts. How many people are vegetarians and how many like desserts?

    d. On the day, not everybody turns up. Anastasia couldn't remember exactly how many turned up, but remembered that they had to reduce the amount of food by 20%. How many people turned up?

19. A pathologist was paid $104 000 last year.

    a. What is the pathologist's fortnightly salary (assume 26 fortnights per year)?

    b. What is the pathologist's annual superannuation payment (assume 11% per year)?

    c. The pathologist is offered a pay increase of 5.5% for next year. What is her new fortnightly salary?

    d. The pathologist can save $20 000 each year towards a new house. What percentage of her new annual salary is this?

    e. Assuming an 80-hour work fortnight, what is the pathologist's new hourly wage?

    f. The pathologist's union also gets the employer to pay 50% extra for any overtime (over 80 hours per fortnight). If the pathologist worked 92 hours last fortnight, what was her pay for that fortnight?

    g. The pathologist pays income tax according to the following table.

Income	Tax rate	Tax payable
0 and $18 200	0%	Nil
$18 201 and $37 000	19%	19 cents for each $1 over $18 200
$37 001 and $90 000	32.5%	$3572 plus 32.5 cents for each dollar over $37 000
$90 001 and $180 000	37%	$20 797 plus 37 cents for each dollar over $90 000
$180 001 and above	45%	$54 097 plus 45 cents for each dollar over $180 000

Calculate the pathologist's annual tax based on her new salary.

20. **a.** If you increase $100 by 30% and then decrease the new amount by 30%, will you end up with more than $100, less than $100 or exactly $100?

   b. Explain your answer to part **a**, using mathematics to support your answer.

   c. If you decrease $100 by 30% and then increase the new amount by 30%, will you end up with more than $100, less than $100 or exactly $100?

   d. Explain your answer to part **c**, using mathematics to support your answer.

   e. Find the percentage change needed in parts **a** and **c** above to get back to the original $100 from the new amount.

   f. Is the percentage change in part **a** greater than or less than the percentage change in part **c**? Why?

Answers

Chapter 4 Percentages

Exercise 4.2 Calculate percentages

1. a. $\dfrac{12}{25}$ b. $\dfrac{67}{250}$

   c. $\dfrac{31}{250}$ d. $\dfrac{1}{125}$

2. a. $\dfrac{23}{25}$ b. $\dfrac{593}{800}$ c. $\dfrac{2}{3}$

3. a. $\dfrac{22}{25}$ b. $\dfrac{1}{4}$ c. $\dfrac{23}{2500}$

   d. $\dfrac{71}{200}$ e. $\dfrac{38}{125}$ f. $\dfrac{291}{400}$

4. a. 0.73   b. 0.943
5. a. 0.024 96   b. 0.0062
6. a. 0.43   b. 0.39   c. 0.80
   d. 0.4725   e. 0.2405   f. 0.0083
7. a. 21%   b. 80%
8. a. 65.2%   b. 32%
9. a. 55%   b. 83%   c. 50%
   d. 87.5%   e. 220%   f. 165%
10. a. 72   b. 80.64 kg
11. a. 157.5   b. $28.81
12. a. 88   b. 327 kg   c. 31 m
   d. 217   e. 753 g   f. 956
13. $59.50
14. $24
15. a. $5.60   b. 12.5 kg
16. a. 55.5   b. 137
17. 56
18. 644 000
19. $495
20. 6 525 litres
21. a. 14 cents   b. $3.24
   c. 15 cents   d. $3.39
22. $131.25
23. a. $1 375   b. $159.50   c. $407
24. a. i. $52.00   ii. $19.60   iii. $71.92
   b. $35.88
25. Save $87.75
26. a. John spends: $420   b. $84
     Jane spends: $336
     John spends more

Exercise 4.3 Express as a percentage

1. a. 64%   b. 65%   c. 30%
2. a. 74%   b. 53%   c. 65%

3. a. 44%   b. 38%   c. 26%
   d. 60%   e. 34%   f. 38%
4. a. 34.1%   b. 74.1%   c. 70.4%
   d. 72.2%   e. 17.0%   f. 72.3%
5. 17.7%
6. 30%
7. a. 34.0%   b. 6.2%   c. 3.4%
   d. 0.1%   e. 27.3%   f. 15.9%
   g. 5.8%   h. 20.0%
8. 46.15%
9. 25%
10. 87.27%
11. 7.8%
12. 88.9%
13. 29.0%
14. a. $152   b. 18.7%
15. Top 1: 40% discount
     Shoes 1: 40% discount
     Top 2: 40% discount
     Shoes 2: 35.5% discount

Exercise 4.4 Percentage decrease and increase

1. a. $446.25   b. $148.75
2. a. $111.96   b. $27.99
3. a. $522.50   b. $495
   c. $467.50   d. $440
   e. $385   f. $275
4. a. 3025 L   b. 51 975 L
5. a. $7   b. $77
6. $2687.50
7. a. $207.90   b. $1815
   c. $167.75   d. $184.64
8. a. i. $465   ii. $240   iii. $652.50
     iv. $1378.50   v. $3250   vi. $3937.50
   b. i. $15 965   ii. $8240   iii. $22 402.50
     iv. $47 328.50   v. $68 250   vi. $82 687.50
9. a. $35   b. $89.70
   c. $27   d. $148.50
10. No you cannot get clothes for free.
     You get 50% off the already reduced price. This means the price is first halved, then another 50% off means it is halved again, so you are paying one quarter of the original price. For example, if something originally costs $160:
     $0.5 \times \$160 = \$80$
     $0.5 \times \$80 = \$40$
11. a. Camera: $86.67
     iPhone: $95
     Speakers: $90
     Bike: $90
     The iPhone is reduced by $95 which is the largest.
   b. Both the speakers and bike are reduced by $90.
   c. $8.33
   d. 30%

12. a. See table bottom of the page.*
    b. $23 039.83
13. a. 2583   b. $255 725
14. a. $36 630.72   b. 87%
15. $85 469.26
16. $224.37
17. 4 years.
18. $92.40
19. a. Store A: $203.50
    Store B: $204
    No, Sarah was not correct, store A was cheaper.
    b. 50 cents
20. Method 1: $60.04
    Method 2: $59.25
    No, it is not the same since taking the 5% discount after the 20% discount is on the reduced value of $63.20, thus this method gives a smaller discount.

Exercise 4.5 Simple interest

1. a. Principal is the amount of money you start with and value of the investment is the combination of the principal and interest.
   b. Interest is a payment from the financial institution and interest rate is the percentage that determines the amount of interest.
2. $28 920
3. D
4. a. $67   b. $134   c. $268
5. $250
6. $18.75
7. C
8. a. $3.33   b. $10   c. $20
9. a. $2000   b. $1100   c. $1180
10. $6860
11. 5.3%
12. Bank A
13. D
14. a. 3.125%   b. 5.6%   c. 18.4%
15. a. $10 187.50
    $10 217.50
    $10 900
    b. i. $10 920.25   ii. $10 939.90
       iii. $10 936.25   iv. $10 930.83

c. Receiving interest payments at the end of each month is the best option. This is because the interest received at the end of each month is reinvested to earn 'interest on interest' more quickly than the other option.
16. a. Bank A
    b. Total interest for Bank A: $867
       Total interest for Bank B: $918
17. a. $2300   b. $12 300
18. $512.50

Exercise 4.6 Review: exam practice

1. B   2. C   3. A   4. C   5. D   6. D
7. A   8. D   9. B   10. D   11. C   12. D
13. a. $559   b. $172   c. $129
14. a. $450   b. $1050
15. a. 28%   b. 5.29%   c. 5.17%
    d. 50%
16. a. $4   b. $3, $2, $1   c. $410
17. a. $1440, $2560, $2800   b. 30%
18. a. 62.5%   b. 37.5%
    c. Vegetarians = 10   d. 32
       Dessert = 30
19. a. $4000   b. $11 440   c. $4220
    d. 18.23%   e. $52.75   f. $5169.50
    g. $28 093.40
20. a. Less than $100
    b. 30% of $100 is $30
       $100 + $30 = $130
       30% of $130 is $39
       $130 − $39 = $91
       This shows the final result is less than $100.
    c. Less than $100
    d. 30% of $100 is $30
       $100 − $30 = $70
       30% of $70 is $21
       $70 + $21 = $91
       This shows the final result is less than $100.
    e. 23%, 43%
    f. The percentage change in part a is less than the percentage change in part c because the new amount in part a is greater than the original so needs to decrease. The new amount in part c is lower than the original so needs to be multiplied by itself (100%) plus the extra percentage (43%) to increase it back to the original amount.

*12. a.

End year	% depreciation	Depreciation ($)	Value ($)
First	10%	$3 500	$31 500
Second	7.5%	$2362.5	$31 500 − 2362.5 = $29 137.50
Third	6%	0.06 × 29 137.5 = $1748.25	$27 389.25
Fourth	5%	$1369.46	$26 019.79
Fifth	4%	$1040.79	$24 979

CHAPTER 5
Data classification, presentation and interpretation

5.1 Overview

LEARNING SEQUENCE

5.1 Overview
5.2 Classifying data
5.3 Categorical data
5.4 Numerical data
5.5 Measures of central tendency and the outlier effect
5.6 Comparing and interpreting data
5.7 Review: exam practice

CONTENT

In this chapter, students will learn to:
- identify examples of categorical data
- identify examples of numerical data
- display categorical data in tables and column graphs
- display numerical data as frequency distribution tables, dot plots, stem-and-leaf plots and histograms
- recognise and identify outliers from a data set
- compare the suitability of different methods of data presentation in real-world contexts [complex].

Fully worked solutions for this chapter are available in the Resources section of your eBookPLUS at www.jacplus.com.au

5.2 Classifying data

5.2.1 Types of data

In today's modern society statistics have become an integral part of our everyday life. Statistical studies are used in all subject areas such as business, education, meteorology and many more.

Statistics is the science of collecting, organising and presenting, analysing and interpreting data. The information collected is called data.

There are two main types of data: **numerical data** and **categorical data**.

5.2.2 Numerical data

Numerical data, also called **quantitative** data, is information collected in the form of numerical values such as number of children in a family, time spent on homework, and shoe sizes.

Numerical data can be classified into two types: discrete and continuous.

Discrete data are observations that can be *counted*. For example, a group of year 11 students was asked to indicate the number of siblings they have. The data collected would be in the form of whole counting numbers: 0, 1, 2, ...

- **Continuous data** are observations that can be *measured*. For example, a mother records her baby's weight daily over a period of two months. Height, length, temperature, time, weight are considered to be continuous data as they represent measurements.

WORKED EXAMPLE 1

Classify the following data as either numerical discrete or numerical continuous.

a. **A group of people was asked the question: 'How many cars does your family have?'.**

b. **Students are asked to complete the statement 'My height is . . .'**

THINK	WRITE/DRAW
a. 1. Ask the question: Is the data countable or a measurement?	The data can be counted.
2. State the type of data.	Numerical discrete
b. 1. Ask the question: Is the data countable or a measurement?	The data is a measurement.
2. State the type of data.	Numerical continuous

5.2.3 Categorical data

Categorical data, also called **qualitative** data, is information collected in the form of categories, such as eye colour (brown, hazel, blue, green), gender (female, male), quality of a service (excellent, very good, good, poor, very poor) etc.

Categorical data can be either nominal or ordinal.

- **Nominal data**, or distinct categories, cannot be ranked. For example, a group of year 11 students were asked to indicate their preference of using a Mac or a PC notebook. As this data cannot be ranked, they will be placed into two categories: Macs and PCs.

- **Ordinal data**, or ordered categories, are recognised by their property of ranking or being placed in order. For example, a group of year 11 students were asked to indicate their level of understanding of the topic of statistics on a rating scale of 1 to 5 represented by: 5 – excellent, 4 – very good, 3 – good, 2 – poor, 1 – very poor.

WORKED EXAMPLE 2

Classify the following data as either categorical nominal or categorical ordinal.

a. A group of people was asked the question 'What is your opinion about the service in your local supermarket?'.

b. Students are asked to complete the statement 'My favourite colour is . . .'

THINK	WRITE/DRAW
a. 1. Ask the question: Can the data be ordered?	Yes
2. State the type of data.	Categorical ordinal
b. 1. Ask the question: Can the data be ordered?	No
2. State the type of data.	Categorical nominal

5.2.4 Categorical data with numerical values

There are instances when categorical data is expressed by numerical values. 'Phone numbers ending in 5' and 'postcodes' are examples of categorical data represented by numerical values.

Classify the following data as either categorical or numerical data.

a. A set of data was collected on 'bank account numbers'.

b. A group of people was asked 'How many minutes do you exercise per day?'

THINK	WRITE/DRAW
a. 1. Ask the question: Is the data countable or a measurement?	Neither
2. State the type of data.	Categorical
b. 1. Ask the question: Is the data countable or a measurement?	Yes, measurable
2. State the type of data.	Numerical

5.2.5 Categorical or numerical data

The diagram shown is a representation of the main types of data.

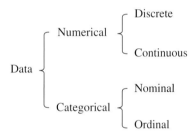

Classify the following data as either categorical or numerical data.

a. A group of people are asked the question: 'What is the area of your backyard?'

b. Students are asked to complete the statement 'The subject I like the most is . . .'

THINK	WRITE/DRAW
a. 1. Ask the question: Is the data countable or a measurement?	Yes, measurable
2. State the type of data.	Numerical
b. 1. Ask the question: Is the data countable or a measurement?	Neither
2. State the type of data.	Categorical

Exercise 5.2 Classifying data

1. **WE1** Classify the following data as either numerical discrete or numerical continuous:
   a. The amount of time spent on homework per night.
   b. The marks in a test.

2. Explain in your own words why 'the number of fruit sold' is a numerical discrete data while 'the quantity of fruit sold' is a numerical continuous data.

3. Classify the following data as either discrete or continuous.

   a. Data is collected on the water levels in a dam over a period of one month.
   b. An insurance company is collecting data on the number of work injuries in ten different businesses.
   c. 350 students were selected by a phone company to answer the question "How many pictures do you take per week using your mobile phone?"
   d. Customers in a shopping centre were surveyed on the number of times they visited the shopping centre in the last 12 months.
   e. 126 people participated in a 5 km charity marathon. Their individual times were recorded.
   f. A group of students were asked to measure the perimeter of their bedrooms. This data was collected and used for a new project.

4. **WE2** Classify the following data as either categorical nominal or categorical ordinal:
   a. The brands of cars sold by a car yard over a period of time.
   b. The opinion about the quality of a website: poor, average or excellent.

5. Explain in your own words why 'favourite movie' is a categorical nominal data while 'the rating for a movie' is a categorical ordinal data.

6. Classify the following data as either ordinal or nominal.
   a. The members of a football club were asked to rate the quality of the training grounds as 'poor', 'average' or 'very good'.
   b. A group of students are asked to state the brand of their mobile phone.
   c. A fruit and vegetable shop is conducting a survey asking the customers 'Do you find our products fresh?' Possible answers: 'Never', 'Sometimes', 'Often', 'Always'.
   d. A chocolate company conducts a survey with the question 'Do you like sweets?'. The possible answers are 'Yes', 'no'.

7. **WE3** Classify the following data as either numerical or categorical:
   a. Data was collected on 'whether the subject outcomes 1, 2, and 3 have been completed or not'.
   b. The number of learning outcomes per subject.

8. Explain in your own words why 'country phone codes' is a categorical data while 'the number of international calls' is a numerical data.

9. **WE4** Classify the following data as either numerical or categorical:
   a. The amount of time spent at the gym per week.
   b. The opinion about the quality of a website: poor, average or excellent.
10. Explain in your own words why 'the length of pencils in a pencil case' is a numerical data while 'the colour of pencils' is a categorical data.
11. **MC** A train company recorded the time trains were late over a period of one month. The data collected is:
   A. numerical discrete.
   B. categorical nominal.
   C. categorical ordinal.
   D. numerical continuous.

12. Classify the following data as either numerical or categorical.
   a. The average time spent by customers in a homeware shop was recorded.
   b. Mothers were asked 'What brand of baby food do you prefer?'.
   c. An online games website asked the players to state their gender.
   d. A teacher recorded the lengths of his students' long jump.

5.3 Categorical data

5.3.1 Displaying categorical data

Visual displays of data are important tools in communicating the information collected. Categorical data can be displayed in tables or graphs.

One-way tables

One-way tables or **frequency distribution tables** are used to organise and display data in the form of frequency counts. The **frequency** of a score means how many times the score occurs in the set of data.

Frequency distribution tables display one variable only.

For large amounts of data a **tally** column helps make accurate counts of the data.

The frequency distribution table shown displays the tally and the frequency for a given data set.

Outcome	Tally	Frequency
0					3
1	ﷻ ﷻ		11
2	ﷻ	5
3						4
Total		23

Alana is organising the year 11 formal and has to choose three types of drinks out of the five available: cola, lemonade, apple juice, orange juice and soda water. She decided to survey her classmates to find their favourite drink. The data collected is shown below.

orange juice	cola	cola	cola	apple juice	lemonade
orange juice	cola	cola	cola	apple juice	lemonade
lemonade	cola	apple juice	lemonade	orange juice	orange juice
lemonade	cola	apple juice	orange juice	cola	soda water
orange juice	cola	soda water	orange juice	cola	cola

By constructing a frequency table, find the three drinks Alana should choose for the formal.

THINK

1. Draw a frequency table with three columns and the headings 'Drink', 'Tally' and 'Frequency'.

2. Write all the possible outcomes in the first column and 'Total' in the last row.

3. Fill in the 'Tally' column.

WRITE/DRAW

Drink	Tally	Frequency

Drink	Tally	Frequency
Apple juice		
Cola		
Lemonade		
Orange juice		
Soda water		
Total		

Drink	Tally	Frequency
Apple juice	|||	
Cola	卌 ||||	
Lemonade	||||	
Orange juice	卌|	
Soda Water	||	
Total		

4. Fill in the 'Frequency' column. Ensure the table has a title also.

Favourite drinks for a Year 11 class

Drink	Tally	Frequency
Apple juice	|||	3
Cola	卌 ||||	9
Lemonade	||||	4
Orange juice	卌|	6
Soda Water	||	2
Total		24

5. Answer the question.

Alana should choose cola, orange juice and lemonade for the Year 11 formal.

Two-way tables

Two-way tables are used to examine the relationship between two related categorical variables. Because the entries in these tables are frequency counts they are also called **two-way frequency tables**.

The table shown is a two-way table that relates two variables. It displays the relationship between gender and handedness.

One variable is **independent** (gender) and the other variable is **dependent** (handedness). The independent variable is the variable that can be manipulated to produce changes in the dependent variable. The dependent variable changes as the independent variable changes. Its outcome is dependent on the input of the independent variable.

Handedness is the dependent variable

Gender is the independent variable

Handedness	Gender		Total
	Boys	Girls	
Right-handed	9	6	15
Left-handed	3	2	5
Total	12	8	20

Total number of right-handed students

Total number of left-handed students

Total number of boys

Total number of girls

Total number students surveyed

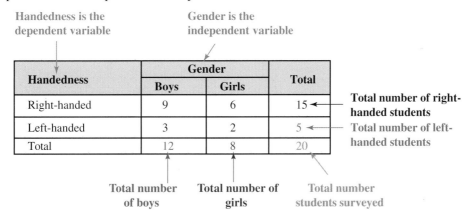

WORKED EXAMPLE 6

A sample of 40 people was surveyed about their income level. The sample consists of 15 men and 25 women. 7 men were on a low level income while 6 women were on a high level income. Display this data in a two-way table and find the number of men on a high level income.

THINK	WRITE/DRAW
1. State the two categorical variables.	Gender: Men and women Income level: Low and high
2. Decide where the two variables are placed.	The independent variable is 'Gender' — columns The dependent variable is 'Income level' — rows

3. Construct the two-way table and label both the columns and the rows.

Income levels for men and women

Income level	Gender		Total
	Men	**Women**	
Low			
High			
Total			

4. Fill in the known data:
  • 7 men are on a low level income
  • 6 women are on a high level income
  • there are 15 men in total and 25 women in total
  • the total number of people in the sample is 40.

Income levels for men and women

Income level	Gender		Total
	Men	**Women**	
Low	7		
High		6	
Total	15	25	40

5. Find the unknown data.

Men on high level income: $15 - 7 = 8$
Women on low level income: $25 - 6 = 19$
Total number of people on a low level income: $7 + 19 = 26$
Total number of people on a high level income: $8 + 6 = 14$

6. Fill in the table.

Income levels for men and women

Income level	Gender		Total
	Men	**Women**	
Low	7	19	26
High	8	6	14
Total	15	25	40

7. Answer the question.

The total number of men on a high level income is 8.

5.3.2 Two-way relative frequency tables

Consider the 'Handedness' table shown. As the number of girls and boys is different, it is difficult to compare the data.

Handedness for boys and girls			
Handedness	Gender		Total
	Boys	Girls	
Right-handed	9	6	15
Left-handed	3	2	5
Total	12	8	20

Nine boys and six girls are right-handed; however, there are more boys surveyed than girls. In order to accurately compare the data, we need to convert these values in a more meaningful way as proportions or percentages.

The frequency given as a proportion is called **relative frequency**.

$$\text{Relative frequency} = \frac{\text{score}}{\text{total number of scores}}$$

The relative frequency table for the data for 'Handedness' is:

Handedness for boys and girls			
Handedness	Gender		Total
	Boys	Girls	
Right-handed	$\frac{9}{20}$	$\frac{6}{20}$	$\frac{15}{20}$
Left-handed	$\frac{3}{20}$	$\frac{2}{20}$	$\frac{5}{20}$
Total	$\frac{12}{20}$	$\frac{8}{20}$	$\frac{20}{20}$

In simplified form using decimal numbers, the relative frequency table for 'Handedness' becomes:

Handedness for boys and girls			
Handedness	Gender		Total
	Boys	Girls	
Right-handed	0.45	0.30	0.75
Left-handed	0.15	0.10	0.25
Total	0.60	0.40	1.00

In a relative frequency table the total value is always 1.

The frequency written as a percentage is called **percentage relative frequency** or **% relative frequency**.

$$\text{Percentage relative frequency} = \frac{\text{score}}{\text{total number of scores}} \times 100\%$$

or

$$\text{Percentage relative frequency} = \text{relative frequency} \times 100\%$$

WORKED EXAMPLE 7

Consider the 'Handedness' data discussed previously with the relative frequencies shown.

Handedness for boys and girls			
Handedness	**Gender**		**Total**
	Boys	**Girls**	
Right-handed	0.45	0.30	0.75
Left-handed	0.15	0.10	0.25
Total	0.60	0.40	1.00

Calculate the percentage relative frequency for all entries in the table out of the total number of students. Discuss your findings.

THINK

1. Write the formula for % relative frequency.

2. Substitute all scores into the percentage relative frequency formula

3. Simplify

WRITE/DRAW

$\text{Percentage relative frequency} = \text{relative frequency} \times 100\%$

Handedness for boys and girls			
Handedness	**Gender**		**Total**
	Boys	**Girls**	
Right-handed	0.45×100	0.30×100	0.75×100
Left-handed	0.15×100	0.10×100	0.25×100
Total	0.60×100	0.40×100	1.00×100

Handedness for boys and girls			
Handedness	**Gender**		**Total%**
	Boys %	**Girls %**	
Right-handed	45	30	75
Left-handed	15	10	25
Total %	60	40	100

4. Discuss your findings.

Of the total number of students, 75% are right-handed and 25% are left-handed. 60% of students are boys and 40% of students are girls. 45% of students are right-handed boys and 30% of students are right-handed girls. 15% of boys are left-handed boys and 10% of students are left-handed girls.

5.3.3 Grouped column graphs

Grouped column graphs are used to display the data for two or more categories. They are visual tools that allow for easy comparison between various categories of the data sets. The frequency is measured by the height of the column. Both axes have to be clearly labelled including scales and units if used. The title should explicitly state what the grouped column graph represents.

Consider the data recorded in the two-way distribution table shown.

Income levels for men and women			
Income level	**Gender**		**Total**
	Men	**Women**	
Low	7	19	26
High	8	6	14
Total	15	25	40

The grouped column graph for this data is shown. Notice that all rectangles have the same width. The colour or pattern of the rectangles differentiates between the two categories, men or women.

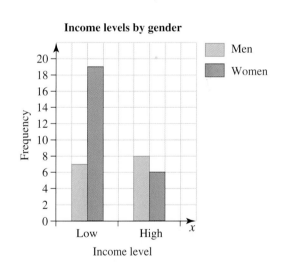

Making comparisons

It is helpful to change the table to a percentage frequency table to make comparisons between the income levels of men and women. We first calculate the percentage relative frequency for all the entries in the two-way table out of the total number of people.

Income levels for men and women			
Income levels	**Gender**		**Total %**
	Men %	**Women %**	
Low	$\dfrac{7}{40} \times 100 = 17.5$	$\dfrac{19}{40} \times 100 = 47.5$	$\dfrac{26}{40} \times 100 = 65.0$
High	$\dfrac{8}{40} \times 100 = 20.0$	$\dfrac{6}{40} \times 100 = 15.0$	$\dfrac{14}{40} \times 100 = 35.0$
Total %	$\dfrac{15}{40} \times 100 = 37.5$	$\dfrac{25}{40} \times 100 = 62.5$	$\dfrac{40}{40} \times 100 = 100$

From these percentages we can conclude that:
- 17.5% of the people surveyed are men on a low income and 47.5% of the people are women on a low income
- 20% of the total number of people surveyed are men on a high income while 15% are women on a high income
- 37.5% of the people surveyed are men while 62.5% are women
- 65% of the people surveyed are on a low income while 35% are on a high income.

Other ways of comparing the results of this survey involve percentages calculated out of the total numbers in columns and in rows.

Row 1: Percentages of men and women on low incomes out of the total number of people on a low income.

$$\text{Men on low income} = \frac{\text{Men on low income}}{\text{Total people on low income}} \times 100\%$$

$$= \frac{7}{26} \times 100\%$$

$$= 26.9\%$$

$$\text{Women on low income} = \frac{\text{Women on low income}}{\text{Total people on low income}} \times 100\%$$

$$= \frac{19}{26} \times 100\%$$

$$= 73.1\%$$

These percentages indicate that 26.9% of the people on a low income are men while 73.1% are women.

Row 2: Percentages of men and women on high incomes out of the total number of people on a high income.

$$\text{Men on high income} = \frac{\text{Men on high income}}{\text{Total people on high income}} \times 100\%$$

$$= \frac{8}{14} \times 100\%$$

$$= 57.1\%$$

$$\text{Women on high income} = \frac{\text{Women on high income}}{\text{Total people on high income}} \times 100\%$$

$$= \frac{6}{14} \times 100\%$$

$$= 42.9\%$$

These percentages indicate that 57.1% of the people on a high income are men while 42.9% are women.

Column 1: Percentages of men on a low and a high income out of the total number of men surveyed.

$$\text{Men on low income} = \frac{\text{Men on low income}}{\text{Total number of men}} \times 100\%$$

$$= \frac{7}{15} \times 100\%$$

$$= 46.7\%$$

$$\text{Men on high income} = \frac{\text{Men on high income}}{\text{Total number of men}} \times 100\%$$

$$= \frac{8}{15} \times 100\%$$

$$= 53.3\%$$

These percentages indicate that 46.7% of the men surveyed are on a low income while 53.3% are on a high income.

Column 2: Percentages of women on a low and a high income out of the total number of women surveyed.

$$\text{Women on low income} = \frac{\text{Women on low income}}{\text{Total number of women}} \times 100\%$$

$$= \frac{19}{25} \times 100\%$$

$$= 76\%$$

$$\text{Women on high income} = \frac{\text{Women on high income}}{\text{Total number of women}} \times 100\%$$

$$= \frac{6}{25} \times 100\%$$

$$= 24\%$$

These percentages indicate that 76% of the women surveyed are on a low income while 24% of women are on a high income.

WORKED EXAMPLE 8

The data shown is part of a random sample of 25 students who participated in a questionnaire run by the Australian Bureau of Statistics through Census@School. The answers were related to how often the students use the internet for school work. Draw a grouped column graph displaying the data sets given.

Use of Internet for school Work			
Time spent	Gender		Total
	Girls	Boys	
Rarely	2	8	10
Sometimes	4	4	8
Often	6	1	7
Total	12	13	25

THINK	WRITE/DRAW

1. Draw a labelled set of axes with accurate scales. Ensure the graph has a title also.

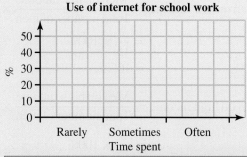

2. Calculate the % relative frequency for the data needed.

Use of internet for school work			
Time spent	Gender		Total %
	Girls %	Boys %	
Rarely	$\frac{2}{25} \times 100 = 8$	$\frac{8}{25} \times 100 = 32$	$\frac{10}{25} \times 100 = 40$
Sometimes	$\frac{4}{25} \times 100 = 16$	$\frac{4}{25} \times 100 = 16$	$\frac{8}{25} \times 100 = 32$
Often	$\frac{6}{25} \times 100 = 24$	$\frac{1}{25} \times 100 = 4$	$\frac{7}{25} \times 100 = 28$
Total	$\frac{12}{25} \times 100 = 48$	$\frac{13}{25} \times 100 = 52$	100

3. Draw the sets of corresponding rectangles. The graph can be drawn vertically.

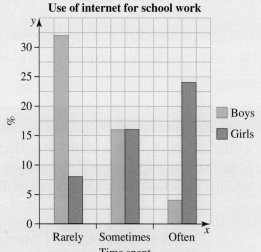

The graph can be drawn horizontally

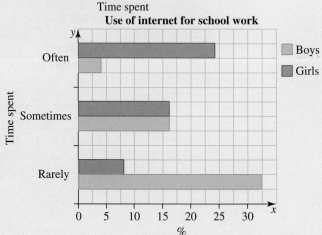

5.3.4 Misuses of grouped column graphs

Column graphs are powerful tools for displaying categorical data. However, if constructed inappropriately, they could be misleading or deceiving.

The graph shown displays three issues that could give a false impression about the data represented:

- the horizontal axis jumps from June 2004 to June 2007 while the rest of the scale represents equal intervals of one year. This gives the impression that the increase from between the first two columns is a lot bigger than the increases between the rest of the columns.
- the overlapping drawing of the woman on the columns gives the impression that the June 2008 column has a higher value than it actually is.

A more accurate representation of this data is represented in the corresponding histogram shown.

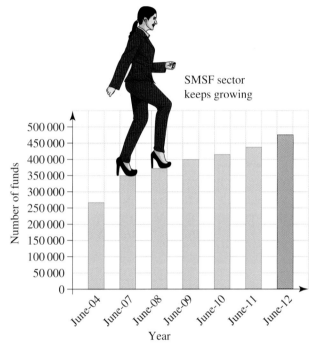

Source: The Age, Money, 27 March, 2013.

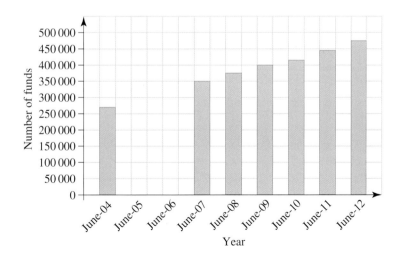

How is the graph shown misleading or deceiving?

Daily sales for Brand A and Brand B

THINK	WRITE/DRAW
1. Check the starting point of the vertical axis.	The vertical axis does not start at zero and does not have a break in it to show this. This gives the impression that the sales for Brand B are a lot higher than the ones for Brand A.
	If we compare the first two columns, it looks like the sales for Brand B are more than double the sales for Brand A. This assumption is incorrect.
	Here is the graph with the scale starting at 0.

Daily sales for Brand A and Brand B

In reality, the sales for Brand B are roughly double the sales of Brand A on day 1.

2. Check the accuracy of the scales on both axes.	The vertical axis has equally spaced scales. The horizontal axis does not have a correct scale.
	Notice that day 4 sales are missing. This makes the jump from the sales on day 3 to the sales on day 5 look a lot higher.
	Here is the graph including day 4 (taken from original data).

Daily sales for Brand A and Brand B

3. Check for axes labels and title.

The title is correct. The vertical axis does not have a label, however its scale is clearly labelled. The horizontal axis is clearly labelled. Add a label to the vertical axis.

Daily sales for Brand A and Brand B

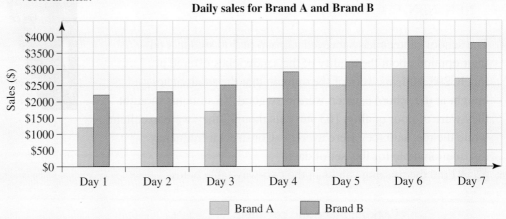

4. Check the colours or patterns of the columns.

Pink is used for one brand and blue for the other brand. Because of this choice of colour, the pink columns tend to stand out more than the blue columns. Care has to be taken when choosing the colours or patterns of the columns so that there is no extreme discrepancy.

 Resources

 Interactivity: Create a bar graph (int-6493)

Exercise 5.3 Categorical data

1. **WE5** The form of transport for a group of 30 students is given below.

car, car, bicycle, tram, tram, tram, car, walk, waik, walk,

tram, walk, car, walk, walk, car, walk, bicycle, bicycle, walk,

bicycle, tram, walk, tram, tram, car, car, walk, tram, tram

Construct a frequency distribution table to display this data.

2. A group of 24 students was asked 'Which social network do you use the most, MyPage or FlyBird?' Their answers were recorded. Draw a frequency table to display this data.

   MyPage, MyPage, FlyBird, MyPage, FlyBird, MyPage, MyPage, MyPage,

   FlyBird, MyPage, MyPage, FlyBird, FlyBird, MyPage, MyPage, FlyBird,

   FlyBird, MyPage, MyPage, MyPage, MyPage, FlyBird, MyPage, FlyBird

3. An internet provider surveyed 35 randomly chosen clients on their preference of internet connection. NBN stands for National Broadband Network, C for Cable and MW for mobile wireless. Display the data shown in a frequency distribution table.

   NBN, NBN, NBN, C, NBN, NBN, NBN, MV, MV,

   MV, MV, MV, NBN, NBN, NBN, MV, MV, MV,

   NBN, MV, MV, NBN, NBN, MV, MV, MV, NBN,

   NBN, NBN, C, C, MV, MV, MV, MV

4. The school canteen has surveyed a group of 500 students on their opinion about the quality of the sandwiches offered at lunch. 298 students were of the opinion that the sandwiches bought at the school canteen were very good, 15 students thought they were of poor quality, 163 students thought they were excellent, while the rest said they were average. Construct a frequency distribution table for this set of data.

5. **WE6** Find the missing values in the two-way table below.

Parking place and car theft			
Theft	**Parking type**		**Total**
	In driveway	**On the street**	
Car theft		37	
No car theft	16		
Total		482	500

6. A survey of 263 students found that 58 students who owned a mobile phone owned a tablet also, while 11 students owned neither. The total number of students who owned a tablet was 174. Construct a complete two-way table to display this data and calculate all the unknown entries.

7. The numbers of the Year 11 students at Senior Secondary College who study a Mathematics subject are displayed in the table shown.

Year 11 students and Mathematics			
Subject	Gender		Total
	Girls	Boys	
Essential Mathematics		23	
General Mathematics		49	74
Mathematical Methods	36	62	
Specialist Mathematics	7		25
Total	80		

   a. Complete the two-way table.
   b. How many Year 11 students study Mathematics at Senior Secondary College?
   c. What is the total number of students enrolled in Essential Mathematics?

8. George is a hairdresser and keeps a record of all the customers and their hairdressing requirements. He made a two-way table for last month and recorded the numbers of customers who wanted a haircut only or a haircut and colour.

   a. Design a two-way table that George would need to record the numbers of male and female customers who wanted a haircut only or a haircut and colour.
   b. George had 56 female customers in total and 2 male customers who wanted a haircut and colour. Place these values in the table from a.
   c. George had 63 customers during last month with 30 female customers requiring a haircut and colour. Fiill in the rest of the two-way table.

9. a. The manager at the local cinema recorded the gender and the ages of the audience at the Monday matinee movie. The total audience was 156 people. A quarter of the audience were children under 10 years old, 21 were teenagers and the rest were adults. There were 82 females, 16 male teenagers, and the number of girls under 10 years old was 11. Construct a complete two-way table.

   b. What is the total number of adult males in the audience?
   c. How many children were in the audience?

10. **WE7** Consider the data given in the two-way table shown.

Cat and dog owners			
Cat owners	Dog owners		Total
	Own	Do not own	
Own	12	78	90
Do not own	151	64	215
Total	163	142	305

    a. Calculate the relative frequencies for all the entries in the two-way table given. Give your answers correct to three decimal places or as fractions.

    b. Calculate the percentage relative frequencies for all the entries in the two-way table given out of the total number of pet owners.

11. Consider the data given in the two-way table shown.

Swimming attendance			
Day of the week	People		Total
	Adults	Children	
Monday	14	31	45
Tuesday	9	17	26
Wednesday	10	22	32
Thursday	15	38	53
Friday	18	45	63
Saturday	27	59	86
Sunday	32	63	95
Total	125	275	400

    a. Calculate the relative frequencies and the percentage relative frequencies for all the entries for adults out of the total number of adults.

    b. Calculate the relative frequencies and the percentage relative frequencies for both adults and children out of the total number of people swimming on Wednesday.

12. Using the two-way table from 7, construct a

    a. relative frequency table, correct to four decimal places

    b. percentage relative frequency table.

13. A ballroom studio runs classes for adults, teenagers and children in Latin, modern and contemporary dances. The enrolments in these classes are displayed in the two-way table shown.

Dance class enrolments				
Dance	**People**			**Total**
	Adults	**Teenagers**	**Children**	
Latin			23	75
Modern		34	49	96
Contemporary	36		62	
Total	80			318

a. Complete the two-way table.

Construct a:

b. relative frequency table for this data set.

c. percentage relative frequency table for this data set.

14. **WE8** Bianca manages a bookstore. She recorded the number of fiction and non-fiction books she sold over two consecutive weeks in the two-way table below. Construct a grouped column graph for this data.

Types of books sold in a fortnight			
Type of book	**Week**		**Total**
	Week 1	**Week 2**	
Fiction	51	73	124
Non-fiction	37	28	65
Total	88	101	189

15. Construct a grouped column graph for the data displayed in the two-way table below.

Type of fitness exercise per gender			
Type of exercise	**Gender**		**Total**
	Girls	**Boys**	
Running	12	10	22
Walking	8	15	23
Total	20	25	45

16. **MC** The horizontal column graph below displays data regarding the school sector enrolments in each of the Australian states collected by the ABC in 2017.

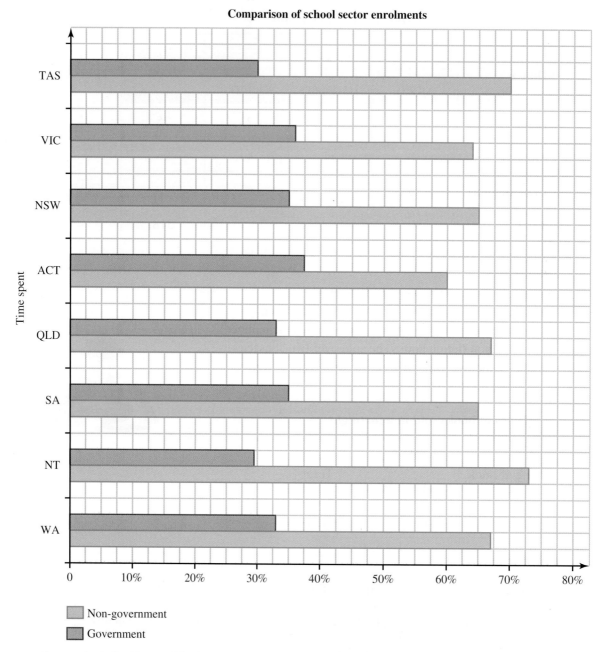

Comparison of school sector enrolments

Non-government
Government

Source: Australian Bureau of Statistics, http://www.abs.gov.au/ausstats/abs@.nsf/Lookup/4221.0main+features22017

Which two states have the highest percentage of students enrolled in non-government schools?
A. VIC and NSW
B. VIC and ACT
C. NSW and QLD
D. VIC and QLD

17. Human blood is grouped in 8 types: O+, O−, A+, A−, B+, B−, AB+ and AB−. The % relative frequency table below shows the percentage of Australians and Chinese that have a particular blood type.

Blood type	Percentage of Australia's population	Percentage of China's population
O+	40	47.7
O −	9	0.3
A+	31	27.8
A−	7	0.2
B+	8	18.9
B−	2	0.1
AB+	2	5.0
AB−	1	0.01

Source: Wikipedia; http://en.wikipedia.org/wiki/ABO_blood_group_system

Use a calculator, a spreadsheet or otherwise to construct a grouped column graph to display the data for both Australia's and China's populations.

18. **WE9** The graph shown represents the relationship between job satisfaction of a group of people surveyed and the years of schooling before conditioning on income (blue columns) and after conditioning on income (pink rectangles). How is this graph misleading or deceiving?

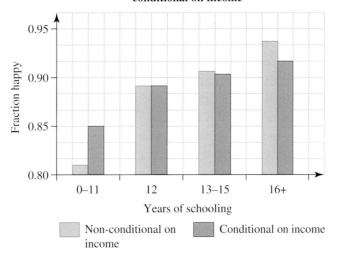

Fraction happy about life by years of completed schooling before and after conditional on income

Source: Oreopoulos, P., Salvanes, K. G., 'Priceless: the nonpecuniary benefits of schooling', *Journal of Economic Perspectives*, vol. 25, no. 1, Winter 2011, pp. 159–84, http://pubs.aeaweb.org/doi/pdfplus/10.1257/jep.25.1.159

19. How is this graph misleading or deceiving?

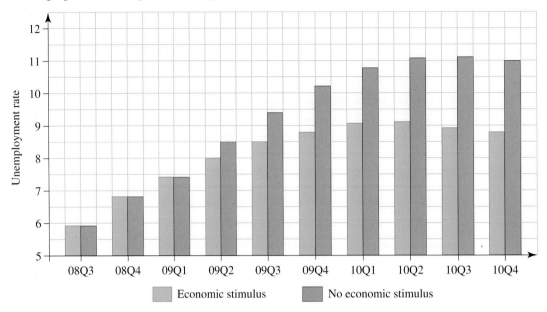

Source: 'The economic impact of the American recovery and reinvestment act', Mark Zandi, Chief Economist, Moody's Economy.com, 21 January 2009. http://www.economy.com/mark-zandi/documents/economic_stimulus_house_plan_012109.pdf

20. **a.** Car insurance companies calculate premiums according to the perceived risk of insuring a person and their vehicle. From historical data, age is a perceived 'risk factor'. In order to determine the premium of their car insurance policies, Safe Drive insurance company surveyed 300 of its customers. They determined three age categories: Category A for young drivers, Category B for mature drivers, and Category C for older drivers, and then recorded whether the customer had had a traffic accident within the last year. Their records showed that 113 drivers from Category B did not have a traffic accident within the last year and 17 drivers out of the 47 in Category C did not have a traffic accident. If 75 drivers had a traffic accident last year and 35 of the drivers were in Category A, construct a complete two-way table for this data set.

   b. How many mature drivers had a traffic accident last year?

   c. What is the total number of drivers who did not have a traffic accident last year?

   d. Use a calculator, a spreadsheet or otherwise to construct a grouped column graph for this data.

5.4 Numerical data

5.4.1 Displaying numerical data

Visual displays of data are important tools in communicating the information collected. Numerical data can be displayed in tables or graphical representations.

Frequency distributions

Frequency distributions or **frequency distribution tables** or just **frequency tables** for numerical data are constructed in the same way as they are constructed for categorical data as they display frequency counts in both cases.

Ungrouped numerical data

For **ungrouped data** each score is recorded separately. For large sets of data it is helpful to use tally marks.

For his statistics project, Radu surveyed 50 students on the number of books they had read over the past 6 months. The data recorded is listed below.

| 3 | 2 | 0 | 6 | 4 | 3 | 3 | 4 | 2 | 2 | 2 | 2 | 1 | 2 | 1 | 1 | 5 | 4 | 4 | 3 | 3 | 2 | 3 | 3 | 3 |
| 5 | 0 | 1 | 1 | 1 | 4 | 2 | 2 | 3 | 2 | 3 | 2 | 2 | 2 | 1 | 2 | 1 | 1 | 5 | 4 | 4 | 3 | 3 | 4 | 2 |

Construct a frequency table for this data.

THINK

1. Draw a frequency table with three columns to display the variable (number of books in this case), tally and frequency.
   - The data has scores between 0 as the lowest score and 6 as the highest score.
   - The last cell in the first column is labelled 'Total'.

WRITE/DRAW

Number of books	Tally	Frequency
0		
1		
2		
3		
4		
5		
6		
Total		

2. Complete the 'Tally' and the 'Frequency' columns.

Number of books	Tally	Frequency
0	II	2
1	JHÍ IIII	9
2	JHÍ JHÍ JHÍ	15
3	JHÍ JHÍ II	12
4	JHÍ III	8
5	III	3
6	I	1
Total		50

3. Check the frequency total with the total number of data.

50

Grouped data

Grouped data is data that has been organised into groups or class intervals. It is used for very large amounts of data and for continuous data.

Care has to be taken when choosing the class sizes for the data collected. The most common class sizes are 5 or 10 units. A class interval is usually written as $5 -\!\!< 10$, meaning that all scores between 5 and 10 are part of this group including 5 but excluding 10.

WORKED EXAMPLE 11

The data below displays the maximum daily temperature in Melbourne for every day in December 2012. The data was downloaded from the Bureau of Meteorology. Construct a frequency table for this data.

Date	Maximum temp (°C)	Date	Maximum temp (°C)	Date	Maximum temp (°C)
1/12/12	27.4	11/12/12	29.1	21/12/12	22.4
2/12/12	20.7	12/12/12	34.3	22/12/12	28.3
3/12/12	21.4	13/12/12	35.1	23/12/12	38.3
4/12/12	18.9	14/12/12	22.9	24/12/12	27.6
5/12/12	18.2	15/12/12	24.8	25/12/12	21.9
6/12/12	21.2	16/12/12	23.1	26/12/12	22.4
7/12/12	30.6	17/12/12	22	27/12/12	31.2
8/12/12	37	18/12/12	27.8	28/12/12	21.7
9/12/12	21.1	19/12/12	32.6	29/12/12	23.3
10/12/12	21.4	20/12/12	23.5	30/12/12	21.9
				31/12/12	24.7

Source: Copyright 2003 Commonwealth Bureau of Meterology

THINK

1. Draw a frequency table with three columns to display the variable (temperature (°C) in this case), tally and frequency.
   - The data has scores between 18.2 °C as the lowest score and 38.3 °C as the highest score. This leads us to five groups of class size 5 °C between 18 °C and 43 °C.
   - The last cell in the first column is labelled 'Total'.

WRITE/DRAW

Temperature (°C)	Tally	Frequency
$18 -\!\!< 23$		
$23 -\!\!< 28$		
$28 -\!\!< 33$		
$33 -\!\!< 38$		
$38 -\!\!< 43$		
Total		

2. Complete the 'Tally' and the 'Frequency' columns.

Temperature (°C)	Tally	Frequency
18 —< 23	ЖІ ЖІ ІІІІ	14
23 —< 28	ЖІ ІІІ	8
28 —< 33	ЖІ	5
33 —< 38	ІІІ	3
38 —< 43	І	1
Total		31

3. Check the frequency total with the total number of data.

31

5.4.2 Dot plots

Dot plots are graphical representations of numerical data made up of dots. They can also be used to represent categorical data with dots being stacked in a column above each category. All data are displayed using identical equally distanced dots where each dot represents one score of the variable. The height of the column of dots represents the frequency for that score.

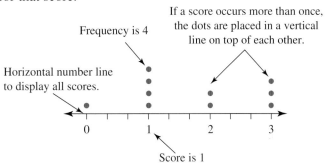

WORKED EXAMPLE 12

At an electronics store customers were surveyed on the number of TV sets they have in their household. The data collected from 20 customers is displayed in the frequency table below. Construct a dot plot for this set of data.

Number of TV sets	Frequency
0	1
1	3
2	5
3	8
4	2
5	1
Total	20

THINK	WRITE/DRAW
1. Draw a labelled horizontal line for the variable 'Number of TV sets'. <ul><li>Start from the minimum score, 0, and continue until the maximum score is placed on the line, 5 in this example.</li><li>The spaces between scores have to be equal.</li></ul>	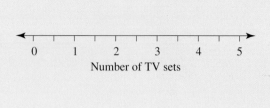
2. Place the dots for each score. <ul><li>The frequency for 0 is 1. Place one dot above score 0.</li><li>The frequency for 1 is 3. Place three dots above score 1.</li><li>Continue until all data is placed above the line.</li><li>Include a title</li></ul>	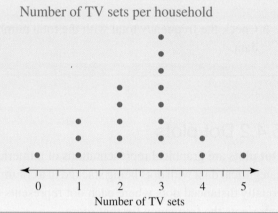

5.4.3 Histograms

Histograms are graphical representations of numerical data. They are very similar to the column graphs used to represent categorical data because the data is displayed using rectangles of equal width. However, although there is a space before the first column, there are no spaces between columns in a histogram. Each column represents a different score of the same variable.

The vertical axis always displays the values of the frequency.

Features

For **ungrouped** data the value of the variable is written in the middle of the column under the horizontal axis.

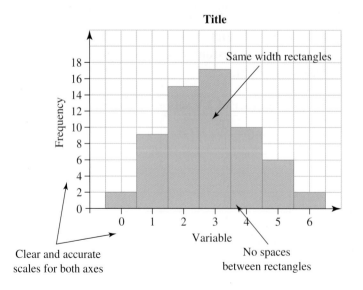

For **grouped data** the ends of the class interval are written under the ends of the corresponding columns.

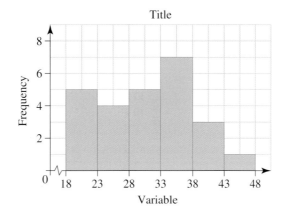

WORKED EXAMPLE 13

Draw a histogram for the data given in the frequency table below.

Number of books read over six months	Frequency
0	2
1	9
2	15
3	12
4	8
5	3
6	1
Total	50

THINK

1. Draw a labelled set of axes:
   - 'Frequency' on the vertical axis
   - the variable 'Number of books' on the horizontal axis
   - ensure the histogram has a title
   - the maximum frequency is 18 so the vertical scale should have ticks from 0 to 18. The ticks could be written by ones (a bit cluttered) or by twos.
   - the data is ungrouped so the ticks are in the middle of each column
   - allow for a space before drawing the columns of the histogram.

WRITE/DRAW

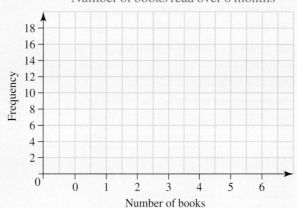

2. Draw the columns.

Number of books read over 6 months

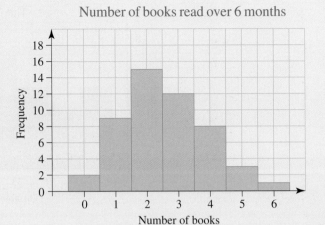

The data displayed in Worked example 13 is numerical discrete data. For numerical continuous data we have to group scores in intervals. The intervals must have to be equal in size and must not overlap.

WORKED EXAMPLE 14

Using the frequency table given, construct a histogram to represent this set of data.

Temperature (°C)	Frequency
18 —< 23	14
23 —< 28	8
28 —< 33	5
33 —< 38	3
38 —< 43	1
Total	31

THINK

1. Draw a labelled set of axes:
   - 'Frequency' on the vertical axis
   - the variable 'Temperature (°C)' on the horizontal axis
   - ensure the histogram has a title
   - the maximum frequency is 14 so the vertical scale should have ticks from 0 to 16. The ticks could be written by ones (a bit cluttered) or by twos.
   - the data is grouped so the ticks should be at the edges of each column
   - allow for a space before drawing the columns of the histogram

WRITE/DRAW

Maximum daily temperatures in Melbourne in December 2012

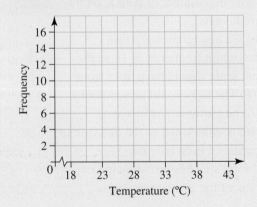

2. Draw the columns.

Maximum daily temperatures in Melbourne in December 2012

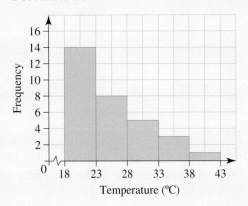

5.4.4 Stem-and-leaf plots

Stem-and-leaf plots are graphical representations of grouped numerical data. The stem-and-leaf plot and the histogram shown are graphical representations of the same set of data.

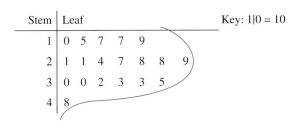

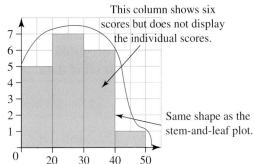

The two graphical representations have similar shapes. However, a stem-and-leaf plot displays all the data collected while a histogram loses the individual scores.

Ten-unit intervals

A stem-and-leaf plot has two columns marked 'Stem' and 'Leaf'. For two-digit numbers, the first digit, the ten, is written in the stem column while the second digit, the unit, in the leaf column. In the plot shown there are four numbers displayed: 31, 37, 35, 48 and 42. The data is displayed in intervals of ten units.

Stem	Leaf			Key: 3|1 = 31
3	1	7	5	
4	8	2		

Ensure that there are equal spaces between all the digits in the leaf column and all numbers line up vertically. Once all scores are recorded, they have to be placed in an increasing order.

Stem	Leaf			Key: 3|1 = 31
3	1	5	7	
4	2	8		

For three-digit numbers, the first two digits are written in the first column and the third digit in the second column. In the plot at right there are four numbers: 120, 125, 126 and 134.

Stem	Leaf			Key: 12|0 = 30
12	0	5	6	
13	4			

For decimal numbers, the whole number is written in the stem column and the decimal digits are written in the leaf column. The plot below displays the data: 3.6, 3.9, 3.0, 2.7, 2.8 and 4.1.

Stem	Leaf	
2	7 8	Key: 2|7 = 2.7
3	0 6 9	
4	1	

Five-unit intervals

A stem-and-leaf plot can also display data in intervals of five units. This situation occurs when the data collected consists of very close values. For example, the set of data 12, 11, 17, 21, 29, 14, 12, 15, 22, 13, 26, 27, 18, 26 is better represented in five-unit intervals. In this case the stem will consist of the values 1, 1∗, 2 and 2∗.

Stem	Leaf	
1	1 2 2 3 4	Key: 2|3 = 23
1∗	5 7 8	2∗|6 = 26
2	1 2	
2∗	6 6 7 9	

The first 1 represents the data values from 10 to 14 inclusive, while the 1∗ represents the data values from 15 to 19 inclusive.

The first 2 represents the data values from 20 to 24 inclusive, while the 2∗ represents the data values from 25 to 29 inclusive.

WORKED EXAMPLE 15

The heights of 20 students are listed below. All measurements are given in centimetres. Construct a stem-and-leaf plot for this set of data:
a. using ten-unit intervals.
b. using five-unit intervals.

$$156 \quad 172 \quad 162 \quad 164 \quad 174 \quad 151 \quad 150 \quad 169 \quad 171 \quad 169$$
$$167 \quad 161 \quad 153 \quad 155 \quad 165 \quad 172 \quad 148 \quad 166 \quad 169 \quad 158$$

THINK

a. 1. Draw a horizontal line and a vertical line as shown and label the two columns stem-and-leaf.

2. Write the values in the stem part of the graph using intervals of 10 units.
   As the values of the data are three-digit numbers, we write the first two digits in the stem column in increasing order. The lowest value is 148 and the highest value is 174. The stem will have the numbers 14, 15, 16 and 17.

WRITE/DRAW

Stem	Leaf

Stem	Leaf
14	
15	
16	
17	

---

3. Write the values in the leaf part of the graph. Start with the first score, 156, and place **6** to the right of 15

The next value is 172. Place 2 to the right of 17.

Stem	Leaf
14	
15	6
16	
17	2

4. Continue until all data is displayed.

Stem	Leaf
14	8
15	6 1 0 3 5 8
16	2 4 9 9 7 1 5 6 9
17	2 4 1 2

5. Rewrite all data in increasing order and a key.

Stem	Leaf
14	8
15	0 1 3 5 6 8
16	1 2 4 5 6 7 9 9 9
17	1 1 2 2 4

Key: 14|8 = 148

b. 1. Rewrite the stem part of the graph using intervals of five units.

Stem	Leaf
14	
14*	
15	
15*	
16	
16*	
17	
17*	

2. Write the values in the leaf part of the graph and a key.

Stem	Leaf
14*	8
15	0 1 3
15*	5 6 8
16	1 2 4
16*	5 6 7 9 9 9
17	1 2 2 4

Key: 15|1 = 151
15*|8 = 158

---

on Resources

Interactivity: Create a histogram (int-6494)
Interactivity: Stem plots (int-6242)
Interactivity: Create stem plots (int-6495)
Interactivity: Dot plots, frequency tables and histograms and bar charts (int-6243)

Exercise 5.4 Numerical data

1. **WE10** A group of 25 Year 11 students was surveyed by a psychologist on the number of hours of sleep they each have per night. The set of scores below shows this data.

   6, 6, 7, 8, 5, 10, 8, 8, 9, 6, 4, 7, 7, 6, 5, 10, 7, 8, 6, 8, 9, 7, 5, 9, 7

   Display this data in a frequency distribution table.

2. 'What is the length, in cm, of your foot without a shoe?' Using the random sampler of 30 students from the Census@School (Australian Bureau of Statistics) website, Eva downloaded the responses of 30 Year 11 students across Australia. Measurements are given to the nearest centimetre. Display this data in a frequency distribution table.

28	27	28	23	27	29	24	26	29	26
23	23	27	25	28	24	26	28	24	28
22	26	23	26	29	21	23	25	24	26

3. **WE11** The times of entrants in a charity run are listed below. The times are given in minutes.

34	29	57	45	26	40	19	28	33	37	39	46
18	52	19	36	28	19	20	19	54	38	38	51
19	21	53	36	37	25	22	30	25	34	18	17

   Display this data in a frequency distribution table.

4. The data set shown represents the number of times last month a randomly chosen group of 30 people spent their weekends away from home. Construct a frequency distribution table for this set of data including a column for the tally.

   4 3 2 0 0 0 1 2 0 1 0 0 2 1 1

   4 1 1 0 0 0 2 1 1 0 2 1 3 3 0

5. For research purposes 36 Tasmanian giant crabs were measured and the widths of their carapaces are given below. Construct a frequency table for this set of data.

35.3,	34.9,	36.0,	35.4,	37.0,	34.2,
35.9,	35.5,	35.0,	36.2,	35.7,	33.1,
36.5,	36.4,	37.2,	35.4,	34.9,	35.1,
34.8,	35.8,	35.2,	35.6,	37.1,	36.7,
35.0,	34.2,	36.3,	33.9,	35.2,	34.1,
36.5,	36.4,	36.6,	34.3,	35.8,	35.7

Source: Naturalise Marine Discovery Centre

6. **WE12** The frequency distribution table below displays the data collected by MyMusic website on the number of songs downloaded by 40 registered customers per week. Construct a dot plot for this data.

Number of songs	Frequency
2	12
3	8
4	7
5	6
6	3
7	1
Total	40

7. Daniel asked his classmates 'How many pairs of shoes do you have?' He collected the data and listed it below. Construct a dot plot for this set of data.

Pairs of shoes	Frequency
3	9
4	5
5	3
6	2
7	3
8	2
9	1
Total	25

8. The frequency table shown displays the ages of 30 people who participated in a spelling contest. Construct a dot plot for this set of data.

Age	Frequency
11	3
12	6
13	8
14	7
15	3
16	2
17	1

9. The data given represents the results in a typing test, in characters per minute, of a group of 24 students in a beginners class.

36 39 40 38 41 39 37 40

41 38 34 35 39 38 37 36

39 41 39 38 42 40 39 37

Construct a dot plot for this set of data.

10. **WE13** The frequency distribution table below displays the data collected by a Year 11 student on the number of emails his friends send per day. Construct a histogram for this data.

Number of emails	Frequency
0	1
1	12
2	17
3	5
4	2
5	3
Total	50

11. 'In how many languages can you hold an everyday conversation?' is one of the questions in the Census@School questionnaire. Using the random sampler from the Census@School (Australian Bureau of Statistics) website, Brennan downloaded the responses of 100 randomly chosen students. He displayed the data in the frequency distribution table at right. Construct a histogram for this data.

Number of languages	Frequency
1	127
2	50
3	13
4	4
5	3
6	2
7	1
Total	200

12. A group of 24 students was given a task that involved searching the internet to find how to construct an ungrouped histogram. The times, in minutes, they spent searching for the answer was recorded. Draw an ungrouped histogram for this set of data.

10 15 18 12 19 13 14 10

16 18 12 15 18 12 13 12

15 12 12 10 15 11 10 16

13. **WE14** A phone company surveyed 50 randomly chosen customers on the number of times they check into their email account on weekends.
   a. Display the data collected in a frequency distribution table using class intervals of 3.
   b. Construct a histogram for this data.

5	10	2	17	8	8	13	6	3	0
11	1	6	10	1	4	0	11	5	12
0	3	4	5	6	5	14	9	4	10
14	9	2	16	3	9	8	10	5	11
7	2	5	6	1	2	2	12	9	17

14. The frequency distribution table below shows the number of road fatalities in Victoria over a period of 12 months to October 2012 per age group involving people between the ages of 30 and 70 years old. Construct a histogram for this data.

Age group	Frequency
30 –< 40	47
40 –< 50	27
50 –< 60	41
60 –< 70	35
Total	210

    Source: TAC road safety statistical summary, page 6

15. The data below displays the responses of 50 year 11 students who participated in a questionnaire related to the amount of money they earned from their part-time jobs over the previous week.

30	103	50	30	0	90	0	80	68	157
0	350	123	70	80	50	0	330	210	26
25	0	0	50	230	50	20	60	0	305
177	126	90	0	0	12	200	120	15	45
90	30	60	0	0	80	0	260	25	150

    a. Display this data in a frequency table.
    b. Draw a grouped histogram to represent this data.

16. **WE15** The BMI or body mass index is a number that shows the proportion between a person's mass and their height. The formula used to calculate the BMI is

$$\text{BMI} = \frac{\text{weight (kg)}}{\text{height (m}^2)}$$

The BMIs of 24 students are listed below. Display this data in a stem-and-leaf plot.

25	24	18	16	29	23	20	21
21	17	19	23	25	24	32	22
22	23	30	18	23	20	19	24

17. The marks out of 50 obtained by a Year 11 Essential Mathematics group of students on their exam are given below. Display this data in a stem-and-leaf plot.

    36   41   29   50   45   23   48   56

    20   12   43   33   35   44   32   49

    39   48   50   18   43   20   38   29

18. Rami is the manager of a timber yard. He started an inventory for the upcoming stocktake sale. The lengths of the first 30 timber planks recorded are given.

    250 220 245 229 260 210 250 261 244 218

    250 251 216 243 232 210 212 227 219 207

    231 243 204 230 265 220 206 253 229 225

Display this data set in an ordered stem-and-leaf plot.

19. Consider the following set of data that represents the heights, in cm, of 40 children under the age of 3.

    58.9 54.6 62.9 45.7 49.6 62.8 43.2 56.7 69.0 65.3

    57.8 59.4 66.7 49.6 72.7 35.2 72.8 75.6 32.9 56.7

    54.8 45.2 69.5 47.3 68.6 67.4 63.4 61.5 52.6 48.0

    56.3 78.4 39.9 75.3 45.2 56.9 66.3 55.4 63.8 72.3

Using a calculator, a spreadsheet or otherwise construct a grouped histogram for this set of data.

20. The reaction times using their non-dominant hand of a group of people are listed below. The times are given in deciseconds (a tenth of a second). Using a calculator, a spreadsheet or otherwise, construct a grouped histogram for this data set.

    48 39 40 36 35 37 46 41 43 39 35

    50 59 37 32 39 42 43 46 33 38 30

    41 29 36 38 51 42 34 39 36 37 41

5.5 Measures of central tendency and the outlier effect

Two key measures are used in statistics to examine a set of data: **measures of central tendency** and **measures of spread**. The focus for this section is on the measures of central tendency.

Central tendency is the centre or the middle of a frequency distribution. The most common measures of central tendency are the **mean, median** and **mode** of a distribution.

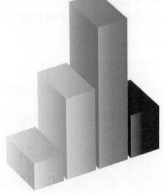

5.5.1 Mean

The **mean**, also called the **arithmetic mean**, is the average value of all the data. It is the most common measure of central tendency. To calculate the mean of a set of data, calculate the sum of all data values and then divide it by the total number of data.

$$\text{Mean} = \frac{\text{Sum of all data values}}{\text{Total number of data values}}$$

$\bar{x} = \dfrac{\sum x_i}{n}$, where x_i = data values (x_1 is the first data score, x_2 is the second data score, ..., x_n is the nth data score)

$$\bar{x} = \text{mean}$$

$\sum x_i$ = sum of all data values. Σ (sigma) is the Greek capital letter standing for the letter S. In Mathematics, Σ stands for the 'sum of ...'

$$n = \text{total number of data values}$$

WORKED EXAMPLE 16

The data below represents the results obtained by 10 students in a statistics test. The maximum number of marks for the test was 20. Calculate the mean for this set of data.

15, 20, 20, 17, 19, 11, 6, 18, 18, 16

THINK	WRITE/DRAW
1. Write the formula for the mean.	$\bar{x} = \dfrac{\sum x_i}{n}$, where $x = 15, 20, 20, 17, 19, 11, 6, 18, 18, 16$ $n = 10$
2. Substitute the data values into the formula for the mean.	$\bar{x} = \dfrac{15 + 20 + 20 + 17 + 19 + 11 + 6 + 18 + 18 + 16}{10}$
3. Determine the answer.	$\bar{x} = \dfrac{160}{10}$ $= 16$

5.5.2 Median

The **median** of a data set is the number in the middle of the distribution when all the scores are arranged in order from the lowest value to the highest value.

The formula $\dfrac{n+1}{2}$ calculates the position of the median for a set n data values.

Care has to be taken when calculating the position of the median for an even number of data and for an odd number of data.

Calculate the median pulse rate for the two sets of data given.

a. Individual pulse rates of a group of 10 adults:

69, 62, 83, 75, 71, 66, 64, 67, 70, 62

b. Individual pulse rates of a group of 11 newborns (0–3 months):

129, 121, 116, 132, 124, 136, 124, 129, 133, 128, 129

THINK	WRITE/DRAW
a. Even number of data	
1. Arrange the data in increasing order.	62, 62, 64, 66, 67, 69, 70, 71, 75, 83
2. Find the position of the median.	$\text{Position} = \dfrac{n+1}{2}$
	$= \dfrac{10+1}{2}$
	$= 5.5 \dots$ between 5th and 6th scores
3. Find the median.	For an even number of data the position of the median is going to be halfway between the two middle scores. The median is the average or the mean of these two scores.

$$62, \quad 62, \quad 64, \quad 66, \quad \underbrace{67, \quad 69,}_{\text{median}} \quad 70, \quad 71, \quad 75, \quad 83$$

$$\text{Mean} = \frac{67 + 69}{2}$$

$$= \frac{136}{2}$$

$$= 68$$

The median of the set of data given is 68.

b. Odd number of data	
1. Arrange the data in increasing order.	116, 121, 124, 124, 128, 129, 129, 129, 132, 133, 136
2. Find the position of the median.	$\text{Position} = \dfrac{n+1}{2}$
	$= \dfrac{11+1}{2}$
	$= 6\text{th score}$
	For an odd number of data the position of the median is exactly in the middle.
3. Find the median.	The median is the score in the middle.

116, 121, 124, 124, 128, (129), 129, 129, 132, 133, 136

The median of the set of data given is 129.

5.5.3 Mode

The **mode** is given by the data value with the highest frequency. Take the data:

1, 1, 2, 3, 3, 4, 5, 5, 5, 6, 9.

The digit that occurs most often is the digit 5. It has the highest frequency, 3. We can conclude that the mode of this set of digits is 5.

WORKED EXAMPLE 18

Michelle is a real estate agent and has listed ten properties for sale in the last fortnight. She listed the bedrooms per house.
3, 2, 4, 5, 2, 2, 1, 2, 3, 2
What is the mode for this data?

THINK	WRITE/DRAW
1. Count the number of each score.	Score 1 has frequency one. Score 2 has frequency five. Score 3 has frequency two. Score 4 has frequency one. Score 5 has frequency one.
2. Find the highest frequency and the mode.	The highest frequency is 5. Therefore, mode = 2.

5.5.4 Mean, median and mode from a frequency distribution table

When working with large amounts of data (i.e. 100 data scores), it is quite tedious to calculate the three measures of central tendency. In this case we have to work with the data from a frequency distribution table.

WORKED EXAMPLE 19

Using the random sampler from the Census@School (Australian Bureau of Statistics) website, Alexandra downloaded the responses of 200 students across Australia who participated in the 2012 questionnaire. She displayed the data values of the students' year level in a frequency table.
a. Calculate the mean of this data.
b. What is the median of this data?
c. Does this set of data have a mode? What is it?

▶

Note: Year level 4 represents '4 or below'. For the purpose of this example we are going to use 4 instead of '4 or below'.

Year level	Frequency
4	8
5	19
6	40
7	23
8	36
9	29
10	27
11	11
12	7
Total	**200**

Source: ABS; http://www.cas.abs.gov.au/cgi-local/cassampler.pl

THINK

a. 1. Add one more column to the frequency table and label it $x \times f$ where x is the variable and f is the frequency.

WRITE/DRAW

Note: the frequency table shows 8 students who were in year 4 in 2012.

$4 + 4 + 4 + 4 + 4 + 4 + 4 + 4 = 4 \times 8$
$= 32$

Year level x	Frequency f	$x \times f$
4	8	32
5	19	95
6	40	240
7	23	161
8	36	288
9	29	261
10	27	270
11	11	121
12	7	84
Total	200	1552

2. Calculate the mean.
Note: Quite often the mean is not one of the scores of the data set. It does not need to be. It can even be an impossible score like in this example.

$\bar{x} = \dfrac{\sum x_i}{n}$, where $\Sigma x = 1552$ and $n = 200$

$\bar{x} = \dfrac{1552}{200}$

$= 7.76$

Mean $= 7.76$

b. 1. Find the position of the median.	Position $= \dfrac{n+1}{2}$
	$= \dfrac{200+1}{2}$
	$= 100.5$ ... between 100th and 101st scores
2. Find the median.	Add up the frequencies until the sum is greater then 100.
	8 (year 4 students) + 19 (year 5 students) = 27
	27 + 40 (year 6 students) = 67
	67 + 23 (year 7 students) = 90
	From 90 to 100 we only need 10 more scores. These scores fall into the year 8 frequency. Therefore, the median is year 8.
c. 1. Find the highest frequency, hence the mode.	The highest frequency occurs for year level 6. Therefore, the mode is year 6.

 Resources

❖ **Interactivity:** Mean, median, mode and quantities (int-6496)

5.5.5 Outliers and their effect on data

A score that is considerably lower or considerably higher than the rest of the data is called an **outlier**. Outliers can influence some of the measures of central tendency. For example, the median house price paid at auction on one particular Saturday in a city could be influenced by a very large price paid for one house. However, for a large number of data, i.e 2000 scores, this influence could be quite insignificant.

WORKED EXAMPLE 20

The marks of 10 students on a test out of a total of 50 marks are:

26, 50, 49, 36, 18, 27, 43, 43, 33, 35

Discuss the effect of adding a very small test mark to this set of data.

THINK	WRITE/DRAW
1. Calculate the mean, median, mode for this set of data.	$\bar{x} = \dfrac{\sum x_i}{n}$, where $\sum x_i = 360$ and $n = 10$
	$= \dfrac{360}{10}$
	$= 36$
	Median 18, 26, 27, 33, 35, 36, 43, 43, 49, 50
	Position $= \dfrac{n+1}{2}$
	$= \dfrac{10+1}{2}$
	$= 5.5 \ldots$ between the 5th and the 6th scores
	Median $= \dfrac{35+36}{2}$
	$= 35.5$
	Mode $= 43$
2. Calculate the mean, median and mode for this set of data after adding a score of 1 mark to the set of data.	1, 18, 26, 27, 33, 35, 36, 43, 43, 49, 50
	$\bar{x} = \dfrac{\sum x_i}{n}$ where $\sum x_i = 361$ and $n = 11$
	$= \dfrac{361}{11}$
	$= 32.8$
	Median position $= \dfrac{11+1}{2}$
	$= 6 \ldots$ 6th score
	Median $= 35$
	Mode $= 43$
3. Describe the differences between the mean, median and mode.	Both the mean and the median have changed while the mode has stayed the same. The score 1 is an outlier for this data and has had an impact on the mean: from 34.5 to 32.8.
4. Discuss the findings from adding a very small test score to the data set.	Adding a very small score or a very large score to a set of data will always change the mean, sometimes changes the median and it rarely changes the mode.
	The most suitable measure for this type of data is the median as it shows that half of the students performed higher than 35.5 while half performed lower than 35.5.
	The mean is less appropriate as it is influenced by outliers and does not give any information about the number of students below or above its value.

Exercise 5.5 Measures of central tendency and the outlier effect

1. **WE16** A postman delivered the following number of letters to 15 households:

   2, 5, 3, 1, 3, 4, 6, 2, 2, 2, 4, 2, 4, 3, 2

   Calculate the mean number of letters per household delivered by the postman.

2. The data set: 171.5, 162.9, 158.4, 156.8, 142.3, 170.3, 163.7, 158.1, 160.6, 162.9 represents the heights in centimetres of 10 students. Calculate the mean height of the 10 students.

3. Calculate the mean for the following sets of data.
   a. 13, 20, 17, 16, 19, 11, 24, 17, 13
   b. 163.2, 219.6, 157.0, 206.4, 211.3, 234.3, 195.8, 179.1

4. Car battery manufacturers have a certain lifetime guarantee for their customers. This means that if something goes wrong with the car battery within the lifetime of the battery, they will replace it for free. To check the lifetime guarantee of their car batteries, CarBat Company checked the lifetime of 200 batteries. This data is displayed in the frequency distribution table shown.

Lifetime (months)	Frequency	fx
21	9	
22	18	
23	44	
24	67	
25	29	
26	20	
27	13	
Total	200	

   a. Calculate the entries in the column labelled fx.
   b. Hence calculate the average lifetime guarantee for this set of car batteries.

5. The principal of Achieve Secondary College wanted to find out the average wage of her teachers. She collected the data and displayed it in the frequency table shown.

Wage ($/annum)	Frequency	fx
57 797	37	
62 880	24	
66 517	16	
70 771	21	
75 248	8	
85 458	5	
90 609	2	
Total	113	

   a. Calculate the entries in the column labelled fx.
   b. Calculate the average teacher wage of the teachers at Achieve Secondary College.

6. **WE17** Amir is a market researcher. He used random numbers software to collect a set of random numbers from 1 to 20.

   12 15 3   19 17 6   11 4 8

   10 14 12 7   13 18 20 9 19

   7   16 1   5   8   10 15 3 2

   a. What is the position of the median?
   b. What is the median of this data set?

7. Richard is a fencer and has to calculate eight quotes for his clients. The data 142.3, 96.2, 57.3, 101.8, 53.4, 81.4, 114.3, 72.6, 81.7, 92.5 represents the lengths of the fences he has to build.
   All measurements are given in metres. Calculate the position and the value of the median length for this data set.

8. Find the median of the following sets of data.
   a. 4.5, 3.6, 5.7, 8.2, 2.9, 6.5, 1.8, 7.4, 9.3
   b. 142, 187, 167, 128, 115, 132, 155, 194, 173, 168

9. A group of 20 students were tested on their knowledge of measures of central tendency by answering a 10 multiple choice test. The numbers of correct answers per student are displayed in the list below.
   What is the mean number of multiple choice questions correctly answered by this group of students?

   7 3 4 5 6 6 7 7 8   8

   9 9 8 8 9 7 9 8 10 7

10. **WE18** Find the mode for the frequency distribution shown below.

    14 14 15 13 17 13 13 14 14 14

    16 16 17 17 16 16 16 13 14 14

    16 17 17 14 14 15 16 16 16 16

11. A group of people aged 20–29 were asked to state their age. What is the mode of this set of data?

    21 25 23 21 23 24 26 22 22 20

    24 27 24 23 29 24 28 24 25 29

12. A shoe store collected the data from 15 people on their shoe size. Calculate the average shoe size for this group of people.

    $8, 8\frac{1}{2}, 10\frac{1}{2}, 9, 9\frac{1}{2}, 10\frac{1}{2}, 10, 9, 8, 8, 10\frac{1}{2}, 8\frac{1}{2}, 9\frac{1}{2}, 9\frac{1}{2}, 9$

13. Find the mode of the following sets of data.

    a. 18, 21, 16, 19, 22, 21, 23, 21, 18, 18, 21, 19

    b. 144.1, 144.8, 144.5, 144.8, 144.9, 144.2, 144.9, 144.8, 144.7, 144.6, 144.3

14. **WE19** The frequency distribution table shown displays the data collected in an airport on the number of return trips 120 passengers booked in the last month.
    a. Add a third column to the frequency distribution table, label it *fx* and fill it in.
    b. Calculate the mean of this data.
    c. What is the median of this data?
    d. Find the mode of this data.

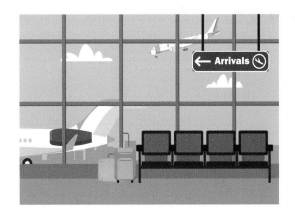

Number of bookings	Frequency
1	27
2	32
3	19
4	26
5	12
6	3
7	1
Total	120

15. Jenny started a new internet business. She recorded the numbers of times the web subscribers logged in over the first month of business. The data is displayed in the frequency distribution below.
    a. Add a third column to the frequency distribution table, label it *fx* and fill it in.
    b. Calculate the mean of this data.
    c. What is the median of this data?
    d. Find the mode of this data.

Number of logins	Frequency
1	23
2	45
3	69
4	51
5	38
6	17
Total	243

16. **WE21** Ling has a school uniform shop and has to order new stock for the year. She collects the dress sizes of a sample of 20 girls from the nearby school.

| 12 | 14 | 8 | 8 | 8 | 10 | 12 | 10 | 8 | 8 |

| 10 | 14 | 8 | 10 | 8 | 8 | 8 | 8 | 24 | 8 |

Identify the outlier and discuss the effect of this outlier on the set of data by calculating the mean, median, and mode for the above set of data

a. without the outlier.

b. with the outlier.

c. Which measure of central tendency is she more interested in: the mean, median, or the mode? Explain.

5.6 Comparing and interpreting data

5.6.1 Back-to-back stem-and-leaf plots

Back-to-back stem-and-leaf plots allow us to compare two sets of data on the same graph. A back-to-back stem-and-leaf plot has a central stem with leaves on either side.

The stem-and-leaf plot at right gives the ages of members of two teams competing in a bowling tournament.

Key: 5 | 5 = 55

Team A	Stem	Team B
1	4	
	5	5 7 8 9
	6	0 2 3 5 7
6 5 4 3	7	0 1 2
8 6 5 4 3 2 1	8	8
	9	0

WORKED EXAMPLE 21

The following data was collected from two Year 11 Maths classes who completed the same test. The total mark available was 100.

Class 1	84	90	86	95	92	81	83	97	88	99	79	100	85	82	97
Class 2	90	55	48	62	70	58	63	67	72	59	60	88	57	65	71

Draw a back-to-back stem-and-leaf plot.

THINK

1. To create a back-to-back stem-and-leaf plot, determine the highest and lowest values of each set of data to help you decide on a suitable scale for the stems.

2. Put the data for each class in order from lowest to highest value.

WRITE

The highest value is 100 and lowest value is 48, so the stems should be 4, 5, 6, 7, 8, 9, 10.

Class 1: 79, 81, 82, 83, 84, 85, 86, 88, 90, 92, 95, 97, 97, 99, 100

Class 2: 48, 55, 57, 58, 59, 60, 62, 63, 65, 67, 70, 71, 72, 88, 90

---

3. Create a back-to-back stem-and-leaf plot.

Key: 4 | 8 = 48

Leaf (Class 1)	Stem	Leaf (Class 2)
	4	8
	5	5 7 8 9
	6	0 2 3 5 7
9	7	0 1 2
8 6 5 4 3 2 1	8	8
9 7 7 5 2 0	9	0
0	10	

5.6.2 Choosing appropriate data representations

Examining multiple data sets is a common practice in statistics. To display similar sets of data, appropriate tabular and/or graphical representations are selected to enable comparisons of similarities and differences.

Side-by-side bar charts or back-to-back bar charts are used when comparing multiple sets of categorical data.

Back-to-back stem-and-leaf plots and dot plots are used to compare distributions of numerical data.

WORKED EXAMPLE 22

A sample of New South Wales residents were asked to state the environmental issue that was most important to them. The results were sorted by the age of the people surveyed. The results were as follows.

	Environmental issue		
Age	**Reducing pollution**	**Conserving water**	**Recycling rubbish**
Under 30	52	63	41
Over 30	32	45	23

a. **What type of data display would you use to compare the data sets?**
b. **Construct a side-by-side bar chart.**
c. **How many people surveyed were over 30?**
d. **What percentage of under 30s surveyed listed conserving water as the most important environmental issue?**

THINK

a. What type of data is being compared?

WRITE

a. Categorical data is being compared. A side-by-side bar chart or back-to-back bar chart would be the most suitable data display.

▶

b. Construct a side-by-side column graph for the data.

b.

Important environmental issues for NSW residents

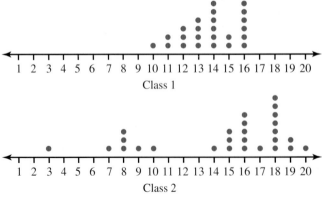

Legend: Under 30, Over 30

c. Total the response of the over 30s.

c. For over-30s there were 32 responses for reducing pollution, 45 for conserving water and 23 for recycling rubbish.

The total number of people over 30 who were surveyed is $32 + 45 + 23 = 100$.

d. To calculate the percentage of under 30s who listed conserving water as an important issue, use

$$\text{percentage} = \frac{\text{frequency}}{\text{total frequency}} \times 100\%.$$

d. Percentage $= \dfrac{63}{52 + 63 + 41} \times 100\%$

$= \dfrac{63}{156} \times 100\%$

$= 40.4\%$

Approximately 40% of residents under 30 listed conserving water as the most important environmental issue.

Exercise 5.6 Comparing and interpreting data

1. The two dot plots below display the latest Maths test results for two Year 11 classes. The results show the marks out of 20.

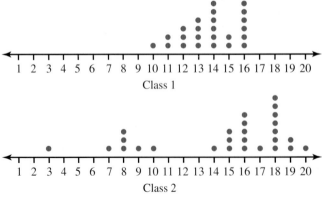

Class 1

Class 2

a. How many students are in each class?

b. For each class, how many students scored 15 out of 20 for the test?

c. For each class, how many students scored more than 10 for the test?

d. Use the dot plots to describe the performance of each class on the test.

2. **WE12** The following data shows the ages of male and female players at a ten-pin bowling centre. Draw a back-to-back stem-and-leaf plot of the data.

Male: 20, 36, 16, 38, 32, 18, 19, 21, 25, 45, 29, 60, 31, 21, 16, 38, 52, 43, 17, 28, 23, 23, 43, 17, 22, 23, 32, 34

Female: 21, 23, 30, 16, 31, 46, 15, 17, 22, 17, 50, 34, 65, 25, 27, 19, 15, 43, 22, 17, 22, 16, 48, 57, 54, 23, 16, 30, 18, 21, 28, 35

3. The comparisons between the battery lives of two mobile phone brands are shown in the back-to-back stem-and-leaf plot below. Which mobile phone brand has the better battery life? Explain.

Key: 6 | 1 = 61 hours

Brand A	Stem	Brand B
8 8 7 5	0	7
9 7 4 1 0	1	0 5 5 5 7 9
2 2 2 1	2	0 2 2 6 7
8 6 4 2 0	3	0 2 4 6 8
	4	
	5	6
1	6	
	7	5

4. **WE14** A city newspaper surveyed a sample of New South Wales residents about an upcoming state election on the issues most important to each age group. The results were as follows:

	Election issue					
Age	Marriage equality	Education	Refugees and immigration	Tax and superannuation	Health	Housing affordability
18–40	18	35	38	43	22	47
Over 40	12	38	19	61	41	32

a. What type of data display would you use to compare the data sets?
b. How many people surveyed were over 40?
c. What issue is most important to the people aged over 40?
d. What percentage of under 40s surveyed listed marriage equality as the most important election issue? Give your answer to the nearest whole number.

5. Ten workers were required to complete two tasks. Their supervisor observed the workers and gave them a score for the quality of their work on each task, where higher scores indicated better quality work. The results are indicated in the following side-by-side bar chart.

a. Which worker had the largest difference between scores for the two tasks?

b. How many workers received a lower score for task B than task A?

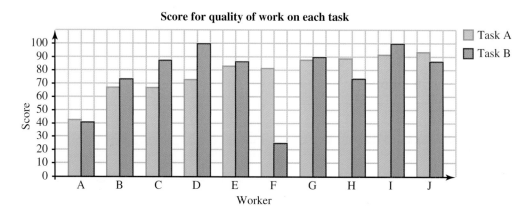

6. The following graphical display summarises the ages at the last birthday of patients seen by two doctors in a medical surgery during one particular day.

a. How many more patients aged under 15 consulted Doctor B compared to Doctor A?

b. Doctor A tends to consult patients aged over 45, whereas Doctor B tends to consult patients aged under 45. True or false?

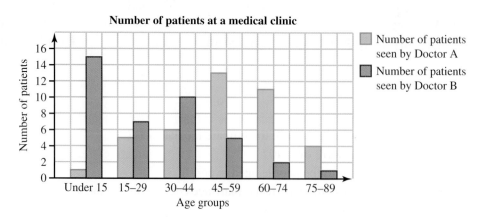

7. Two households estimate the electricity consumed by different household appliances and devices as follows.

Appliance or device	Household 1 electricity consumption (%)	Household 2 electricity consumption (%)
Water heater	29	32
Refrigerator	18	21
Stove/cooktop	21	24
Washing machine	6	10
Lighting	8	7
Computer	7	2
Audiovisual equipment	2	2
Air-conditioner	6	2
Heating	3	3

Assuming that both households use the same amount of electricity overall, answer true or false to the following statements.

a. Household 1's computer uses more electricity than household 2's computer.

b. Household 2 uses less electricity in heating and air-conditioning than household 1.

c. Household 2 watches less TV than household 1.

8. The daily number of hits a fashion blogger gets on her new website over 3 weeks are:

126   356   408   404   420   425   176

167   398   433   446   419   431   189

120   431   390   495   454   215   117

At the same time, the daily number of hits a healthy lifestyle blogger gets on his new website over 3 weeks are:

240   156   462   510   420   474   520

225   402   426   563   621   339   195

320   621   340   495   700   415   371

a. Compare the two data sets using an appropriate graphical display.

b. Comment on the two data sets.

9. The winning times in seconds for the women's and men's 100 metre sprint in the Olympics are shown below.

Year	Women's 100 m sprint	Men's 100 m sprint
1928	12.2	10.8
1932	11.9	10.3
1936	11.5	10.3
1948	11.9	10.3
1952	11.5	10.4
1956	11.5	10.5
1960	11.0	10.2
1964	11.4	10.0
1968	11.0	9.9
1972	11.07	10.14
1976	11.08	10.06
1980	11.60	10.25
1984	10.97	9.99
1988	10.54	9.92
1992	10.82	9.96
1996	10.94	9.84
2000	10.75	9.87
2004	10.93	9.85
2008	10.78	9.69

a. Display the winning times for women and men using a stem-and-leaf plot.
b. Is there a large difference in winning times? Explain your answer.

10. The following data sets show the rental price (in $) of two-bedroom apartments in two different suburbs of Wollongong.

Suburb A

215   225   211   235   244   210   215   210   256   207

200   200   242   225   231   205   240   205   235   200

Suburb B

235   245   231   232   240   280   280   270   255   275

275   285   245   265   270   255   260   258   251   285

a. Draw a back-to-back stem-and-leaf plot to compare the data sets.
b. Compare and contrast the rental price in the two suburbs.

11. A side-by-side bar chart shows the distribution of New South Wales road fatalities from March 2016 to March 2017.

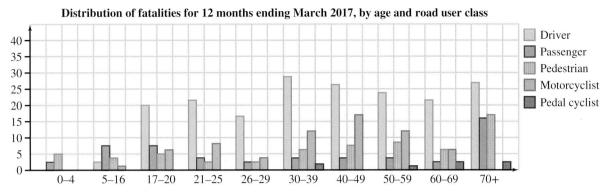

Distribution of fatalities for 12 months ending March 2017, by age and road user class

a. Which age group had the most passenger and pedestrian fatalities? Give a reason why this might be.
b. The New South Wales government wants to introduce a road campaign. Which age groups should the government focus on? Explain your answer.
   Source: Transport for NSW, Centre for Road Safety

12. The horizontal side-by-side bar chart at right shows a monthly comparison of the New South Wales road fatalities.

a. Using the data from 2015 and 2016, which year had the most fatalities on New South Wales roads?
b. Which year has had the most fatalities from January to March?
c. Why do you think there were 44 fatalities in the month of April 2016?

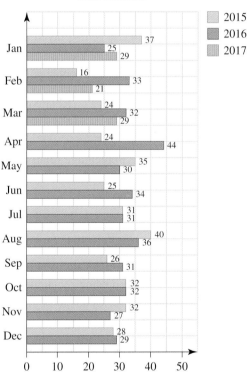

Monthly comparison of New South Wales road fatalities

Source: Transport for NSW, Centre for Road Safety

13. A coffee bar serves either skim, reduced-fat or whole milk in coffees. The coffees sold on a particular day are shown in the table below, sorted by the type of milk and the genders of the customers.

Type of milk	Gender	
	Male	Female
Skim	87	124
Reduced fat	55	73
Whole	112	49

a. How many coffees were sold on this day?
b. Represent this data in a graphical display.
c. What percentage of males used skim milk for their coffees? Give your answer to the nearest whole number.
d. What percentage of coffees sold contained reduced-fat milk? Give your answer to the nearest whole number.
e. If this was the daily trend of sales for the coffee bar, what percentage of the coffee bar's customers would you expect to be female? Give your answer to the nearest whole number.

5.7 Review: exam practice

5.7.1 Data classification, presentation and interpretation: summary

Classifying data

- There are two main types of data: numerical data and categorical data.
- Numerical data can be either discrete or continuous.
- Discrete data are countable observations such as number of siblings or number of pets.
- Continuous data are measurable observations such as height, length and temperature.
- Categorical data can be either nominal or ordinal.
- Nominal data, or distinct categories, cannot be ranked.
- Ordinal data, or ordered categories, can be ranked or placed in order.
- There are instances when a categorical data is expressed by numerical values. 'Phone numbers ending in 5' and 'postcodes' are examples of categorical data represented by numerical values.

Categorical data

- One-way tables or frequency distribution tables organise and display data in the form of frequency counts for one variable only.
- Two-way tables or two-way frequency distribution tables are tables that display the relationship between two variables.
- Categorical data can be displayed as proportions or percentages. Proportions are called relative frequencies and percentages are called percentage relative frequency. Percentage relative frequency tables display data as percentages.

$$\text{Relative frequency} = \frac{\text{score}}{\text{total number of scores}}$$

$$\text{Percentage relative frequency} = \frac{\text{score}}{\text{total number of scores}} \times 100\%$$

or

$$\text{Percentage relative frequency} = \text{Relative frequency} \times 100\%$$

- Grouped column graphs are used to display the data for two or more categories.
  - Data is displayed using rectangles of equal width.
  - The same data series has to be represented by identical rectangles: scale, colour, and texture.
  - The columns representing the same category have no gaps.
  - Different categories are separated by spaces.
  - There is a half space before the first set of rectangles.
  - Both axes are clearly labelled including units if used.
  - The title should be explicitly stating what the grouped column graph represents.
  - Clear and accurate scales are required for both axes.
  - The frequency is measured by the height of the column.
  - Add a key if necessary.
- To avoid misleading grouped column graphs look for the following.
  - Title and labels
  - Starting value for the vertical and/or horizontal axis
  - Clear and accurate scales
  - Distortions in the area or volume of columns
  - Bright colours that exaggerate columns

- Patterns and textures that cause optical illusions
- Perspective that obstruct values.

Numerical data

- Numerical data, also called quantitative data, is information collected in the form of numerical values. Numerical data is discrete data for countable observations and continuous data for measurable observations.
- Frequency distributions organise and display data in the form of frequency counts for one variable.
- Dot plots are graphical representations of numerical data made up of dots where each dot represents one score of the variable. The graph has one horizontal line labelled with the scores of the variable. If a score occurs more than once, the dots are placed in a vertical line on top of each other. The height of the vertical line of dots represents the frequency for that score.
- Histograms are graphical representations of numerical data using rectangles of equal width with no spaces in between. However, there is a space before the first rectangle. Each column represents a different score of the same variable. The vertical axis always displays the values of the frequency. For ungrouped data the value of the variable is written in the middle of the column under the horizontal axis. For grouped data the ends of the class interval are written under the ends of the corresponding columns.
- Stem-and-leaf plots are graphical representations of grouped numerical data with two columns marked 'stem' and 'leaf'. Once all scores are recorded, they have to be placed in an increasing order. A key is essential in the interpretation of the graph.

Measures of central tendency and the outlier effect

- Central tendency is the centre or the middle of a frequency distribution and is defined by the mean, median or mode of the set of data
- The mean or the average of a set of data is calculated using the formula
$\bar{x} = \dfrac{\sum x_i}{n}$ where $\sum x_i$ is the sum of all scores and n is the total number of data scores
- The median of a set of data is the middle score when the scores are arranged in order of size.
- The mode is the score with the highest frequency.
- The measures of central tendency for grouped data are mean, median class, and modal class.
- An outlier is a score that has a value that is far removed from the rest of the data.
- When outliers are present in the data set, the median is a better measure of central tendency than the mean.

Comparing and interpreting data

- Data sets can be compared using back-to-back stem-and-leaf plots and other graphical displays as appropriate.

Exercise 5.7 Review: exam practice

Simple familiar

1. Categorical data can be graphically represented using a:
   A. histogram.
   B. dot plot.
   C. stem-and-leaf plot.
   D. column graph.
2. An example of numerical discrete data is the:
   A. depth of water in a dam over a period of one month.
   B. number of children enrolled in 20 high schools.
   C. daily average temperatures in April 2015 at a meteorological station.
   D. weekly average time spent doing household chores for a group of 46 students.

3. The Australian Bureau of Statistics collects data on the price of unleaded petrol. What type of data is being collected?

    A. nominal   **B.** ordinal   **C.** discrete   **D.** continuous

4. An outlier is a score:

    A. with an extreme value far less than the minimum score or far greater than the maximum score.

    B. outside the interquartile range.

    C. less than Q_3 but greater than Q_1.

    D. that lies between the median and the maximum score of the data set.

5. Which graph represents the data as a grouped vertical column graph?

    A.

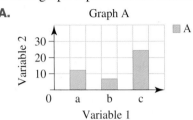

    B.

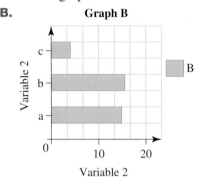

    C.

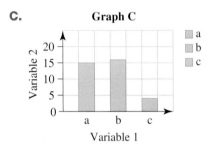

    D.

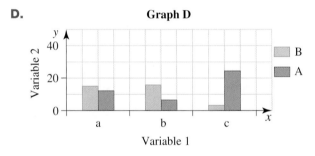

6. Choose the incorrect statement.

    A. Dot plots are used to represent categorical data.

    B. Each dot in a dot plot represents a score.

    C. A dot plot requires a number line to show the scores.

    D. The spaces between dots in a dot plot are equal.

7. Consider the following set of data that represents the heights, in cm, of 40 children under the age of 3.

58.9	54.6	62.9	45.7	49.6	62.8	43.2	56.7	69.0	65.3
57.8	59.4	66.7	49.6	72.7	35.2	72.8	75.6	32.9	56.7
54.8	45.2	69.5	47.3	68.6	67.4	63.4	61.5	52.6	48.0
56.3	78.4	39.9	75.3	45.2	56.9	66.3	55.4	63.8	72.3

A suitable graphical representation for the data set given is:

    A. a histogram with ungrouped data.

    B. a histogram with grouped data.

    C. a stem-and-leaf plot.

    D. both **B** and **C**.

8. Sara is training for the next 50 m freestyle swimming race. Her coach has recorded Sara's times in seconds over 20 consecutive days.

26.5   26.3   25.8   25.6   25.7   25.6   25.5   25.3   24.8   25.0

25.6   25.1   25.5   25.4   25.6   25.5   25.6   25.7   24.6   25.2

The mode of this set of data is:
A. 26.5.   B. 25.5.   C. 24.6.   D. 25.6.

9.

Outcome	Frequency	fx
1	12	12
2	31	62
3	15	45
4	24	96
5	19	95
6	8	48

The mean, median and mode respectively for the data set displayed in the frequency table shown are:
A. 5.19, 3 and 2.   B. 2, 3.28, and 3.   C. 3, 3.28 and 4.
D. 3.28, 3 and 2.

10. In a group of 165 young people aged 16–20 years, 28 females have a learner's driver license while 15 males have a P driver license. Out of the 117 young people who have a license, 59 are learners. If the total number of males surveyed was 73, the number of females who have a P driver license is:

Type of driver license for young people aged 16–20 years:			
	Gender		
Driver license	**Male**	**Female**	**Total**
L	31	28	59
P	15	43	58
None	27	21	48
Total	73	92	165

A. 92.   B. 58.   C. 64.   D. 43.

11. The data sets displayed in the back-to-back stem-and-leaf plot shown can be described by:
A. the data set on the right is symmetrical while the data set on the left is positively skewed.
B. the data set on the left is symmetrical while the data set on the right is negatively skewed.
C. the data set on the left is symmetrical while the data set on the right is positively skewed.
D. the data set on the right is symmetrical while the data set on the left is negatively skewed.

Left	Stem	Right
	10	03445
52	11	012223589
887511	12	014669
75543210	13	33789
9886420	14	3568
9662	15	168
733	16	02
	17	04
	18	6

Stem units: 10

12. As part of his preparation for the Essential Mathematics exam, Andrew completes a weekly test. The results, in percentages, for his last 9 tests are:

67, 61, 69, 66 68, 66, 71, 69, 69

How will the mean, median and mode of this set of data be affected if he scored 100% in the 10th test?

A. The mean increases, the median increases and the mode does not change.

B. The mean increases, the median does not change and the mode does not change.

C. The mean does not change, the median increases and the mode does not change.

D. The mean does not change, median does not change and the mode increases.

Complex familiar

13. Classify the following data as either discrete or continuous.

a. Data is collected on the water levels in a dam over a period of one month.

b. An insurance company is collecting data on the number of work injuries in ten different businesses.

c. 350 students were selected by a phone company to answer the question 'How many pictures do you take per week using your mobile phone?'

d. A group of students were asked to measure the perimeter of their bedrooms. This data was collected and used for a new project.

e. 126 people participated in a 5 km charity marathon. Their individual times were recorded.

f. Customers in a shopping centre were surveyed on the number of pairs of shoes they bought in the last 12 months.

14. A road worker recorded the number of vehicles that passed through an intersection in the morning and in the afternoon and summarised the data in a two-way table. A total of 47 heavy vehicles passed through the intersection during the day and 126 light vehicles in the afternoon. Of the 238 vehicles that passed through the intersection in the morning, 199 were light vehicles.

a. Construct a two-way table for this set of data.

b. How many vehicles passed through the intersection during the whole day?

c. How many heavy vehicles passed through the intersection in the afternoon?

15. The numbers of the immediate family members for the teachers in a small school is given.

5 6 3 8 2 2 3 4 3 4 6 7

2 5 4 5 4 6 1 3 4 5 5 4

a. Present this set of data in a frequency table.

b. Present this data as a dot plot.

16. A car company surveyed a group of 250 people on the number of cars per household. The data is displayed in the frequency distribution table below.

    a. Calculate the entries in the column labelled *fx*.
    b. Calculate the mean of this data set correct to two decimal places.
    c. What is the median of this data set?
    d. Does this data set have a mode? What is it?

Number of cars	Frequency	*fx*
1	36	
2	107	
3	89	
4	13	
5	5	
Total	250	

17. A small family business has 6 fulltime employees with the following salaries:

    $54 283.10, $55 681.70, $64 709.30, $69 281.50, $63 458.30, $54 283.10

    As the business is growing, there is a need to employ a new manager on a salary of $93 591.80. How is this salary going to affect the mean and the median of the given data set?

Complex unfamiliar

18. Calculate the percentage relative frequency, correct to one decimal place, for all the entries in the two-way table given.

Eye colour and hair colour				
Hair colour	Eye colour			Total
	Hazel	Blue	Green	
Light	49	38	26	113
Dark	106	12	29	147
Total	155	50	55	260

19. Consider the grouped bar graph shown that displays the numbers of people lodging visa applications in Australia, by sex and age, in 2011–12.

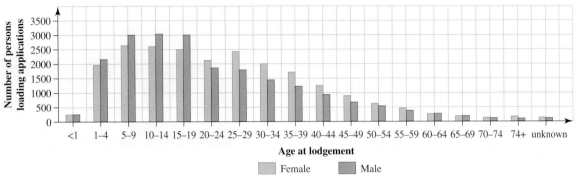

Notes:
1. Visas counted include subclass 200 (Refugee), 201 (In-Country Special Humanitarian Program), 202 (Global Special Humanitarian Program), 203 (Emergency Rescue) and 204 (Woman at Risk).
2. Data was extracted from Departmental systems on 06 July 2012.

a. How many males in the age bracket 30–34 year of age lodged a visa application in 2011–12?
b. How many children aged between 1–4 years of age lodged a visa application in 2011–12?
c. In what age brackets was the number of female applicants equal to the number of male applicants?
d. What age bracket has the highest proportion of female applicants?

20. The local childcare centre, HappyTodd is attended by 37 children aged between 2–4 years. In order to provide an appropriate diet for these toddlers, their weights, in kilograms, were recorded at enrolment.

14.5 12.4 16.3 13.0 13.6 12.9 15.4 14.7 14.3 14.9

13.8 15.2 15.4 16.1 13.2 13.8 13.5 13.4 15.1 12.7

14.2 13.6 14.4 14.1 14.7 14.8 14.2 14.3 13.5 13.1

14.6 15.3 14.2 15.5 15.7 15.8 16.5

a. Display this data in a grouped frequency table using intervals of 0.5 kg including a column for the midpoint of the class intervals.
b. Construct a histogram to represent this data set.
c. Estimate the average weight at enrolment of the toddlers at HappyTodd childcare centre by first adding a column for fx in the frequency distribution table.
d. State the minimum and the maximum weights, hence calculate the range of this data set.

Answers

Chapter 5 Data classification, presentation and interpretation

Exercise 5.2 Classifying data

1. a. Numerical continuous
   b. Numerical discrete

2. The number of fruit is discrete as it is represented by counting numbers. The quantity of fruit sold is categorical because it represents a measurement; there is always another value between any two values.

3. a. Continuous   b. Discrete
   c. Discrete   d. Discrete
   e. Continuous   f. Continuous

4. a. Categorical nominal
   b. Categorical ordinal

5. 'Favourite movies' identifies a category; it cannot be ordered. 'The opinion about a movie' can be ordered in categories like poor, average, very good and excellent.

6. a. Ordinal   b. Nominal
   c. Ordinal   d. Nominal

7. a. Categorical nominal
   b. Numerical discrete

8. Although 'Country phone codes' are numerical, they are neither countable nor measurements. They represent categories. 'The number of international calls' are numerical data as they are countable data.

9. a. Numerical continuous
   b. Categorical ordinal

10. Data collected on 'the length of pencils' is a measurement while data collected on 'the colour of pencils' can be represented in colour categories.

11. D

12. a. Numerical   b. Categorical
    c. Categorical   d. Numerical

Exercise 5.3 Categorical data

1.

Mode of transport to school		
Mode of transport	Tally	Frequency
Car	ﬀﬀ ﬀ II	7
Bicycle	IIII	4
Walk	ﬀﬀ ﬀﬀ	10
Tram	ﬀﬀ IIII	9
Total		30

2.

Favourite social network		
Social network	Tally	Frequency
MyPage	ﬀﬀ ﬀﬀ ﬀﬀ	15
FlyBird	ﬀﬀ IIII	9
Total		24

3.

Internet connection		
Type of internet connection	Tally	Frequency
C	III	3
NBN	ﬀﬀ ﬀﬀ ﬀﬀ	15
MW	ﬀﬀ ﬀﬀ ﬀﬀ II	17
Total		35

4.

Opinion	Frequency
Excellent	163
Very good	298
Average	24
Poor	15
Total	500

5.

Parking place and car theft			
Theft	Parking type		Total
	In driveway	On the street	
Car theft	2	37	39
No car theft	16	445	461
Total	18	482	500

6.

Mobile phones and tablets			
Mobile phones	Tablets		Total
	Own	Do not own	
Own	58	78	136
Do not own	116	11	127
Total	174	89	263

7. a.

Year 11 students and mathematics			
Subject	Gender		Total
	Girls	Boys	
Essential mathematics	12	23	35
General mathematics	25	49	74
Mathematical methods	36	62	98
Specialist mathematics	7	18	25
Total	80	152	232

   b. 232 students
   c. 35 students

8. a.

Hairdressing requirements per gender			
Hairdressing requirement	Gender		Total
	Males	Females	
Haircut			
Haircut and colour			
Total			

b.

Hairdressing requirements per gender			
Hairdressing requirement	Gender		Total
	Males	Females	
Haircut			
Haircut and colour	2		
Total		56	

c.

Hairdressing requirements per gender			
Hairdressing requirement	Gender		Total
	Males	Females	
Haircut only	5	26	31
Haircut and colour	2	30	32
Total	7	56	63

9. a.

Monday matinee audience per gender			
Audience type	Gender		Total
	Females	Males	
Adults	66	30	96
Teenagers	5	16	21
Children	11	28	39
Total	82	74	156

b. 30

c. 39

10. a. Relative frequencies

Cat and dog owners			
Cat owners	Dog owners		Total
	Own	Do not own	
Own	0.039	0.256	0.295
Do not own	0.495	0.210	0.705
Total	0.534	0.466	1.000

b. Percentage relative frequencies

Cat and dog owners			
Cat owners	Dog owners		Total (%)
	Own (%)	Do not own (%)	
Own (%)	3.9	25.6	29.5
Do not own (%)	49.5	21.0	70.5
Total	53.4	46.6	100.0

11. a. Relative frequencies

Swimming attendance			
Day of the week	People		Total
	Adults	Children	
Monday	0.0350	0.0775	0.1125
Tuesday	0.0225	0.0425	0.0650
Wednesday	0.0250	0.0550	0.0800
Thursday	0.0375	0.0950	0.1325
Friday	0.0450	0.1125	0.1575
Saturday	0.0675	0.1475	0.2150
Sunday	0.0800	0.1575	0.2375
Total	0.3125	0.6875	1.000

b. Percentage relative frequencies

Swimming attendance			
Day of the week	People		Total
	Adults (%)	Children (%)	
Monday (%)	3.50	7.75	11.25
Tuesday (%)	2.25	4.25	6.50
Wednesday (%)	2.50	5.50	8.00
Thursday (%)	3.75	9.50	13.25
Friday (%)	4.50	11.25	15.75
Saturday (%)	6.75	14.75	21.50
Sunday (%)	8.00	15.75	23.75
Total (%)	31.25	68.75	100.00

12. a. Relative frequency table

Year 11 students and mathematics			
Subject	Gender		Total
	Girls	Boys	
Essential mathematics	0.0517	0.0991	0.1509
General mathematics	0.1078	0.2112	0.3190
Mathematical methods	0.1552	0.2672	0.4224
Specialist mathematics	0.0302	0.0776	0.1078
Total	0.3448	0.6551	1.0000

b. Percentage relative frequency table

Year 11 students and mathematics

Subject	Gender		Total (%)
	Girls (%)	Boys (%)	
Essential mathematics (%)	5.17	9.91	15.09
General mathematics (%)	10.78	21.12	31.90
Mathematical methods (%)	15.52	26.72	42.24
Specialist mathematics (%)	3.02	7.76	10.78
Total (%)	34.48	65.52	100.00

Note: Due to rounding, some of the percentages written do not add up to the actual value in some of the total columns. To have the correct values in the total columns ensure that the actual values (no rounding) are summed up.

13. a.

Dance class enrolments

Dance	People			Total
	Adults	Teenagers	Children	
Latin	31	21	23	75
Modern	13	34	49	96
Contemporary	36	49	62	147
Total	80	104	134	318

b. Relative frequency table

Dance class enrolments

Dance	People			Total
	Adults	Teenagers	Children	
Latin	0.10	0.07	0.07	0.24
Modern	0.04	0.11	0.15	0.30
Contemporary	0.11	0.15	0.19	0.45
Total	0.25	0.33	0.41	1.00

Note: Due to rounding, some of the percentages written do not add up to the actual value in some of the total columns. To have the correct values in the total columns ensure that the actual values (no rounding) are summed up.

c. Percentage relative frequency table

Dance class enrolments

Dance	People			Total (%)
	Adults (%)	Teenagers (%)	Children (%)	
Latin (%)	10	7	7	24
Modern (%)	4	11	15	30
Contemporary (%)	11	15	19	45
Total (%)	25	33	41	100

14.

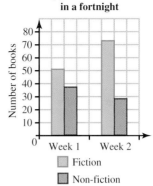

Types of books sold in a fortnight

15.

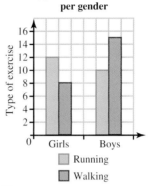

Types of fitness exercise per gender

16. B

17. See figure at the foot of the page*.

18. The vertical axis starts at 0.80 with no break shown. The differences between columns look exaggerated by choosing a non-zero base line. The percentage difference is quite minimal, approximately 7% between the people who never finished high school and the people who have a postgraduate degree. Here is the graph with a base line of zero:

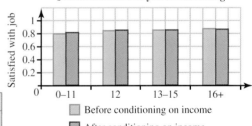

Job satisfaction and years of schooling

*17.

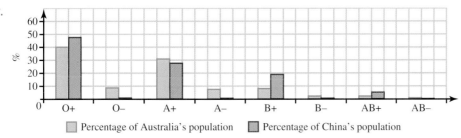

19. The vertical axis starts at 5 with no break shown. The differences between columns look exaggerated by choosing a non-zero baseline. This makes the difference between the columns look greater than they actually are.

20. **a.** See table at foot of the page.*
b. 10
c. 225
d.

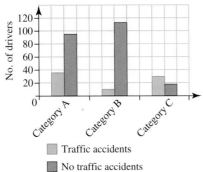

Traffic accidents per age categories

☐ Traffic accidents
■ No traffic accidents

Exercise 5.4 Numerical data

1.

Number of hours of sleep	Tally	Frequency
4	|	1
5	|||	3
6	卌	5
7	卌 |	6
8	卌	5
9	|||	3
10	||	2
Total		25

2.

Foot length	Tally	Frequency
21	|	1
22	|	1
23	卌	5
24	||||	4
25	||	2
26	卌 ||	6
27	|||	3
28	卌 |	5
29	|||	3
Total		30

3.

Time	Tally	Frequency
17−<22	卌 卌	10
22−<27	||||	4
27−<32	||||	4
32−<37	卌	5
37−<42	卌 |	6
42−<47	||	2
47−<52	|	1
53−<57	|||	3
57−<62	|	1
Total		36

4.

Weekends	Tally	Frequency
0	卌 卌 |	11
1	卌 ||||	9
2	卌	5
3	|||	3
4	||	2
Total		30

*20.a.

Drivers and traffic accidents				
	Drivers			
Event	Category A Young drivers	Category B Mature drivers	Category Older drivers	Total
Traffic accident	35	10	30	75
No traffic accident	95	113	17	225
Total	130	123	47	300

5.

Width	Tally	Frequency
33.0−<34.0	||	2
34.0−<35.0	⊪⊬ ||	7
35.0−<36.0	⊪⊬ ⊪⊬ ⊪⊬	15
36.0−<37.0	⊪⊬ ||||	9
37.0−<38.0	|||	3
Total		36

6. **Number of songs downloaded**

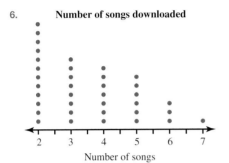

7. **Number of pairs of shoes**

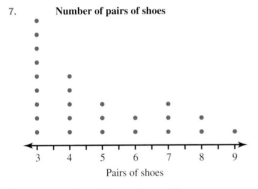

8. **Ages of participants in a spelling contest**

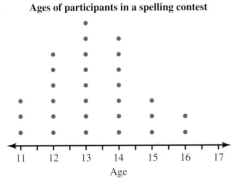

9. **Typing test results**

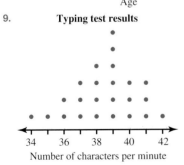

10. **Number of emails sent per day by Year 11 students**

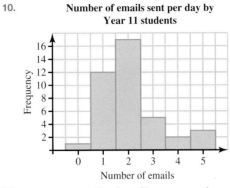

11. **Number of languages spoken**

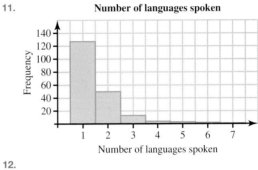

12. **Time spent searching the internet**

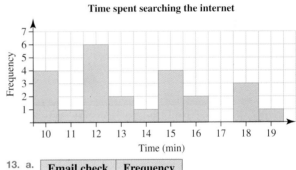

13. a.

Email check	Frequency
0−<3	11
3−<6	12
6−<9	8
9−<12	11
12−<15	5
15−<18	3

Email check per weekend

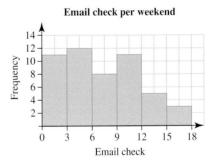

14.

Road fatalities in Victoria, 12 Months to October 2012

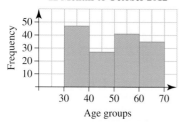

15. a.

Money earned	Frequency
0 —<50	22
50 —<100	14
100 —<150	4
150 —<200	3
200 —<250	3
250 —<300	1
300 —<350	2
350 —<400	1
Total	50

b.

Weekly wage from part-time jobs

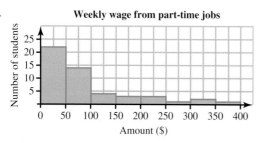

16.

Stem	Leaf
1*	6 7 8 8 9 9
2	0 0 1 1 2 2 3 3 3 3 4 4 4
2*	5 5 9
3	0 2

key: 1|6 = 16

17.

Stem	Leaf
1	2 8
2	0 0 3 9 9
3	2 3 5 6 8 9
4	1 3 3 4 5 8 8 9
5	0 0 6

key: 1|2 = 12

18.

Stem	Leaf
20	4 6 7
21	0 0 2 6 8 9
22	0 0 5 7 9 9
23	0 1 2
24	3 3 4 5
25	0 0 0 1 3
26	0 1 5

key: 20|4 = 204

19.

Heights of children under the age of 3

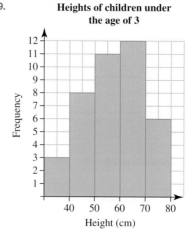

20.

Reaction time with non-dominant hand

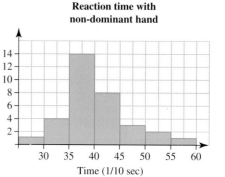

Exercise 5.5 Measures of central tendency and the outlier effect

1. Mean = 3 letters

2. Mean = 160.75 cm

3. a. Mean = 16   **b.** Mean = 195.8

4. a.

Lifetime (months)	Frequency	fx
21	9	189
22	18	396
23	44	1012
24	67	1608
25	29	725
26	20	520
27	13	351
Total	200	4801

b. Mean = 24.0 months

5. a.

Wage ($ /annum)	Frequency	fx
57 797	37	2 138 489
62 880	24	1 509 120
66 517	16	1 064 272
70 771	21	1 486 191
75 248	8	601 984
85 458	5	427 290
90 609	2	181 218
Total	113	7 408 564

b. Average = $65 562.5

6. a. Position 14.   **b.** Median = 10

7. Median = 87.1m

8. a. Median = 5.7m   **b.** Median = 161

9. Mean = 7.25

10. Mode = 16

11. Mode = 24 years

12. Mean = 9.2

13. a. 21   **b.** 144.8

14. a.

Number of bookings	Frequency	fx
1	27	27
2	32	64
3	19	57
4	26	104
5	12	60
6	3	18
7	1	1
Total	120	337

b. Mean = 2.81 trips per given month
c. Median = 3 trips per given month
d. Mode = 2 trips per given month

15. a.

Number of logins	Frequency	fx
1	23	23
2	45	90
3	69	207
4	51	204
5	38	190
6	17	102
Total	243	816

b. Mean = 3.36 subscribers
c. Median = 3 subscribers
d. Mode = 3 subscribers

16. a. Mean = 9.5, median = 8, mode = 8
b. Mean = 10.2, median = 8, mode = 8. When size 24 is included, this large score increases the mean from 9.5 to 10.2. The median and the mode are not affected by this score.
c. It is more appropriate to use the median or the mode for this data set as extreme sizes highly affect the value of the mean.

Exercise 5.6 Comparing and interpreting data

1. a. Class 1: 25; Class 2: 27
b. Class 1: 2; Class 2: 3
c. Class 1: 24; Class 2: 20
d. Eleven students in class 2 scored better than class 1 but 6 students scored worse than class 1. Class 2's results are more widely spread than class 1's results.

2. Key: 1 | 6 = 16 years

Leaf (female)	Stem	Leaf (male)
9 8 7 7 6 6	1	5 5 6 6 6 7 7 7 8 9
9 8 5 3 3 3 2 1 1 0	2	1 1 2 2 2 3 3 5 7 8
8 8 6 4 2 2 1	3	0 0 1 4 5
5 5 3	4	3 6 8
2	5	0 4 7
0	6	5

3. Brand B seems to have a better battery life, as Brand A has more batteries that have a battery life of less than 10 hours. Brand B has a battery life of up to 75 hours, which is higher than Brand A's battery life maximum of 61 hours.

4. a. Side-by-side bar chart
b. 203
c. Tax and superannuation
d. 9%

5. a. Worker F   **b.** 4 (A, F, H & J)

6. a. 14.   **b.** True

7. a. True   **b.** True   **c.** False

8. a.

Number of hits	Fashion blogger	Healthy lifestyle blogger
100 —<150	3	0
150 —<200	3	2
200 —<250	1	2
250 —<300	0	0
300 —<350	0	3
350 —<400	3	1
400 —<450	9	4
450 —<500	2	3
500 —<550	0	2
550 —<600	0	1
600 —<650	0	2
650 —<700	0	0
700 —<750	0	1

See figure at foot of the page.*

b. Answers will vary. Example answer:
The fashion blogger had 400 —<450 hits on her website on 9 days in the 3-week period. The healthy lifestyle blogger got a wider spread of hits. He received over 400 hits for 11 days in the 3-week period. His website seems to be gaining in popularity as its spread falls over a higher maximum and minimum than the spread of the fashion blog.

9. a. Key: 9 | 69 = 9.69

Leaf: Female	Stem	Leaf: Male
	9	69
	9	84 85 87 90 92 96 99
	10	00 06 14 20 25 30 30 30 40
97 94 93 82 75 54	10	50 80
40 08 07 00 00	11	
90 90 60 50 50 50	11	
20	12	

b. Answers will vary. There is not a large difference within each gender, but there is a large difference in time between the two genders.

10. a. Key: 23 | 1 = $231

Leaf: Suburb *A*	Stem	Leaf: Suburb *B*
7 5 5 0 0 0	20	
5 5 1 0 0	21	
5 5	22	
5 5 1	23	1 2 5
4 2 0	24	0 5 5
6	25	1 5 5 8
	26	0 5
	27	0 0 5 5
	28	0 0 5 5

b. Answers will vary. Suburb A has lower rent than Suburb B. Suburb B is a more expensive suburb to rent an apartment.

11. a. 70+ age group. Answers about the reason will vary. Elderly people don't drive as much and tend to walk or be passengers.

b. Answers will vary. The road campaign could focus on:
- elderly people as pedestrians — being alert when crossing a road
- motorcyclists in the 30–49 age group, not on probationary licence — risk taking
- driving for all age groups — awareness and safety for all drivers.

12. a. 2016
b. 2016
c. Answers will vary but could include that more people are on the roads during the Easter break and school holidays.

*8.a.

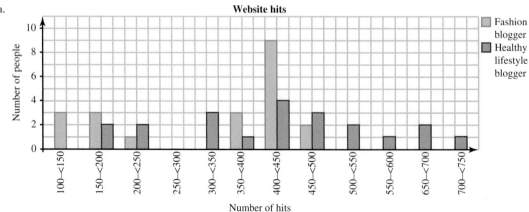

13. a. 500

b.

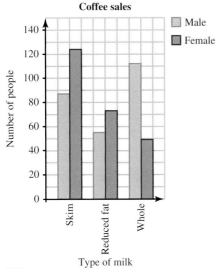

c. 34%

d. 26%

e. 49%

Exercise 5.7 Review

Simple familiar

1. D	**2.** B
3. D	**4.** A
5. D	**6.** A
7. D	**8.** D
9. D	**10.** E
11. C	**12.** A

Complex familiar

13. a. Numerical continuous

  b. Numerical discrete

  c. Numerical discrete

  d. Numerical continuous

  e. Numerical continuous

  f. Numerical discrete

14. a.

Number of vehicles			
	Type of vehicle		
Time of the day	**Heavy vehicles**	**Light vehicles**	**Total**
Morning	39	199	238
Afternoon	8	126	134
Total	47	325	372

b. 372 vehicles

c. 8 heavy vehicles

15. a.

Family members	Frequency
1	1
2	3
3	4
4	6
5	5
6	3
7	1
8	1

b.

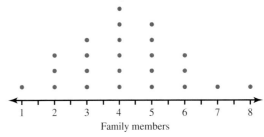

16. a.

Number of cars	Frequency	fx
1	36	36
2	107	214
3	89	267
4	13	52
5	5	25
Total	250	594

b. Mean = 2.38

c. Median = 2

d. The data set has a mode of 2.

17. The mean increases from $60 282.80 to $65 041.30. The median increases from $59 570 to $63 458.30.

Complex unfamiliar

18.

Eye colour and hair colour				
	Eye colour			
Hair colour	**Hazel**	**Blue**	**Green**	**Total**
Light	18.8%	14.6%	10%	43.5%
Dark	40.8%	4.6%	11.1%	56.5%
Total	59.6%	19.2%	21.1%	100%

19. a. 1400

b. 4200

c. Babies under 1 year of age and 65–69 year olds.

d. 5–9 years of age

20. a.

Weight (kg) class	Midpoint (kg)	Frequency
12.0 —<12.5	12.25	1
12.5 —<13.0	12.75	2
13.0 —<13.5	13.25	4
13.5 —<14.0	13.75	6
14.0 —<14.5	14.25	7
14.5 —<15.0	14.75	6
15.0 —<15.5	15.25	5
15.5 —<16.0	15.75	3
16.0 —<16.5	16.25	2
16.5 —<17.0	16.75	1
Total		37

b.

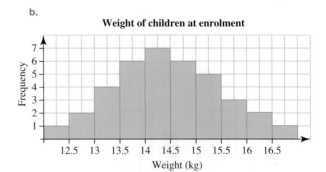

Weight of children at enrolment

c.

Weight class (kg)	Midpoint (kg)	Frequency	fx (kg)
12.0—<12.5	12.25	1	12.25
12.5—<13.0	12.75	2	25.50
13.0—<13.5	13.25	4	53.00
13.5—<14.0	13.75	6	82.50
14.0—<14.5	14.25	7	99.75
14.5—<15.0	14.75	6	88.50
15.0—<15.5	15.25	5	76.25
15.5—<16.0	15.75	3	47.25
16.0—<16.5	16.25	2	32.50
16.5—<17.0	16.75	1	16.75
Total		37	534.25

Mean = 14.4 kg

d. Minimum weight = 12.4 kg, maximum weight = 16.5 kg, range = 4.1 kg

PRACTICE ASSESSMENT 1

Essential Mathematics: Problem solving and modelling task

Unit
Unit 1: Money, measurement and relations

Topic
Topic 1: Consumer arithmetic

Conditions

Duration	Mode	Individual/group
5 weeks	Written report, up to 8 pages (maximum 1000 words) excluding appendix	Individual
Resources permitted		
The use of technology required, for example: • calculator • spreadsheets • Internet • other mathematical software		

Criterion*	Grade
Formulate *Assessment objectives 1, 2, 5	
Solve *Assessment objectives 1, 6	
Evaluate and verify *Assessment objectives 4, 5	
Communicate *Assessment objective 3	
Milestones	
Week 1	
Week 2	
Week 3	
Week 4	
Week 5 (assessment submission)	

*Queensland Curriculum & Assessment Authority, *Essential Mathematics Applied Senior Syllabus 2019 v1.1*, Brisbane, 2018.
   For the most up to date assessment information, please see www.qcaa.qld.edu.au/senior.

Context

Years 11 and 12 are seen to be very important years of your schooling as it is during these years that you will decide whether to continue with further study at university or TAFE, take up a trade, or go into the workforce. In these year levels there is an increased work load and hence an increase in the number of hours per week that you need to spend doing homework. This task looks at the number of hours students spend doing homework per week.

Data can be misleading. You need to analyse the data given, and then compare it to data you will collect in your school.

Task

Part A

A class of 20 Year 11 students were asked to record how many hours a week they spent on homework. The results are shown below:

Hours per week spent on homework									
9	12	8	9	11	18	21	14	0	5
7	7	10	14	20	3	17	14	19	24

Analyse the data and complete the following:
- Determine measures of central tendency.
- Create a graphical representation of the data.
- Create a stem-and-leaf plot of the data.

Part B

Data can be misleading. The data above is from one class, but in this section of the task you need to consider if it is representative of your Year 11 class.
To determine if this data is representative of your class you will need to complete the following:
- collect data from students in your own class
- analyse the data from your class and compare it to the data given by completing the same analysis as you conducted on the original data set
- after making the comparison, what can you conclude about the data given and the data collected? Use mathematical terms and graphs to justify your conclusions.

To complete this task you must:

- Present your findings as an investigative report based on the problem-solving and mathematical modelling approach outlined in the Essential Mathematics syllabus, and on the flow chart on the following page of this instrument.
- Provide a response that highlights the real-life application of mathematics and demonstrates your knowledge of the Unit 1 subject matter.
- Develop a unique response that is appropriately structured.
- Show all working to support your response.

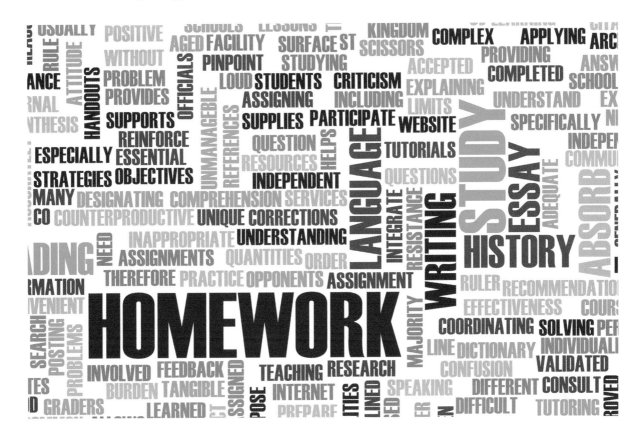

Approach to problem-solving and modelling

In this task you will investigate the number of hours per week spent on homework for a Year 11 student. Your report should consider:

- measures of central tendency
- graphical representation of the data
- analysis and comparison of the two data sets.

Design a detailed plan, identifying the mathematical procedures required to solve this problem. Remember to state the necessary assumptions, variables and observations. You must also explain how you will make use of technology.

Solve

Construct graphs to graphically represent the data so it can be analysed and conclusions can be drawn on any patterns or observations. Link the analysis back to the context (weekly study hours). You will make further refinements and comparisons as necessary.

You must use technology efficiently and show detailed calculations demonstrating the procedures used.

Is it solved?

Evaluate and verify

Evaluate the reasonableness of your original solution.

Based on your representation of the data, consider whether it could be represented better to emphasise your main discussion points.

Justify and explain all procedures you have used and decisions you have made. Considering the original task, how valid is your solution?

Is the solution verified?

Communicate

Once you have completed all necessary work, you should consider how you have communicated all aspects of your report. Communicate using appropriate language that refers to the calculations, tables and graphs included in previous sections. Your response should be coherently and concisely organised.

Ensure you have:

- used mathematical, statistical and everyday language
- considered the strengths and limitations of your solution
- drawn conclusions by discussing your results
- included recommendations.

CHAPTER 6
Reading, interpreting and using graphs

6.1 Overview

CONTENT

In this chapter, students will learn to:
- interpret information presented in two-way tables
- interpret information presented in graphs, such as step graphs, column graphs, pie graphs, picture graphs, conversion graphs of calories ↔ kilojoules, line graphs using units of energy to describe consumption of electricity, including kilowatt hours
- discuss and interpret tables and graphs, including misleading graphs found in the media and in factual texts [complex]
- use graphs in practical situations
- interpret graphs in practical situations [complex].

Fully worked solutions for this chapter are available in the Resources section of your eBookPLUS at www.jacplus.com.au

6.2 Two-way tables

6.2.1 Reading and interpreting two-way tables

Two-way tables were introduced in chapter 5. In this section, we consider in more detail how to read and interpret two-way tables.

The table shown is a two-way table that relates two variables, the ability to swim and gender.

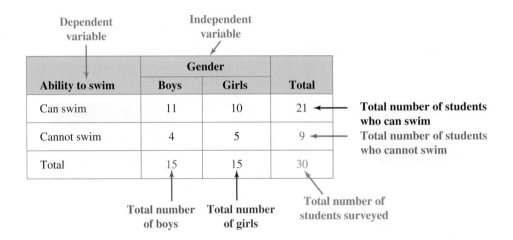

How do we interpret this information? If we want to find the number of boys who can swim, draw an imaginary line through the row 'Can swim' and an imaginary line through the column 'Boys'. The cell where the two imaginary lines meet has the answer: 11 boys can swim.

| | Gender | | |
Ability to swim	Boys	Girls	Total
Can swim	11	10	21
Cannot swim	4	5	9
Total	15	15	30

The other cells of the table represent:

| | Gender | | |
Ability to swim	Boys	Girls	Total
Can swim	11 boys who can swim	10 girls who can swim	21 students who can swim
Cannot swim	4 boys who cannot swim	5 girls who cannot swim	9 students who cannot swim
Total	15 boys surveyed	15 girls surveyed	30 students surveyed

Two new school policies were introduced at Burumin High School:
School policy 1.
The school uniform has to be worn on the way to school as well as on the way home from school.

School policy 2.
Mobile phones are not to be used at school between 8:30 am and 3:30 pm.
The two-way table shown displays the responses to a survey conducted on 200 students.

School policy 2	School policy 1		Total
	Agree	Disagree	
Agree	68	14	82
Disagree	103	15	118
Total	171	29	200

a. How many students agree with both new school policies?
b. How many students agree with School policy 1 but disagree with School policy 2?
c. What is the total number of students who disagree with School policy 2?
d. How many students disagree with School policy 1 but agree with School policy 2?

THINK

a. 1. Draw a horizontal imaginary line through the row 'Agree' with School policy 2 and a vertical imaginary line through the column 'Agree' with the School policy 1.

2. Read the number of students in the box where the two lines meet.

b. 1. Draw a horizontal imaginary line through the row 'Disagree' with School policy 2 and a vertical imaginary line through the column 'Agree' with School policy 1.

WRITE/DRAW

School policy 2	School policy 1		Total
	Agree	Disagree	
~~Agree~~	68	~~14~~	~~82~~
Disagree	103	15	118
Total	171	29	200

68 students agree with both new school policies.

School policy 2	School policy 1		Total
	Agree	Disagree	
Agree	68	14	82
~~Disagree~~	103	~~15~~	~~118~~
Total	171	29	200

2. Read the number of students in the box where the two lines meet.

103 students agree with School policy 1 but disagree with School policy 2.

c. 1. Draw a horizontal imaginary line through the row 'Disagree' with School policy 2 and a vertical imaginary line through the column 'Total'.

School policy 2	School policy 1		Total
	Agree	Disagree	
Agree	68	14	82
Disagree	103	15	118
Total	171	29	200

118 students disagree with School policy 2.

2. Read the number of students in the box where the two lines meet.

d. 1. Draw a horizontal imaginary line through the row 'Agree' with School policy 2 and a vertical imaginary line through the column 'Disagree' with School policy 1.

School policy 2	School policy 1		Total
	Agree	Disagree	
Agree	68	14	82
Disagree	103	15	118
Total	171	29	200

2. Read the number of students in the box where the two lines meet.

14 students disagree with School policy 1 but agree with School policy 2.

Exercise 6.2 Two-way tables

1. **WE1** The students in a school were surveyed on whether they wear glasses for reading or not. The data is displayed in the two-way table shown.

Reading glasses	Gender		
	Girls	Boys	Total
Wear glasses for reading	26	23	49
Do not wear glasses for reading	291	313	603
Total	317	335	652

a. How many students were surveyed altogether?
b. How many girls wear glasses for reading?
c. How many students do not wear glasses for reading?
d. How many boys were surveyed altogether?

2. The two-way table shown displays the way a group of 150 students check the weather.

Students	Weather app		Total
	Weather app	TV news	
Junior (years 7–10)	21	47	68
Senior (years 11 and 12)	75	7	82
Total	96	54	150

a. How many students check the weather using a weather app?
b. How many senior students check the weather watching the TV news?
c. How many junior students check the weather using a weather app?
d. How many junior students were surveyed?

3. The contingency table shown below displays the information gained from a medical test screening for a virus. A positive test indicates that the patient has the virus.

Virus	Test results		Total
	Accurate	Not accurate	
With virus	45	3	48
Without virus	922	30	952
Total	967	33	1000

a. How many patients were screened for the virus?
b. How many positive tests were recorded? (That is, in how many tests was the virus detected?)
c. What percentage of test results were accurate?
d. Based on the medical results, if a positive test is recorded what is the percentage chance that you actually have the virus?

4. The two-way table below indicates the results of a radar surveillance system. If the system detects an intruder, an alarm is activated.

Detection	Test results		Total
	Alarm activated	Not activated	
Intruders	40	8	48
No intruders	4	148	152
Total	44	156	200

a. Over how many nights was the system tested?
b. On how many occasions was the alarm activated?
c. If the alarm is activated, what is the percentage chance that there actually is an intruder?
d. If the alarm was not activated, what is the percentage chance that there was an intruder?
e. What was the percentage of accurate results over the test period?
f. Comment on the overall performance of the radar detection system.

5. **MC** For the two-way table shown, the number of male nurses is:

Career	Gender		Total
	Female	Male	
Teachers	58	26	84
Nurses	77	39	116
Total	135	65	200

A. 77   **B.** 65   **C.** 39   **D.** 116

6. The two-way table shown displays the relationship between two categorical variables: gender and pierced ears.

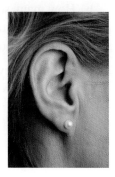

Pierced ears	Gender		Total
	Boys	Girls	
Yes	2	35	37
No	41	7	48
Total	43	42	85

a. How many girls have pierced ears?
b. What is the total number of students surveyed?
c. How many boys were surveyed?
d. What is the total number of students that do not have their ears pierced?

The information below is to be used in questions **7** to **9**.
A test for a medical disease does not always produce the correct result. A positive test indicates that the patient has the condition. The table indicates the results of a trial on a number of patients who were known to either have the disease or known not to have the disease.

Disease status	Test results		Total
	Accurate	Not accurate	
With disease	57	3	60
Without disease	486	54	540
Total	543	57	600

7. **MC** The overall accuracy of the test is:
   A. 9.5%   **B.** 90%   **C.** 90.5%   **D.** 92.5%
8. **MC** Based on the table, what is the percentage of patients with the disease who have had it detected by the test?
   A. 9.5%   **B.** 90%   **C.** 90.5%   **D.** 95%

9. **MC** Which of the following statements is correct?
    A. The test has a greater accuracy with positive tests than with negative tests.
    B. The test has a greater accuracy with negative tests than with positive tests.
    C. The test is equally accurate with positive and negative test results.
    D. The test is equally inaccurate with positive and negative test results.
10. Airport scanning equipment is tested by scanning 200 pieces of luggage. Prohibited items were placed in 50 bags and the scanning equipment detected 48 of them. The equipment detected prohibited items in five bags that did not have any forbidden items in them.
    a. Use the above information to complete the contingency (or two-way frequency) table below.

Bags	Test results		Total
	Accurate	**Not accurate**	
With prohibited items			
With no prohibited items			
Total			

    b. Use the table to answer the following:
        i. What percentage of bags with prohibited items were detected?
        ii. What was the percentage of false positives among the bags that had no prohibited items?
        iii. What percentage of prohibited items pass through the scanning equipment undetected?
        iv. What is the overall percentage accuracy of the scanning equipment?

6.3 Line graphs, conversion graphs and step graphs

6.3.1 Graphical representations of data

Data is often presented in the form of graphs. Graphical representations of data are easier to read and interpret. In this section we are going to discuss line graphs, conversion graphs and step graphs.

Line graphs

Line graphs are used to represent the relationship between two numerical continuous datasets.

These graphs consist of individually plotted points joined with a straight line.

The graph shown represents the relationship between two numerical continuous datasets: the temperature readings (°C) in Canberra recorded hourly on 7 August 2018.

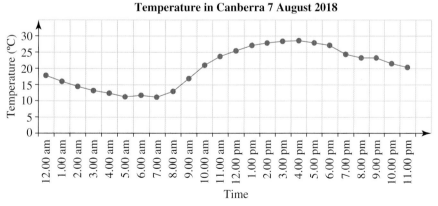

Temperature in Canberra 7 August 2018

Source: Adapted from Bureau of Meteorology.

Information we can gather from this graph:
- the minimum (lowest) temperature of the day and the time when it occurred (10.9 °C at 7:00 am)
- the maximum (highest) temperature of the day and the time when it occurred (28.5 °C at 4:00 pm)
- the temperature at any hour of the day
- the time of the day when a certain temperature occurred
- the times between the temperature was increasing (from 7:00 am to 4:00 pm)
- the times between the temperature was decreasing (from 4:00 pm to 8:00 pm).

WORKED EXAMPLE 2

The graph shown is a representation of the temperature readings (°C) in Sydney on 13 March 2018 recorded hourly.

a. State the maximum temperature in Sydney on 13 March 2018.

b. State the minimum temperature in Sydney on 13 March 2018.

c. State the times between which the temperature in Sydney was increasing.

d. What was the temperature in Sydney at 12:00 pm?

e. At what times was the temperature in Sydney 24 °C?

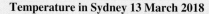

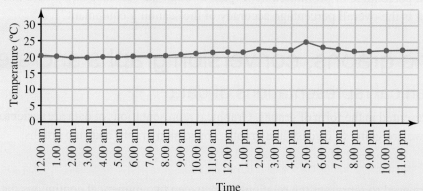

Source: Adapted from Bureau of Meteorology, http://www.bom.gov.au/products/IDN60901/IDN60901.94768.shtml

THINK

a. 1. The maximum temperature occurs at the highest point on the graph.

WRITE/DRAW

2. State the temperature shown on the graph.

Approximately 25 °C at 5.00 pm

b. 1. The minimum temperature occurs at the lowest point on the graph.

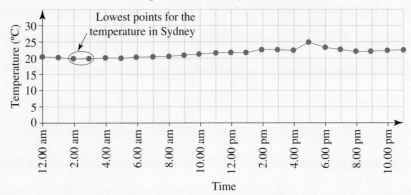

2. State the temperature shown on the graph.

Approximately 20 °C between 2.00 am and 3:00 am

c. 1. The increase in temperature is shown by a line that slopes upwards from left to right.

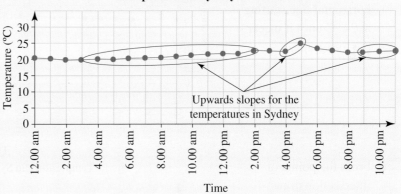

2. State the times required.

The temperature increase in Sydney on 13 March 2018 occurred between 2.00 am and 2:00 pm, between 4:00 pm and 5:00 pm and between 9:00 pm and 11:00 pm.

d. 1. To find the temperature at 12:00 pm, draw a vertical line from 12:00 pm until it meets the line graph.

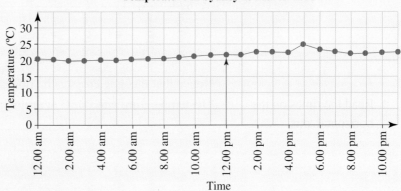

2. From the point of intersection draw a horizontal line to the left until it crosses the vertical axis.

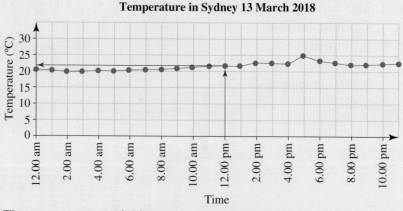

3. Read the temperature on the vertical axis.

The temperature at 12:00 pm was approximately 22 °C.

e. 1. To find the time when the temperature is 24 °C, draw a horizontal line starting at 24 °C until it meets the line graph.

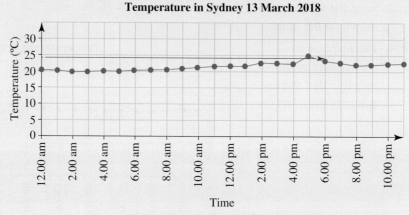

2. From the points of intersection draw vertical lines down until they cross the horizontal axis.

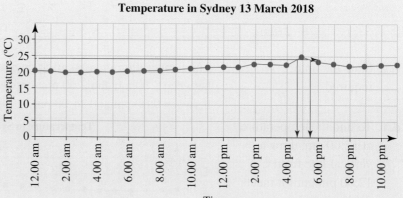

Notice that this line crosses the line graph in two points.

3. Read the times on the horizontal axis.

The temperature was 24 °C at around 4:45 pm and 5:30 pm.

6.3.2 Conversion graphs

As the name explicitly states, **conversion graphs** are just that: line graphs that convert one quantity into another.

Conversion graphs are often used to convert between foreign currencies or between measurements. These graphs are always represented by a straight line because the relationship between the two variables does not change.

Temperature can be measured in either degrees Celsius (°C) or degrees Fahrenheit (°F). In Australia we measure temperature using degrees Celsius. Other countries, such as the USA, measure temperature using degrees Fahrenheit. The graph shown represents this relationship.

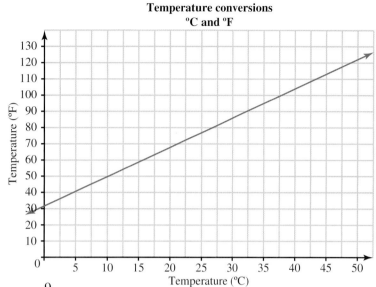

The relationship between the two measurements is given by the formula $°F = 32 + \dfrac{9}{5}°C$.

WORKED EXAMPLE 3

The conversion table shown displays the relationship between speed given in km/h and speed given in m/s. Use the conversion graph to convert

a. 180 km/h into m/s

b. 100 m/s into km/h.

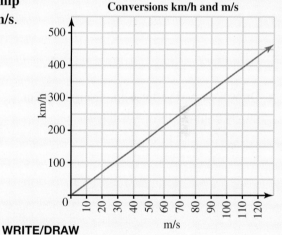

THINK

a. 1. Draw a dashed horizontal line starting at 180 km/h until it touches the graph.

WRITE/DRAW

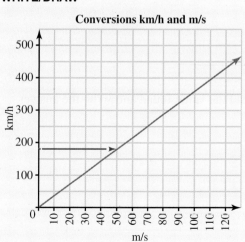

2. Draw a dashed vertical line from the point on the graph to the horizontal axis.

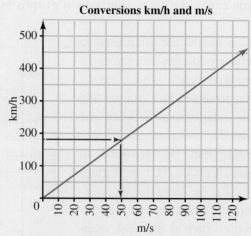

3. Read the value found and write the answer.

A speed of $180\,\text{km/h} = 50\,\text{m/s}$

b. 1. Draw a dashed vertical line starting at $100\,\text{m/sec}$ until it touches the graph.

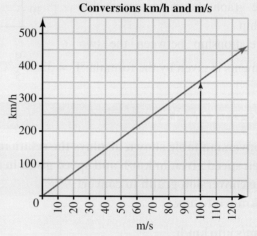

2. Draw a dashed vertical line from the point on the graph to the vertical axis.

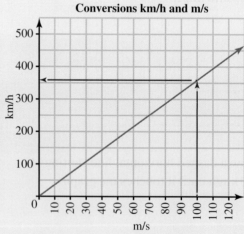

3. Read the value found and write the answer.

A speed of $100\,\text{m/s} = 360\,\text{km/h}$

6.3.3 Step graphs

Step graphs are made up of horizontal straight lines. These graphs are used when the value remains constant over intervals. Postage and parking costs are examples of step graphs.

The first step represents a cost of $0.60 for envelopes with a weight of less than 20 g. This means that regardless of whether the envelope weighs 10 g or 19 g, the postage cost is $0.60.

The open circle at the end of the interval means that the postage cost for 20 g is not $0.60, it jumps to the next step which is $1.20. This is why the next step has a closed dot at the beginning of the interval.

- The closed dot, ●, means that the interval includes that value.
- The empty circle, ○, means that the interval does not include that value.

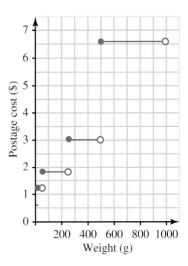

WORKED EXAMPLE 4

The step graph shown displays the parking cost in a car park. Describe the costs involved if you were to use this car park.

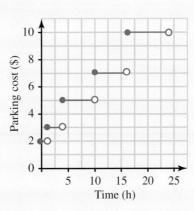

a. What is the parking cost for 4 hours?

b. What is the parking cost for 15 hours?

c. What is the parking cost for 23 hours?

THINK	WRITE/DRAW
a. 1. Describe the first step of the graph.	The cost for parking for one hour or less is $2.
2. Describe the second step of the graph.	As soon as an hour has passed, the cost for between 1 hour and up to 4 hours of parking is $3.
3. State the parking cost.	The four-hour mark is on the second step so the cost will be $3
b. 1. Describe the third step of the graph.	Step three shows a cost of $5 for between 4 hours and up to 10 hours .

2. Describe the fourth step of the graph.	Step four shows a cost of $7 for parking between 10 hours and up to 16 hours.
3. State the parking cost.	The 15 hour mark is on the fourth step so the cost will be $7.
c. 1. Describe the fifth step of the graph.	Step five shows a cost of $10 for parking between 16 hours and up to 24 hours.
2. State the parking cost.	The 23-hour mark is on the fifth step so the cost will be $10.

6.3.4 Comparing line graphs

Line graphs are very useful to help compare similar types of data. The line graph shown displays the proportion of ongoing employees working part-time by gender for the period 1998 to 2012.

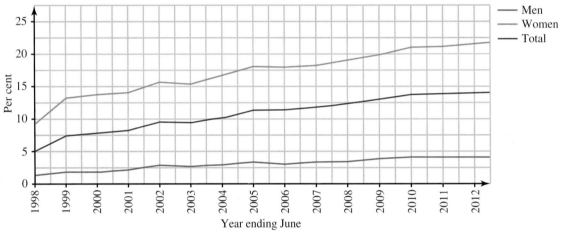

Source: Australian Public Service Commission (APSCD), http://www.apsc.gov.au/about-theapsc/ parliamentary/state-of-the-service/new-sosr/06-diversity

This graph clearly shows that the percentage of women who work part-time is higher than the percentage of men working part-time for the period 1998 to 2012.

WORKED EXAMPLE 5

The line graph shown displays the total number of employees by gender for the period 1998 to 2012.

a. When was the number of employed women equal to the number of employed men?

b. What was the number of employed women and the number of employed men in 2008?

WRITE/DRAW

a. 1. Mark the point where the two lines graphs intersect.

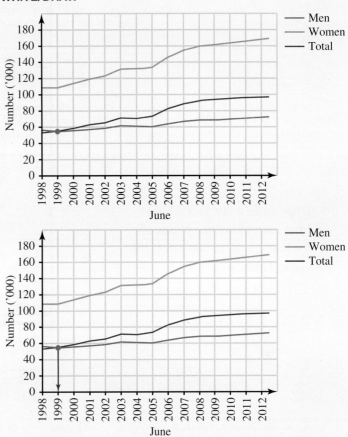

2. Draw a dotted vertical line from this point until it meets the horizontal axis.

3. Answer the question.

In 1999 the numbers of employed women and employed men are equal.

b. 1. Draw a dashed vertical line starting at 2008 until it touches the two line graphs.

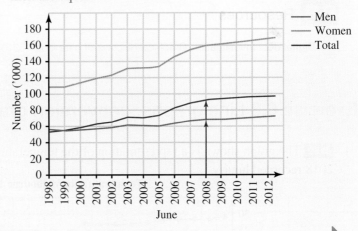

2. Mark the two points.

3. Draw dashed horizontal lines from each of the two points on the line graphs until it meets the vertical axis.

4. Read the values required.

The number of female employees in 2008 was 95 000 and the number of male employees was 70 000.

---

on Resources

📄 **Digital doc: Investigation** Graphical displays of data (doc-15910)

Exercise 6.3 Line graphs, conversion graphs and step graphs

1. **WE2** The graph shown is a representation of the temperature readings (°C) in Melbourne on 13 March 2018 recorded hourly.

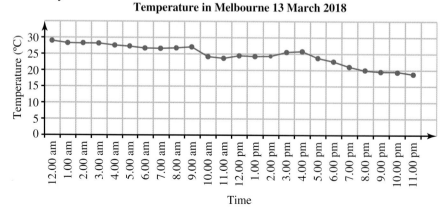

Temperature in Melbourne 13 March 2018

Source: Adapted from Bureau of Meteorology, http://www.bom.gov.au/products/IDV60901/IDV60901.94868.shtml

a. State the maximum temperature in Melbourne on 13 March 2018.
b. State the minimum temperature in Melbourne on 13 March 2018.
c. State the times between which the temperature in Melbourne was decreasing.
d. What was the temperature in Melbourne at 9.00 am?
e. At what times was the temperature in Melbourne 21 °C?

2. The line graph shown displays the average rainfall levels (mm) in Australia for the period 2006–2018.

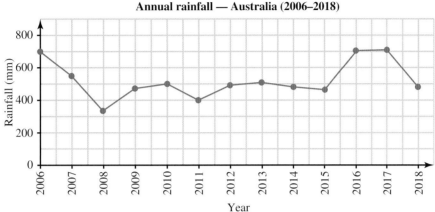

Annual rainfall — Australia (2006–2018)

Source: Adapted from Bureau of Meteorology, http://www.bom.gov.au/products/IDV60901/IDV60901.94868.shtml

a. When did the maximum average annual rainfall occur?
b. State the average annual rainfall in Australia for 2011.
c. State the years between which the average annual rainfall in Australia was increasing.
d. In what years was the average annual rainfall in Australia 500 mm?
e. When did the minimum average annual rainfall in Australia occur and what was it?

3. The line graph shown displays the daily maximum temperature in Launceston for the month of August 2018.

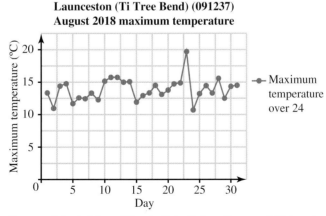

Launceston (Ti Tree Bend) (091237)
August 2018 maximum temperature

Source: Adapted from Climate Data Online, Bureau of Meteorology Copyright Commonwealth of Australia, 2013

Use the line graph given to answer the following questions:
a. What day had the lowest maximum temperature for the month and what was this temperature?
b. What day had the highest maximum temperature for the month and what was this temperature?
c. What is the difference between the maximum temperatures on 14 and 19 August 2018?
d. On what day was the maximum temperature 15.5 °C?
e. For how many days was the maximum temperature less than 12 °C?

4. The line graph shown displays the mean maximum temperature in Antarctica at the station Mawson for the years 1954–2018.

Location: 300001 MAWSON

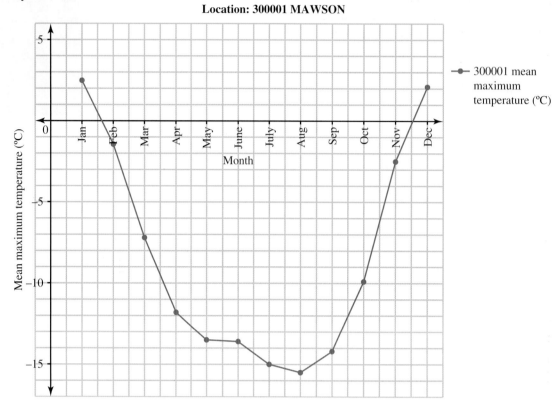

— 300001 mean maximum temperature (°C)

Source: Bureau of Meteorology

Use the line graph given to answer the following questions.

a. What month had the lowest mean maximum temperature in Antarctica and what was this temperature?

b. What month had the highest mean maximum temperature in Antarctica and what was this temperature?

c. What is the difference between the two temperatures found in **a** and **b**?

d. In what month was the mean maximum temperature in Antarctica –2.1°C?

e. For how many months was the mean maximum temperature in Antarctica greater than –8°C?

5. **WE3** The conversion table shown displays the relationship between the Australian dollar (A$) and the American dollar (US$).

Use the conversion graph to convert

a. US$80 into A$.

b. A$100 into US$.

Conversions A$ and US$

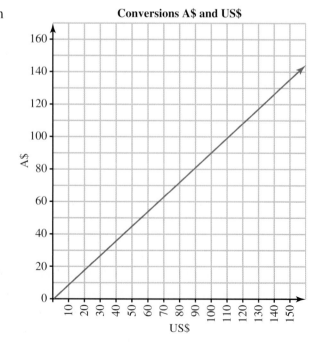

6. The conversion table shown displays the relationship between centimetres and inches.
   Use the conversion graph to convert
   a. 10 cm, 15 cm, 46 cm and 51 cm into inches
   b. 20 inches, 35 inches, 80 inches and 102 inches into centimetres.

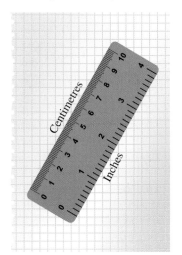

Conversions centimetres and inches

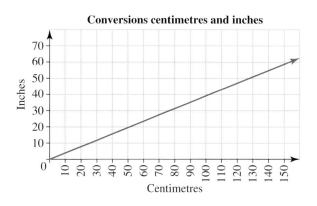

7. Use the temperature conversion graph given to answer the following questions:
   a. What is the temperature in °F equivalent to 35 °C?
   b. What is the temperature in °F equivalent to 10 °C?
   c. What is the temperature in °C equivalent to 41 °F?
   d. What is the temperature in °C equivalent to 104 °F?

Temperature conversions °C and °F

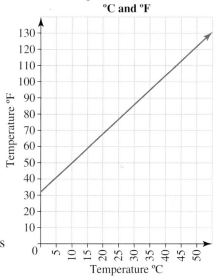

8. The graph shown represents a conversion graph between miles and kilometres.

Conversions miles to kilometres

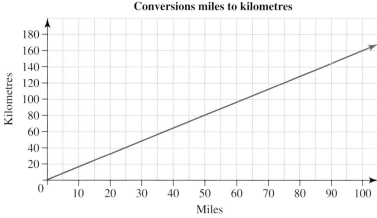

   a. Approximately how many kilometres equal 80 miles?
   b. Approximately how many kilometres equal 50 miles?
   c. Approximately how many miles equal 30 kilometres?
   d. Approximately how many miles equal 140 kilometres?

9. **WE4** The step graph shown represents the cost of parking at the domestic terminal at the Brisbane Airport.

a. What is the parking cost for 45 minutes?
b. What is the parking cost for 2 hours?
c. What is the parking cost for 3.5 hours?

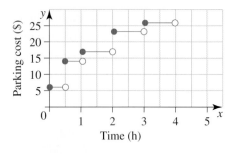

10. The cost of an international call is displayed in the step graph shown.

a. How much will a 5 minute call cost?
b. How much will a 50 minute call cost?
c. For how long would a call last if the cost of the call is $3.50?

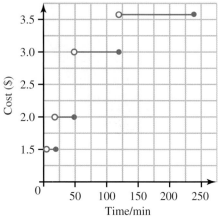

11. **WE5** The line graph shown displays the percentage of full-time Aboriginal and Torres Strait Islander secondary school students by gender in 2007.

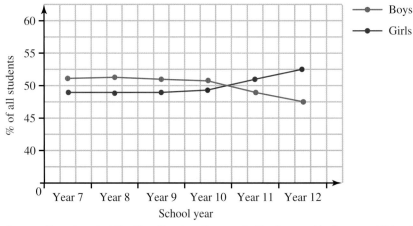

Source: Department of Families, Housing, Community Services and Indigenous Affairs, Australian Government http://www.fahcsia.gov.au/our-responsibilities/women/publications-articles/general/women-in-australia/women-in-australia-2009?HTML
Source: Australian Bureau of Statistics 2008, Schools, Australia, 2007, Catalogue No.4221.0, ABS, Canberra, Table 9. Figures do not include students categorised as 'ungraded'.

a. What was the percentage of Aboriginal and Torres Strait Islander boys and the percentage of Aboriginal and Torres Strait Islander girls enrolled in Year 7 in 2007?
b. What was the percentage of Aboriginal and Torres Strait Islander boys and the percentage of Aboriginal and Torres Strait Islander girls enrolled in Year 12 in 2007?
c. In what year level was the percentage of Aboriginal and Torres Strait Islander boys equal to the percentage of Aboriginal and Torres Strait Islander girls?

12. The line graph shown displays the gender wage gap in Australia from 1990 to 2009.

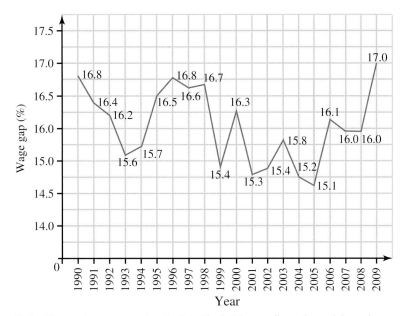

Note: The gender wage gap is calculated for full time, ordinary time adult employees, using original data. The reference period for data used in this figure is February for each year.
Source: ABS Average Weekly Earnings, Date cube, 2009, Cat No. 6302.0
http://www.fahcsia.gov.au/sites/default/files/documents/05_2012/gender_wage_gap.pdf

a. In what year was the wage gap between men and women at its lowest percentage and what was its value?

b. In what year was the wage gap between men and women at its highest percentage and what was its value?

c. What is the percentage increase in wage gap from 1999 to 2000?

d. Between what years was the highest decrease in the wage gap?

e. In what year was the wage gap 16.5%?

13. The line graph shown displays the percentages of male and female daily smokers aged 14 and over for the period 1995 to 2015.

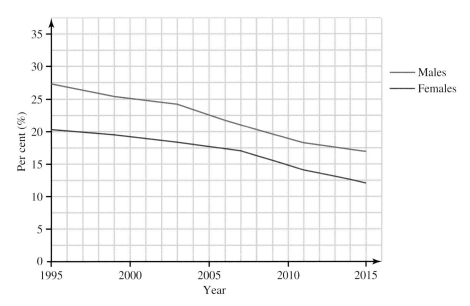

Source: http://www.abs.gov.au/ausstats/abs@.nsf/Lookup/by%20Subject/4364.0.55.001
~2014−15~Main%20Features~Smoking~24

a. What were the percentages of female and male daily smokers in 2013?
b. When was the percentage of female daily smokers maximum and what was the maximum?
c. When was the percentage of male daily smokers 25?

6.4 Column graphs and picture graphs

6.4.1 Some other types of graphs

There are many other types of graphs that can be used to display data. In this section we are going to look at column graphs and picture graphs.

Column graphs

Column graphs are made up of columns that can be either vertical or horizontal. These graphs are used to represent categorical data.

The column graph shown displays the power consumed by a household over a period of one year.

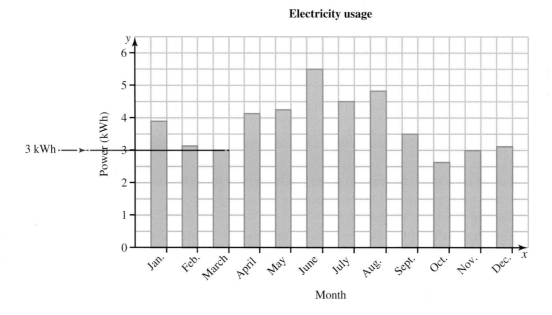

This column graph can be interpreted in the following ways.
- The use of electricity can be read on the vertical axis.
- The units are given in kWh (kilowatt hour), which means the power used over a period of one hour.
- Electricity use increased from the summer months (January and February) to the winter months (June, July and August).
- Electricity use decreased from the winter months (June, July and August) to the summer month (December).

Water bills display the water use in a household using column graphs. Describe the water consumption for the household represented by the column graph shown.

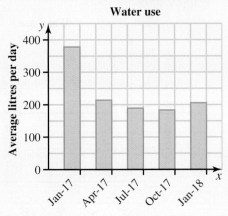

a. **What is the average number of litres used per day in the January 2017 bill?**
b. **What is the average number of litres used per day in the July 2017 bill?**
c. **What is the average number of litres used per day in the January 2018 bill?**
d. **Interpret the graph.**

THINK

a. 1. Draw a dashed horizontal line at the top of the column labelled Jan-17. Read the corresponding value on the vertical axis.

2. State the answer.

b. 1. Draw a dashed horizontal line at the top of the column labelled Jul-17. Read the corresponding value on the vertical axis.

WRITE/DRAW

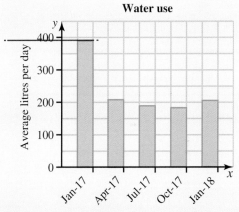

The average number of litres per day used in the first quarter of 2017 (January, February and March) is approximately 370 L.

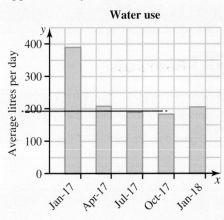

2. State the answer.

The average number of litres per day used in the third quarter of 2017 (July, August and September) is approximately 190 L.

c. 1. Draw a dashed horizontal line at the top of the column labelled Jan-18. Read the corresponding value on the vertical axis.

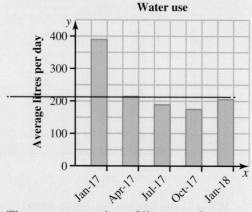

2. State the answer.

The average number of litres per day used in the first quarter of 2018 (January, February and March) is approximately 210 L.

d. 1. Interpret the graph.

This household decreased their usage of water from the first quarter 2017 to the last quarter of 2017.

More water is used in the quarter starting in January 2018 compared to the second half of 2017.

However, the average number of litres used per day in the quarter starting in January 2018 is a lot lower than in the quarter starting in January 2017.

Possible reasons include:
- there are fewer people in the household
- the garden was watered less because of water restrictions
- higher rainfall than average.

Resources

Interactivity: Column and bar graphs 1 (int-4058)

Interactivity: Column and bar graphs 2 (int-4059)

Digital doc: SpreadSHEET Column graphs (doc-3441)

6.4.2 Picture graphs

Picture graphs or **pictographs** are statistical graphs that use pictures to display categorical data.

The picture graph shown displays the average sales of four categories of fruit by a retailer on Mondays and Saturdays.

In order to be able to read the sales of the fruits, read the key underneath the graph.

The fruit shop sells an average of 5 kg of apples on Mondays, and 26 kg of apples on Saturdays.

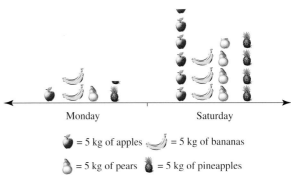

Average fruit sales on Mondays and Saturdays

Parts of a picture represent parts of the full quantity represented by the full picture.

5 kg   4 kg   3 kg   2 kg   1 kg

Pictographs are typically easy to read and show trends in data clearly. However, sometimes they are hard to read, especially when parts of the picture are shown.

Misleading picture graphs

There are times when picture graphs can be misleading. This can happen either on purpose or by mistake.

Purchasing power of the Canadian dollar, 1980 to 2000

1980 = $1.00

1985 = $0.70

1990 = $0.56

1995 = $0.50

2000 = $0.46

Source: Statistics Canada, http://www.statcan.gc.ca/edu/power-pouvoir/ch9/picto-figuratifs/5214825-eng.htm

The picture graph shown has the intention to express the idea that the value of the Canadian dollar decreased from $1.00 to less than a half, $0.46, over the 20 years from 1980 to 2000 due to inflation. However, the visual impact is a lot stronger because the difference between the area of the dollar representing $1.00 in 1980 is about three times larger than the area that represents $0.46 in 2000.

The picture graph shown below is a more accurate representation of this situation.

Canadian dollar

Using three-dimensional shapes can give the same misleading visual representations. The picture graph shown displays the number of plum puddings two friends Katie and Julia baked for their family and friends. Julia baked twice as many puddings as Katie.

In the first picture graph the quantity of plum puddings baked by Julia looks a lot greater than the quantity of plum puddings baked by Katie.

The visual impact of the first picture graph is a lot stronger than the second picture graph. This happens because all the dimensions of the picture have been doubled: length, width and height.

This picture graph is a more realistic representation of the situation.

WORKED EXAMPLE 7

A group of 50 students was asked to state which of the following foods they liked the most: pizza, hamburgers or sandwiches. The results are displayed in the picture graph shown.

a. How many students prefer hamburgers?

b. If 30 students preferred pizza, how many slices of pizza would have to be drawn?

c. If 16 students preferred sandwiches, how many sandwiches would have to be drawn?

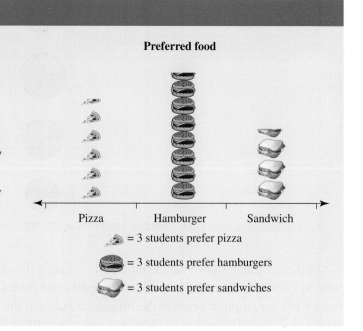

THINK	WRITE/DRAW

THINK

a. 1. Count the number of pictures that represent hamburgers.

2. Multiply the number of pictures by the corresponding number given in the key.

3. State the answer.

b. 1. Check the key for pizzas.

2. Find the number of pictures required and work out the answer.

3. State the answer.

c. 1. Check the key for sandwiches.

2. Find the number of pictures required and work out the answer.

3. State the answer

WRITE/DRAW

There are $7 + \frac{1}{3}$ hamburgers.

Each 🍔 represents 3 students

$$\left(7 + \frac{1}{3}\right) \times 3 = 7 \times 3 + \frac{1}{3} \times 3$$

$$= 21 + 1$$

$$= 22 \text{ students}$$

22 students prefer hamburgers.

One 🍕 represents 3 students who prefer pizza.

One 🍕 = 3 students

We need to find the number of pizza slices required for 30 students. Multiply both sides of the equation by 10.

10 slices of pizza = 30 students

30 students who prefer pizza would be represented in the graph by 10 slices of pizza.

One 🥪 represents 3 students who prefer sandwiches.

One 🥪 = 3 students

We need to find the number of 🥪 required for 16 students.

Step 1

Divide both sides of the equation by 3 to find the number of 🥪 required to represent one student.

$$\frac{1}{3} \text{🥪} = 1$$

Step 2

Multiply both sides of the equation by the number of students required

$$16 \times \frac{1}{3} \text{🥪} = 1 \text{ student} \times 16$$

$$\frac{16}{3} \text{🥪} = 16 \text{ students}$$

Step 3

Convert the improper fraction into a mixed number.

$$\frac{16}{3} \text{🥪} = 5\frac{1}{3} \text{🥪}$$

16 students who prefer sandwiches would be represented by $5\frac{1}{3}$ 🥪.

Exercise 6.4 Column graphs and picture graphs

1. **WE6** The column graph shown displays the average daily usage of gas for a household over a period of one year.

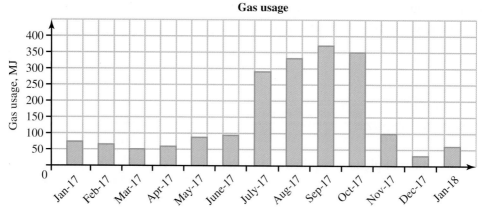

Gas usage

a. What is the average gas usage per day in June 2017?
b. What is the average gas usage per day in September 2017?
c. In what months was the average gas usage 60 MJ?
d. Interpret the graph.

2. The horizontal column graph shown displays the percentage of petrol volume sold by brand over the period 2013–14.

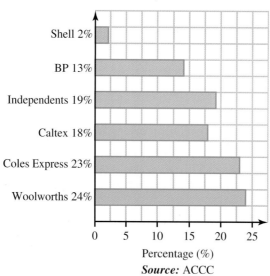

**Market shares: petrol volume
sold by brand 2013–14**

Source: ACCC

Source: Australian Institute of Petroleum (AIP), http://www.aip.com.au/
pricing/facts/Facts_About_the_Australian_Retail_Fuels_Market_and_Prices.htm

a. Which brand of petrol sold in 2013–14 had the highest volume percentage?
b. Which brand of petrol sold in 2013–14 had the lowest volume percentage?
c. What company sold 18% of the total volume of petrol in 2013–14?

3. The column graph shown displays the average daily usage of electricity for a household over a period of one year.

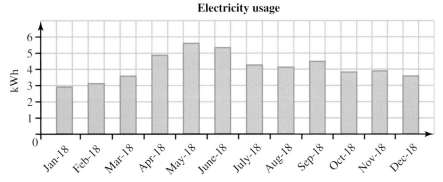

a. What is the average power usage per day in April 2018?
b. What is the average power usage per day in August 2018?
c. In what month was the minimum average power usage per day and what was its value?
d. Interpret the graph.

4. The column graph shown displays the top 10 countries of birth for the female overseas-born population.
a. What percentage of women were born in the Middle East?
b. From what country did the highest percentage of women come?
c. What percentage of women came from South Eastern Europe?

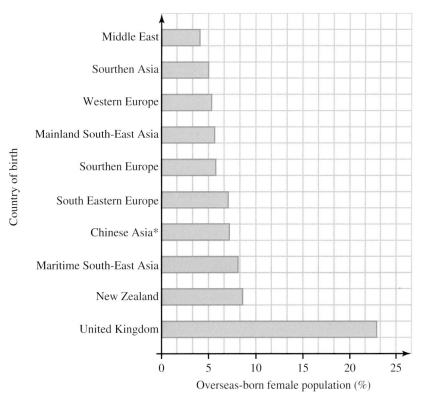

Source: Department of Families, Housing and Community Services and Indigenous Affairs,
http://www.abs.gov.au/ AUSSTATS/abs@.nsf/Lookup/3201.0Main+Features1Jun%202010?OpenDocument

5. The graph shown represents the population estimate for states and territories in Australia in 2007. State the approximate population of each of the Australian states and territories in 2007.

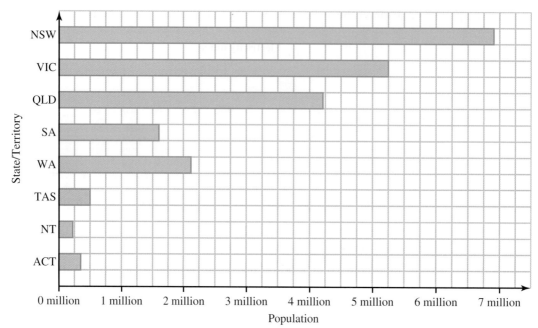

Source: ABS http://www.abs.gov.au/websitedbs/d3310114.nsf/Home/Animated+Historical+Population+Chart

6. The column graph shown displays the average daily usage of gas for a household over a 31 day month.

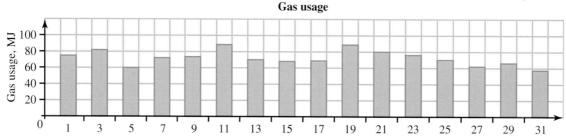

a. What is the gas usage on the thirteenth of the month?

b. What day had the maximum gas usage for the month and what was the gas usage?

c. What day had the minimum gas usage for the month and what was the gas usage?

7. **WE7** The average prices of houses in three suburbs are represented in the picture graph shown.

a. What is the average price of houses in Suburb 3?

b. What is the average price of houses in Suburb 1?

c. By how much are the houses in Suburb 2 more expensive than the houses in Suburb 4?

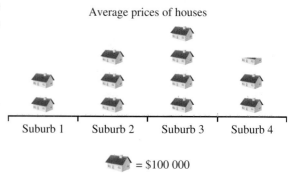

8. The population of the Australian states and territories at the end of December 2017 is represented in the picture graph shown.

Australia's population per states and territories December 2017

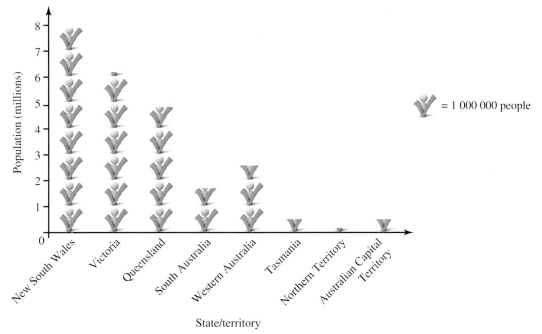

Source: Adapted from Australian Bureau of Statistics, http://www.abs.gov.au/ausstats/abs@.nsf/mf/3101.0

a. What was the population of Tasmania in December 2017?
b. What was the population of Queensland in December 2017?
c. What was the total population of Australia in December 2017?

9. The picture graph shown displays the blood types of a group of people. How many people were surveyed altogether?

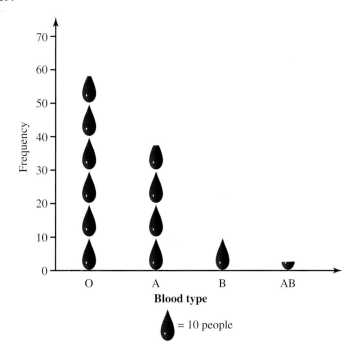

10. A group of 48 students were asked whether they have been regularly bullied at school or not. Their answers are displayed in the picture graphs shown.
    a. How many students were regularly bullied at school?
    b. How many students were surveyed altogether?

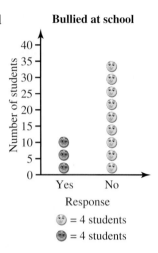

Bullied at school

= 4 students
= 4 students

6.5 Review: exam practice

6.5.1 Reading, interpreting and using graphs: summary

Two-way tables

- The independent variable is the variable that can be manipulated to produce changes in the dependent variable.
- The dependent variable changes as the independent variable changes. Its outcome is dependent on the input of the independent variable.

Dependent variable	Independent variable		Total
	A	B	
C	Number of data with the property of both A and C	Number of data with the property of both B and C	Total number of C
D	Number of data with the property of both A and D	Number of data with the property of both B and C	Total number of D
Total	Total number of A	Total number of B	Total number data

Line graphs, conversion graphs and step graphs

- **Line graphs** are used to represent the relationship between two numerical continuous datasets.
- **Conversion graphs** are used to convert one quantity into another. These graphs are always represented by a straight line because the relationship between the two variables does not change.
- **Step graphs** are made up of horizontal straight lines. These graphs are used when the value remains constant over intervals.
- The closed dot, • , means that the interval includes that value.
- The empty circle, ○ , means that the interval does not include that value.
- Line graphs are often used to compare similar types of data.

Column graphs and picture graphs

- **Column graphs** are made up of columns that can be either vertical or horizontal. These graphs are used to represent categorical data.
- **Picture graphs** or **pictographs** are statistical graphs that use pictures to display categorical data.

Exercise 6.5 Review: exam practice

Simple familiar

1. **MC** What is missing from the graph shown?

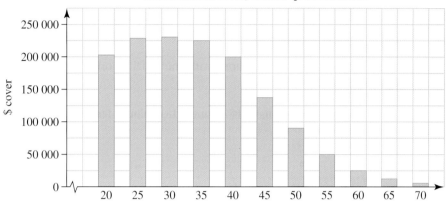

Maximum life cover, with no questions asked

A. Title
B. Vertical axis
C. Horizontal axis
D. Horizontal axis label

2. **MC** The literacy and numeracy skills of 142 primary school students are recorded in the table shown.

	Literacy skills		
Numeracy skills	**Good**	**Poor**	**Total**
Good	101	12	113
Poor	26	3	29
Total	127	15	142

The number of students who have good numeracy skills but poor literacy skills is:

A. 3.
B. 15.
C. 29.
D. 12.

3. **MC** A skip can be hired for $120 for the first 2 days, $160 for more than 2 days and up to and including 4 days, $210 for more than 4 days up to and including 7 days, and $280 for more than 7 days and up to and including 14 days.

The graph that represents the relationship between the cost of hiring the skip and the number of days hired is:

A.

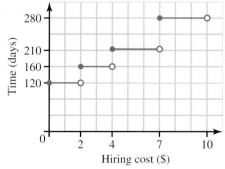

B.

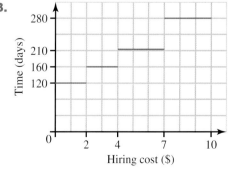

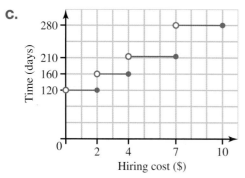

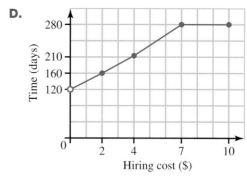

4. **MC** The daily minimum temperatures for the month of June 2018 recorded at Mount Wellington station are displayed in the graph shown.

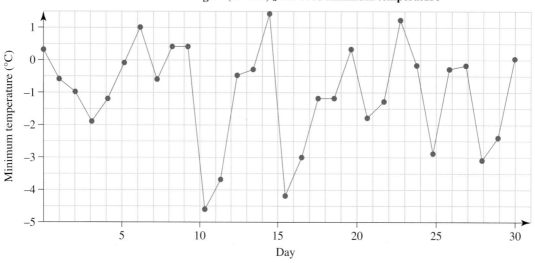

Mount Wellington (094087) June 2018 minimum temperature

— Minimum temperature over 24h —— Minimum temperature period uncertain ● No data

Source: Adapted from http://www.bom.gov.au/jsp/ncc/cdio/weatherData/av?p_display_type=dataDGraph&p_stn_num=094087&p_nccObsCode=123&p_month=06&p_startYear=2006.

The highest minimum temperature in June 2018 was recorded on

A. 1 June. **B.** 11 June. **C.** 15 June. **D.** 30 June.

5. **MC** The graph shown displays the exchange rates between A$ and US$ between 13 February 2013 and 15 March 2013.

Which of the following statements is not correct?

A. The value of the A$ decreased between 5 March and 6 March.

B. A$0.980 = US$1 on 1 and 4 March.

C. The value of the A$ decreased between 17 February and 19 February.

D. The value of the A$ increased between 22 February and 3 March.

Australian dollars (A$) to 1 US dollar (US$)

min = 0.96068 (March 15)   avg = 0.97292   max = 0.98122 (March 3)

Source: Exchange-Rates.com http://www.exchange-rates.org/history/AUD/USD/G/30.

6. **MC** The step graph shown displays the income tax rates for the financial year 2012–13.

The tax rate is 37 cents/$ for:

A. all amounts less than $80 000.
B. all amounts greater than $180 001.
C. amounts between $80 000 and $180 000.
D. amounts between $80 001 and $180 000.

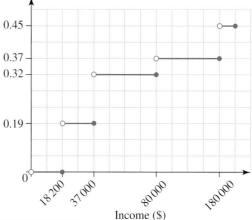

7. **MC** The column graph shown displays the labour force participation rate for mothers aged between 20–54 years by age of their youngest child in 2018.

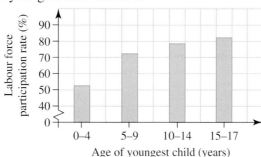

Source: Adapted from Department of Families, Housing and Community Services and Indigenous Affairs, http://www.abs.gov.au/AUSSTATS/abs@.nsf/ Lookup/3201.0Main+Features1Jun%202010?OpenDocument

The labour force participation rate of approximately 80% represents

A. women aged 20 to 54 with the youngest child aged 0–4.
B. women aged 20 to 54 with the oldest child aged 15–17.
C. women aged 20 to 54 with the oldest child aged 10–14.
D. women aged 20 to 54 with the youngest child aged 10–14.

8. **MC** The column graph shown was drawn using a sample population of:

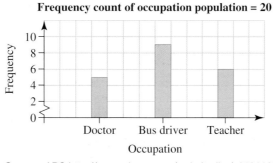

Source: ABS http://www.abs.gov.au/websitedbs/a3121120.nsf/ 89a5f3d8684682b6ca256de4002c809b/e200e8e572a 2ae52ca25794900127f4f!OpenDocument

A. 9 people.   B. 20 people.   C. 7 people.   D. 10 people.

9. **MC** A group of students collected clothing and shoes to donate to a charity shop.

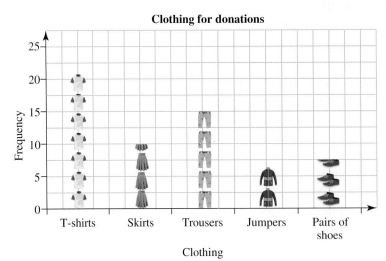

If each symbol represents 3 items, the number of T-shirts, skirts, trousers, jumpers and pairs of shoes respectively are:

A. 20, 9, 15, 10, 6.

B. 21, 10, 14, 10, 8.

C. 20, 10, 15, 5, 8.

D. 21, 10, 15, 6, 8.

A group of Year 11 music students were asked which musical instrument they were learning. The results are summarised in the graph below. Use this graph to answer questions **10–12**.

Year 11 students who learn a musical instrument

10. **MC** The most popular instrument is

A. piano.   **B.** guitar.   **C.** drums.   **D.** flute.

11. **MC** The least popular instrument is

A. piano.   **B.** guitar.   **C.** drums.   **D.** clarinet.

12. **MC** The total number of music students in this group is

A. 80.   **B.** 83.   **C.** 87.   **D.** 89.

13. A transport company charges different fees depending on the weight of the luggage. Luggage less than or equal to 5 kg is free of charge. Luggage over 5 kg and less than or equal to 10 kg has a fee of $15, and luggage over 10 kg and less than or equal to 20 kg has a fee of $25. Luggage heavier than 20 kg and less than or equal to 30 kg incurs a charge of $30. Construct a step graph to represent this data.

14. The picture graph shown displays five types of goods sold at a shop during a given month.

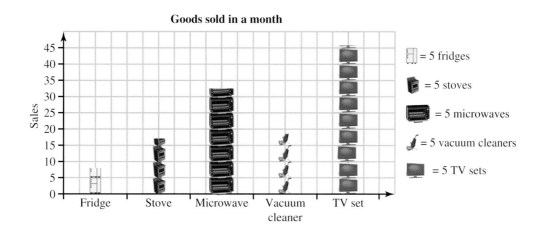

Goods sold in a month

a. How many of each product were sold during the month given?
b. What is the total number of products sold?

15. The two-way table shown displays the extracurricular activities 90 Year 11 students undertake outside school hours.

| Sport played | Other | | Total |
	Singing	Playing an instrument	
Soccer	4	12	16
Netball	23	17	40
Basketball	8	26	34
Total	35	55	90

a. How many students play soccer and learn to sing?
b. How many students play basketball?
c. How many students learn to play an instrument?

16. Redraw the two-way table from question 15 to add 30 more students who learn to dance in their spare time. Two of these students play soccer, 15 play netball and 13 play basketball.

Complex unfamiliar

Use the following data collected from the 2006 census for questions 17 and 18.

It details age groups and education details for residents of Townsville.

	Males	**Females**	**Persons**
Total persons	48 396	47 068	95 464
Age groups:			
0–4 years	3 003	2 850	5 853
5–14 years	6 415	5 978	12 393
15–19 years	3 749	3 819	7 568
20–24 years	4 783	4 266	9 049
25–34 years	7 306	6 900	14 206
35–44 years	6 873	6 846	13 719
45–54 years	6 630	6 295	12 925
55–64 years	4 919	4 422	9 341
65–74 years	2 723	2 797	5 520
75–84 years	1 576	2 073	3 649
85 years and over	419	822	1 241
Age of persons attending an educational institution			
0–4 years	357	344	701
5–14 years	5 662	5 293	10 955
15–19 years	2 171	2 623	4 794
20–24 years	1 054	1 487	2 541
25 years and over	1 351	2 219	3 570

Source: Australian Bureau of Statistics, 2006 Census of Population and Housing: Townsville.

17. a. Construct a contingency table displaying the number of male and female 15–19 year olds in Townsville compared with all other age groups there. Show all totals.
   b. Is it correct to claim that:
      i. 50.5% of 15–19 year olds are female?
      ii. the percentage of males who are 15–19 years of age is greater than the percentage of females who are 15–19 years old?
      Provide calculations to support your answers to each of these.

18. a. Construct a contingency table displaying males and females of 15–19 years 'Attending an educational institution' and 'Not attending an educational institution'. Show all totals.
   b. From your contingency table calculate:
      i. the percentage of males in this age group in an educational institution.
      ii. the percentage of those in an educational institution who are 15–19 years old and male.
   c. Would it be correct to say that only 55% of 15–19 year old females attend an educational institution?

The line graphs shown display the percentages of water stored at Samson Brook during 2015, 2016, 2017 and 2018.

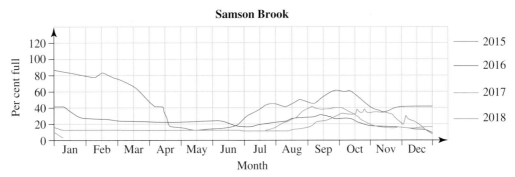

Samson Brook

Source: Adapted from Australian Government Bureau of Meteorology http://water.bom.gov.au/waterstorage/awris/ #urn:bom.gov.au:awris:common:codelist:feature:samsonbrook

19. a. State the month for the highest percentage of stored water in the four years.
    b. In what months was the amount of water stored in 2015 the same as the amount of water stored in 2018?
20. a. Between what months was the water stored in 2017 greater than the amount stored in 2018?
    b. Which year had the highest variation in the percentage of the water stored?

Answers

Chapter 6 Reading, interpreting and using graphs

Exercise 6.2 Two-way tables

1. **a.** 652 **b.** 26 **c.** 603 **d.** 335
2. **a.** 96 **b.** 7 **c.** 21 **d.** 68
3. **a.** 1000 **b.** 75 **c.** 96.7% **d.** 60%
4. **a.** 200
   b. 44
   c. 90.9%
   d. 5.1%
   e. 94%
   f. Sample responses can be found in the worked solutions in the online resources.
5. C
6. **a.** 35 **b.** 85 **c.** 43 **d.** 48
7. C
8. D
9. A
10. **a.**

	Test results		
Bags	**Accurate**	**Not accurate**	**Total**
With prohibited items	48	2	50
With no prohibited items	145	5	150
Total	193	7	200

   b. **i.** 96% **ii.** 3.3%
   iii. 4% **iv.** 96.5%

Exercise 6.3 Line graphs, conversion graphs and step graphs

1. **a.** 29 °C
   b. 19 °C
   c. The temperature is decreasing between 12:00–7:00 am, 9.00–11:00 am, 12:00–2:00 pm, 4:00–11:00 pm.
   d. 27 °C
   e. 7 pm
2. **a.** 2017
   b. 400 mm
   c. 2008–10, 2011–13 and 2015–17
   d. 2010 and 2013
   e. 2008, 330 mL
3. **a.** 24th, 10.6°C **b.** 23rd, 19.6°C
   c. 2°C **d.** 28th
   e. 4 days

4. **a.** August, −15.5 °C **b.** January, 2.5 °C
   c. 18 °C **d.** In February and in November
   e. 5
5. **a.** A\$72 **b.** US\$110
6. **a.** 4 inches, 6 inches, 18 inches, 20 inches
   b. 51 cm, 89 cm, 203 cm, 259 cm
7. **a.** 95 °F **b.** 50 °F **c.** 5 °C **d.** 40 °C
8. **a.** 128 km **b.** 80 km
   c. 19 miles **d.** 87 miles
9. **a.** \$14 **b.** \$17 **c.** \$26
10. **a.** \$1.50
    b. \$2.00
    c. Between 2 hours and 4 hours.
11. **a.** 51% Year 7 boys and 49% Year 7 girls
    b. 48% Year 12 boys and 52% Year 12 girls
    c. Year 10
12. **a.** 2005, 15.1%
    b. 2009, 17.0%
    c. 1%
    d. From 1998 to 1999, 1.3%
    e. 1995
13. **a.** Females ≈ 16%
    Males ≈ 20%
    b. 1995. Approximately 20.5%
    c. 2000

Exercise 6.4 Column graphs and picture graphs

1. **a.** 95 MJ
   b. 370 MJ
   c. April 2014, February 2014, or January 2015
   d. The daily gas usage is higher in the winter months, July, August, September and October. This usage decreases to very low values for the rest of the year. This could mean that they used gas for heating the house during the winter months.
2. **a.** Woolworths **b.** Shell **c.** Caltex
3. **a.** 4.9 kWh
   b. 4.1 kWh
   c. In January 2014, 2.9 kWh
   d. The highest usage of electricity is in the period April–June 2014. The lowest usage of electricity is in January 2014.
4. **a.** 4% **b.** United Kingdom
   c. 7%
5. NSW 7 000 000; VIC 5 200 000; QLD 4 200 000; SA 1 600 000; WA 2 100 000; TAS 500 000; NT 250 000; ACT 350 000.
6. **a.** 70 MJ **b.** 11th, 89 MJ **c.** 31st, 58 MJ
7. **a.** \$400 000 **b.** \$200 000 **c.** \$ 50 000
8. **a.** 500 000 approx. **b.** 4 900 000 approx.
   c. 24 700 000
9. 110
10. **a.** 12 **b.** 48

Exercise 6.5 Review: exam practice

1. D
2. D
3. C
4. C
5. A
6. C
7. D
8. B
9. D
10. A
11. D
12. C
13.

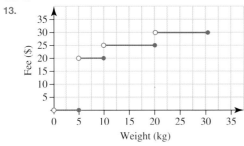

14. a. 9 fridges, 17 stoves, 33 microwaves, 18 vacuum cleaners and 41 TV sets.
    b. 118

15. a. 4   b. 34   c. 55

16.

Sport played	Other			Total
	Singing	Playing an instrument	Learn to dance	
Soccer	4	12	2	18
Netball	23	17	15	55
Basketball	8	26	13	47
Total	35	55	30	120

17. a.

	Male	Female	Total
15–19 years	3 749	3 819	7 568
All other ages	44 647	43 249	87 896
Total	48 396	47 068	95 464

   b. i. Yes
      ii. No — 7.7% for males and 8.1% for females

18. a.

Educational institution	Gender		Total
	Male	Female	
Attending	2171	2623	4794
Not attending	1578	1196	2774
Total	3749	3819	7568

   b. i. 58%   ii. 45%
   c. No — 69% attend

19. a. January.   b. May and November.

20. a. Between July and October.   b. 2015.

CHAPTER 7
Drawing and using graphs

7.1 Overview

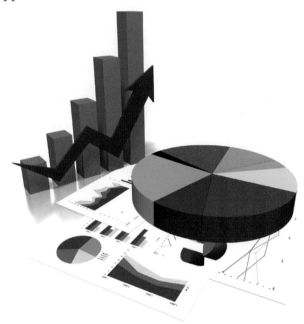

CONTENT

In this chapter, students will learn to:
- draw a line graph to represent any data that demonstrates a continuous change, such as hourly temperature [complex]
- use graphs in practical situations
- interpret the point of intersection and other important features (*x*- and *y*-intercepts) of given graphs of two linear functions drawn from practical contexts [complex]
- draw graphs from given data to represent practical situations [complex]
- discuss and interpret tables and graphs, including misleading graphs found in the media and in factual texts [complex]
- use spreadsheets to tabulate and graph data [complex]
- draw graphs from given data to represent practical situations [complex]
- determine which type of graph is best used to display a dataset

Fully worked solutions for this chapter are available in the Resources section of your eBookPLUS at www.jacplus.com.au.

7.2 Drawing and using line graphs

7.2.1 Line graphs

Line graphs are the most popular representations of datasets because they are easy to draw and easy to read. They clearly show trends in the data.

Line graphs show the relationship between two variables:
- The values of the independent variable are represented on the horizontal axis (the x axis).
- The values of the dependent variable are represented on the vertical axis (the y axis).

7.2.2 Numerical continuous data

The line graphs shown display the relationships between two numerical continuous variables.

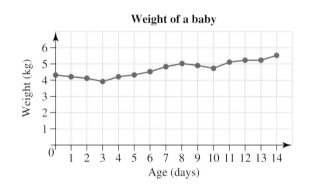

Weight of a baby

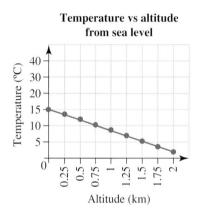

Temperature vs altitude from sea level

The first line graph displays the change of the weight of a baby over a period of 14 days.

'Age' is the independent variable and 'weight' is the dependent variable.

The second line graph displays the change in the temperature of the atmosphere in relationship to altitude (the vertical distance from sea level).

'Altitude' is the independent variable and 'temperature' is the dependent variable.

WORKED EXAMPLE 1

A worker in a greenhouse recorded the temperature in the greenhouse every hour for a period of 24 hours. The data were recorded in the table of values shown. Draw a continuous line graph for this dataset.

Time, hours	0	1	2	3	4	5	6	7	8	9	10	11	12
Temperature, °C	9	8	7	9	10	12	15	17	20	21	23	25	26

Time, hours	13	14	15	16	17	18	19	20	21	22	23	24
Temperature, °C	24	22	20	17	13	12	10	9	9	11	13	16

THINK	WRITE/DRAW
1. Draw the two axes and label them with the independent variable and the dependent variable.	The independent variable is the time and the dependent variable is the temperature in the greenhouse. The temperature is the dependent variable because its values depend on the time of reading. It does not affect the time. 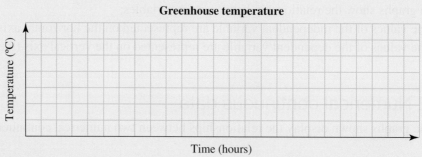
2. Choose the scales of the two axes.	The first variable takes values from 0 to 24 hours. The scale will have 12 ticks from 0 to 24 hours. Each tick will represent 2 hours. The second variable takes values from 7 °C to 26 °C. The scale of the vertical axis will have 6 ticks from 0 °C to 30 °C. Each tick will represent 5 °C. 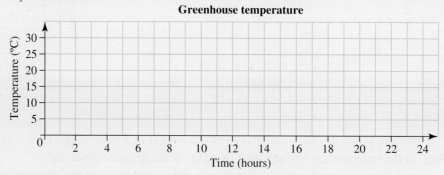
3. Plot the first data point.	The first data point is: at time 0 hours the temperature is 9 °C. 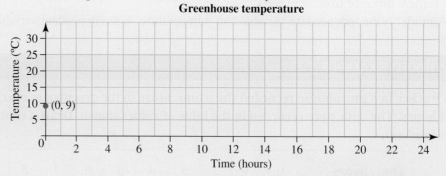
4. Plot the remaining values.	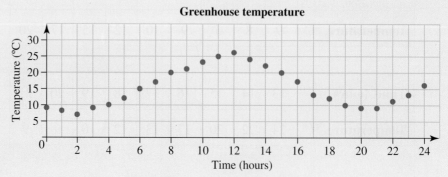

5. Connect all the points with a continuous line.

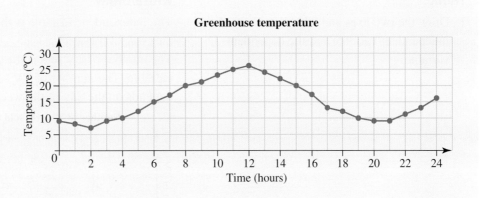

Greenhouse temperature

7.2.3 Discrete numerical data and categorical data

Numerical discrete data such as number of marks in a test or final score in a footy game can be represented by line graphs showing trends over time.

The line graphs shown display the relationships between a numerical discrete variable and a categorical data. The first line graph displays the results in the first five Mathematics tests of a student. The second line graph displays the final scores of a football team in 15 consecutive games.

Note: These graphs represent a set of discrete data so information cannot be read from the graphs at intervals between the points. The points are joined with a broken line only to illustrate any trend.

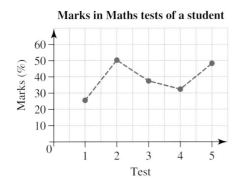

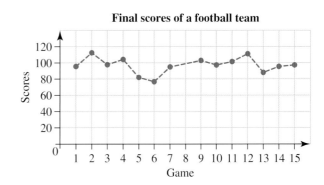

The categorical data, the test and the game, are the independent variables and the numerical discrete data, marks and scores, are the dependent variables.

The horizontal axis has a variable that shows a category. However, it is important to notice that the category is related to time. The discrete data are related to a sequence of chronological events.

WORKED EXAMPLE 2

The manager of an electronics shop recorded the number of TV sets sold per day over a period of one week. The data were recorded in the table of values shown. Draw a discrete line graph for this dataset.

Day	Monday	Tuesday	Wednesday	Thursday	Friday	Saturday	Sunday
TV sets	7	10	8	13	19	25	18

THINK	WRITE/DRAW
1. Draw the two axes and label them with the independent variable and the dependent variable.	The independent variable is the categorical data day of the week. The number of TV sets is the dependent variable. The number of TV sets sold depends on the day of the week. 
2. Choose the scales for the two axes.	The categorical variables are equally spaced on the horizontal axis. As there are 7 days in a week, the horizontal axis will have 7 categories. The scale of the vertical axis will have 6 ticks from 0 to 30. Each tick will represent 5 TV sets. 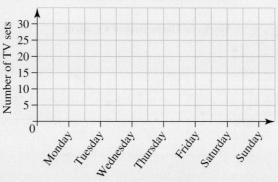
3. Plot the first data point.	The first data point is: on Monday the store sold 7 TV sets. 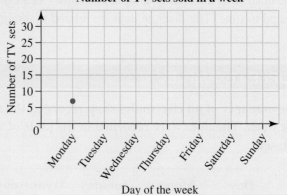

4. Plot the remaining values.

Number of TV sets sold in a week

Number of TV sets (vertical axis: 0, 5, 10, 15, 20, 25, 30)
Day of the week (horizontal axis: Monday, Tuesday, Wednesday, Thursday, Friday, Saturday, Sunday)

5. Connect all the points with a dotted line because the data is discrete.

Number of TV sets sold in a week

Number of TV sets (vertical axis: 0, 5, 10, 15, 20, 25, 30)
Day of the week (horizontal axis: Monday, Tuesday, Wednesday, Thursday, Friday, Saturday, Sunday)

7.2.4 Interpreting the point of intersection of two line graphs

Line graphs can be used to compare two different options related to the same situation.

WORKED EXAMPLE 3

The costs of hiring two electricians, Ben and Belinda, are illustrated on the graph at right. Use the information from the graph to answer the following questions.

a. How much does Ben charge for the initial call out?

b. How much does Belinda charge for the initial call out?

c. How much would Ben charge for a job that would take 2 hours?

d. How much would Belinda charge for a job that would take 2 hours?

e. If a job was estimated to take 3 hours, which electrician would complete it for the least amount of money?

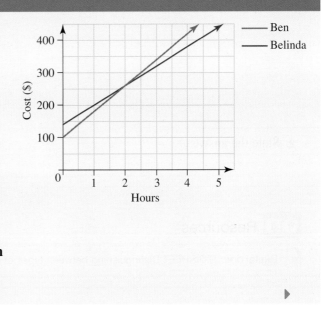

THINK	WRITE/DRAW
a. 1. Read the information on the horizontal and vertical axis	The independent variable is hours. The dependent variable is cost.
2. A call out charge is the amount of money that needs to be paid before any hours of work have been completed. Check the cost for Ben for zero hours of work, which is the intersection of the blue line with the vertical axis.	Ben's call out fee is $100.
b. Check the cost for Belinda for zero hours of work, which is the intersection of the red line with the vertical axis.	Belinda's call out fee is $140.

c. Locate two hours on the horizontal axis and then find the cost for the blue line

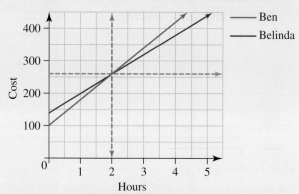

Ben charges $260 for a 2 hour job.

Belinda also charges $260 for a 2 hour job.

d. Locate two hours on the horizontal axis and then find the cost for the red line

e. 1. Locate 3 hours on the horizontal axis and find the cost for Ben and Belinda

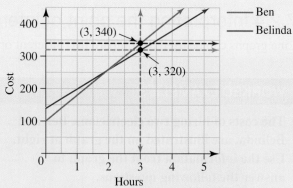

Ben charges $340 for a 3 hour job, Belinda charges $320.

Belinda would complete the job for a cheaper price.

2. State the answer.

---

on Resources

Digital doc: SkillSHEET Distinguishing between types of data (doc-5339)

Exercise 7.2 Drawing and using line graphs

1. **WE1** A mother recorded the yearly height of her son over a period of 10 years. She recorded the data in the table of values given. Draw a line graph for the data given.

Time, years	0	1	2	3	4	5	6	7	8	9	10
Height, cm	39	58	71	80	92	105	110	112	125	142	156

2. The cooling temperature of a drink placed in the fridge was recorded in the table of values shown. Draw a line graph for the data given.

Time, min	0	5	10	15	20	25	30	35	40	45	50	55	60
Temperature, °C	30	25	21	18	15	14	12	11	10	10	9	9	9

3. Draw a line graph for the data given in the table shown.

Time (hours)	Temperature (°C)
0	37
1	39
2	42
3	43
4	41
5	38
6	39
7	35
8	32
9	27
10	24

4. The percentages of five students in their Mathematics tests are recorded in the table shown.

Student \ Test	Test 1	Test 2	Test 3	Test 4	Test 5	Test 6	Test 7
Helen	94	85	87	86	99	91	95
Ahim	59	49	52	59	58	61	65
Lilly	63	56	58	68	71	70	72
Scott	88	67	77	82	81	82	85
Elaine	77	62	64	72	75	82	79

a. On the same set of axes draw line graphs to represent the results of the five students.
b. In which test did the students obtain the lowest scores?
c. In what test did each student score the highest percentage?

5. **WE2** The manager of a car yard recorded the number of cars sold per month over a period of one year. The data was recorded in the table of values shown. Draw a line graph for the data given.

Time, months	Jan	Feb	Mar	Apr	May	Jun	Jul	Aug	Sep	Oct	Nov	Dec
Cars sold	56	47	42	39	75	91	48	53	57	54	68	84

6. The manager of a small business recorded the amount of GST paid every month over a period of one year. The data was recorded in the table of values shown. Draw a line graph for the data given.

Time, month	Jan	Feb	Mar	Apr	May	Jun	Jul	Aug	Sep	Oct	Nov	Dec	Jan
GST, $1000	7	5	4	6	7	8	7	5	6	9	10	9	8

7. **WE3** The costs of hiring two furniture removal companies, Rent-a-truck and Do-it-today, are illustrated on the graph on the right.

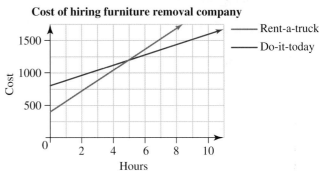

Cost of hiring furniture removal company

Use the information from the graph to answer the following questions.
   a. How much does Rent-a-truck charge for the hire of the truck before adding an hourly charge?
   b. How much does Do-it-today charge for the hire of the truck before adding an hourly charge?
   c. How much would Rent-a-truck charge for a job that would take 5 hours?
   d. How much would Do-it-today charge for a job that would take 5 hours?
   e. If a job was estimated to take 7 hours, which company would complete it for the least amount of money?

8. The population of Australia from 1860 to 2010 is recorded in the table given. Draw a line graph for this set of data.

Year	1860	1870	1880	1890	1900	1910	1920	1930
Population (million)	1.2	1.6	2.2	3.2	3.8	4.4	5.4	6.5

Year	1940	1950	1960	1970	1980	1990	2000	2010
Population (million)	7.0	8.3	10.4	12.7	14.8	17.2	19.3	22.3

9. Use the table from 8 and calculate the population growth in Australia for each decade from 1860 to 2010. Draw a new line graph for the two variables 'decade' and 'population growth'.

10. The average rainfall over a week is recorded in the table shown. Draw a line graph to represent this dataset.

Day	1	2	3	4	5	6	7
Rainfall (mm)	130	155	200	50	0	125	100

11. The share prices of two stocks over five days are recorded in the table shown. On the same set of axes draw a line graph to represent these datasets.

Day	1	2	3	4	5
Stock A ($)	1.21	1.25	1.20	1.22	1.29
Stock B ($)	0.59	0.65	0.64	0.61	0.62

12. A manager has recorded the number of people in the shop every hour for seven hours. The table shown displays this data. Draw a line graph for this dataset.

Time (hours)	1	2	3	4	5	6	7
People	5	16	24	28	26	17	3

13. Water is boiled from room temperature to 100°C. The temperatures over two minutes are recorded in the table shown. Draw a line graph for this dataset.

Time, sec	15	30	45	60	75	90	105	120
Temperature, °C	18	21	26	33	49	61	88	100

14. The monthly mean maximum temperatures for the month of December recorded at the Perth Metro weather station from 2000 to 2018 are displayed in the table shown. Using a calculator, a spreadsheet or otherwise, construct a line graph to represent this dataset.

Year	Temperature, °C	Year	Temperature, °C
2000	29.8	2010	30.0
2001	27.7	2011	23.7
2002	27.2	2012	30.4
2003	30.2	2013	27.7
2004	29.2	2014	27.8
2005	31.5	2015	30.8
2006	29.7	2016	29.3
2007	26.6	2017	30.5
2008	30.4	2018	31.2
2009	28.2		

15. The monthly mean maximum temperatures for the month of December recorded at the Perth Metro weather station and Adelaide (Kent Town) station, from 2000 to 2018 are displayed in the table shown. Using a calculator, spreadsheet or other method, construct a line graph to represent these datasets.

Year	Temperature, °C Perth Metro	Temperature, °C Adelaide (Kent Town)
2000	29.8	28.9
2001	27.7	25.7
2002	27.2	25.7
2003	30.2	26.7
2004	29.2	27.6
2005	31.5	25.7
2006	29.7	28.3
2007	26.6	23.3
2008	30.4	27.8
2009	28.2	29.0
2010	30.0	27.3
2011	23.7	27.3
2012	30.4	28.5
2013	27.7	28.9
2014	27.8	25.5
2015	30.8	28.5
2016	29.3	26.7
2017	30.5	27.9
2018	31.2	28.3

16. Two full water tanks, A and B, have developed cracks and are leaking. There has not been any rain to replenish them. Their water storage information is illustrated in the graph below.

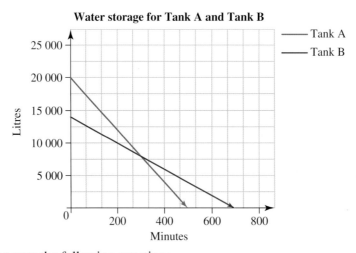

Using the graph, answer the following questions.
a. How much water does Tank A contain when it is full?
b. How much water does Tank B contain when it is full?
c. How long did it take Tank A and Tank B to empty?
d. At what time did Tank A and Tank B contain the same amount of water?

7.3 Drawing and using column and pie graphs

7.3.1 Column and pie graphs

In this section you will learn to draw both column graphs and pie charts. Recall that column graphs are used to represent categorical data. Pie graphs are circular graphical representations of data.

7.3.2 Drawing column graphs

Column graphs consist of vertical or horizontal bars of equal width. The frequency is measured by the height of the column. Both axes have to be clearly labelled and appropriate and accurate scales are required. The title should explicitly state what the column graph represents.

For the vertical column graphs, the frequency is always shown on the vertical axis and the categories of the data are shown on the horizontal axis. The axes of the horizontal column graphs are labelled the other way around. Horizontal column graphs are usually drawn when the category names are long.

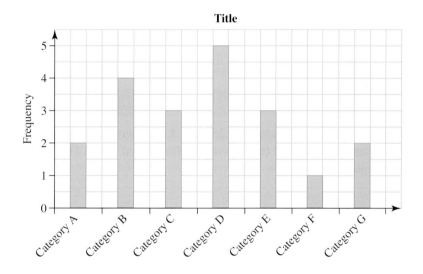

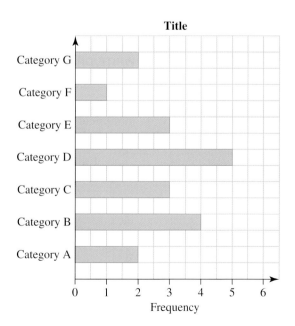

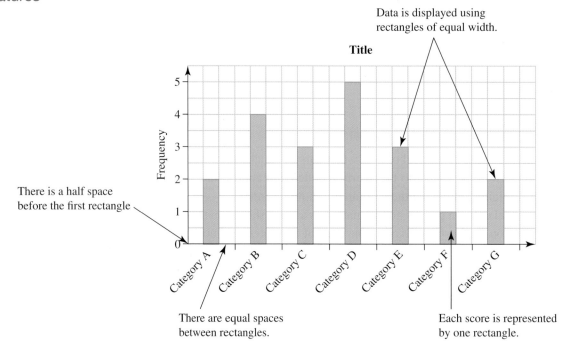

Data is displayed using rectangles of equal width.

Title

There is a half space before the first rectangle

There are equal spaces between rectangles.

Each score is represented by one rectangle.

WORKED EXAMPLE 4

The data shown is part of a random sample of 25 Year 11 students. The answers were related to how often the students use the Internet to do research for school work. Draw a column graph displaying the data for girls only.

Use of Internet for school work	Gender		Total
	Girls	Boys	
Rarely	2	8	10
Sometimes	4	4	8
Often	6	1	7
Total	12	13	25

THINK

1. Draw the two axes and label them with the two variables.

WRITE/DRAW

The categorical variable, Internet usage, is displayed on the horizontal axis and the numerical discrete variable, number of girls, on the vertical axis.

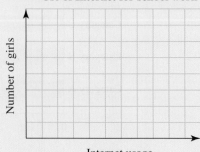

Use of Internet for school work

Number of girls

Internet usage

2. Choose the scales of the two axes.

The categorical variables are equally spaced on the horizontal axis. The scale of the vertical axis will have 6 ticks from 0 to 6.

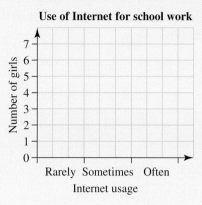

3. Draw a rectangle for each category.

The category 'rarely' is 2 units high because there are 2 girls using the Internet for school work rarely.
The category 'sometimes' is 4 units high and the category 'often' is 6 units high.

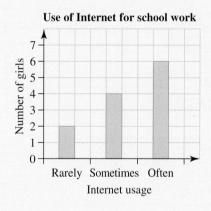

7.3.3 Drawing pie graphs

Pie graphs are circles divided into sectors. Each sector represents one item of the data.

The pie chart shown at right has four sectors: Net wage, Tax, Medicare levy and Superannuation contribution.

Each sector represents the percentage of each item from the gross wage. This pie chart is based on a gross wage (before tax) of $65 000 per year.

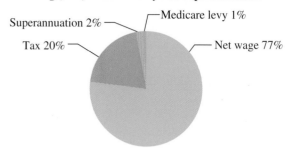

Wage, tax, Medicare levy and superannuation

Superannuation 2% — Medicare levy 1%

Tax 20% — Net wage 77%

%	Angle
1% is the Medicare levy: 1% of \$65 000 $= \dfrac{1}{100} \times 65\,000$ $= \$650$	The corresponding angle on the circle is 1% of $360° = \dfrac{1}{100} \times 360°$ $= 3.6°$
2% is the superannuation contribution: $2\% = \dfrac{2}{100} \times 65\,000$ $= \$1300$	The corresponding angle on the circle is 2% of $360° = \dfrac{2}{100} \times 360°$ $= 7.2°$
20% is the tax: $20\% = \dfrac{20}{100} \times 65\,000$ $= \$13\,000$	The corresponding angle on the circle is 20% of $360° = \dfrac{20}{100} \times 360°$ $= 72°$
77% is the net wage (after tax): $77\% = \dfrac{77}{100} \times 65\,000$ $= \$50\,050$	The corresponding angle on the circle is 77% of $360° = \dfrac{77}{100} \times 360°$ $= 277.2°$

Features of pie graphs

1. Each sector has to be coloured differently.
2. All labels and other writing have to be horizontal and equally distanced from the circle.

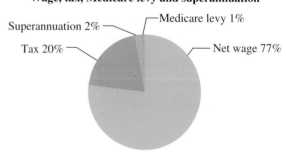

Wage, tax, Medicare levy and superannuation

Superannuation 2% — Medicare levy 1%

Tax 20% — Net wage 77%

A recipe for baked beans contains the following ingredients:

300 g dry beans, 50 g onion, 750 g diced tomatoes, 30 g mustard and 870 g smoked hock.

a. Construct a pie chart to represent these ingredients.

b. How many grams of dry beans are required to make 100 g of this dish?

c. How many grams of dry beans are required for a serve of 250 g of this dish?

THINK	WRITE/DRAW
a. 1. Establish the number of sectors required for the pie chart.	The pie chart requires 5 sectors, one for each ingredient: dry beans, onion, diced tomatoes, mustard, smoked hock
2. Calculate the total quantity.	$300 + 50 + 750 + 30 + 870 = 2000 \text{ g}$
3. Calculate the percentage for each quantity.	Dry beans: $\dfrac{300}{2000} \times 100 = 15\%$ Onion: $\dfrac{50}{2000} \times 100 = 2.5\%$ Diced tomatoes: $\dfrac{750}{2000} \times 100 = 37.5\%$ Mustard: $\dfrac{30}{2000} \times 100 = 1.5\%$ Smoked hock: $\dfrac{870}{2000} \times 100 = 43.5\%$
4. Check that the percentage sum is 100%.	$15\% + 2.5\% + 37.5\% + 1.5\% + 43.5\% = 100\%$
5. Calculate the angles for each sector out of the 360°. To calculate the angle for each sector use the formula: angle $= x\% \times 360°$ $= \dfrac{x}{100} \times 360°$	Dry beans: $\dfrac{15}{100} \times 360° = 54°$ Onion: $\dfrac{2.5}{100} \times 360° = 9°$ Diced tomatoes: $\dfrac{37.5}{100} \times 360° = 135°$ Mustard: $\dfrac{1.5}{100} \times 360° = 5.4°$ Smoked hock: $\dfrac{43.5}{100} \times 360° = 156.6°$
6. Check that the sum of the angles is 360°.	$54 + 9 + 135 + 5 4 + 156 6 = 360°$
7. Draw a circle for the pie chart and a vertical radius from the centre of the circle to the top of the circle.	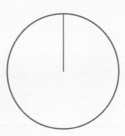

8. Draw the first sector of the pie graph.

The first ingredient is dry beans with an angle of 54°. Measure 54° from the vertical line in a clockwise direction.

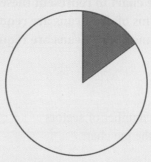

9. Draw the second sector of the pie graph.

The second ingredient is onion with an angle of 9°. Measure 9° from the right-hand side of the sector in a clockwise direction.

10. Draw the rest of the sectors of the pie graph.

Continue to draw the remaining sectors going in a clockwise direction around the circle.

11. Label all sectors.

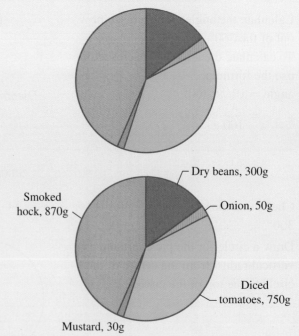

b. 1. Calculate the quantity required.

Use the formula:

$$\frac{\text{quantity of the item}}{\text{total quantity}} \times \text{new quantity}$$

Dry beans:

$$\frac{300}{2000} \times 100$$

$$= 15 g$$

c. 1. Calculate the quantity required.

Use the formula:

$$\frac{\text{quantity of the item}}{\text{total quantity}} \times \text{new quantity}$$

Dry beans:

$$\frac{300}{2000} \times 250$$

$$= 37.5 g$$

---

on Resources

Interactivity: Pie graphs (int-4061)

Digital doc: SkillSHEET Finding the size of a sector (doc-30355)

Digital doc: SkillSHEET Drawing pie graphs (doc-30356)

Exercise 7.3 Drawing and using column and pie graphs

1. **WE4** The ingredients on a package are: green beans 70 g, baby corn spears 63 g, asparagus 67 g, butter 4 g and water 6 g. Draw a column graph to display this data.

2. Vlad was interested in the water consumption in Victoria. He researched the Australian Bureau of Statistics to determine the consumption of water for various industries. He downloaded the Water Account released on 27 November 2018 and recorded the water consumption for Victoria in the table below. Draw a column graph to display this data.

Victoria	
Industry	**ML (million litres)**
Agriculture	1 234
Forestry and fishing	13
Mining	10
Manufacturing	144
Electricity and gas	117
Water supply	310
Other industries	220
Household	3 11
Total	**2359**

3. A group of students was surveyed about their favourite type of chocolate and the results have been recorded in the table shown. Draw a column graph for this dataset.

Type of chocolate	Frequency
Dark chocolate	35
Milk chocolate	47
White chocolate	18

4. A new program to improve school attendance was introduced at Diramu College. A survey was conducted at the end of the program to seek teachers' opinion about the program's effect on attendance. These opinions were recorded in the table shown. Using this data draw a vertical column graph.

Opinion	Frequency
No improvement at all	9
Some improvement	15
Moderate improvement	56
High improvement	28
Total	108

5. For the data given in question 4
   a. calculate the percentages for each category.
   b. construct a pie graph to represent this data.

6. Fiona has been the manager of a toy shop for six months. She recorded the sales over this period of time in the table shown. Construct a column graph to represent this data.

Month	Jan	Feb	Mar	Apr	May	Jun
Sales ($)	24 000	26 000	30 000	35 000	32 000	27 000

7. The top 12 countries based on the gold medal tally at the 2012 London Olympics are recorded in the table shown. Construct a column graph to represent this data.

Country	Gold medals
United States of America	46
People's Republic of China	38
Great Britain	29
Russian Federation	24
Republic of Korea	13
Germany	11
France	11
Italy	8
Hungary	8
Australia	7
Japan	7
Kazakhstan	7
Total	209

8. **WE5** The ingredients on a package are 70 g green beans, 63 g baby corn spears, 67 g asparagus, 4 g butter and 6 g water.
   a. Draw a pie graph to display this data.
   b. How many grams of asparagus are in a 1 kg package?

9. Nicole has a gross wage of $95 000. She pays $40 000 in tax, $1425 for the Medicare levy and deposits $1575 in her superannuation fund.
   a. How much money does she actually receive?
   b. Draw a pie graph to display this data, including the amount of money she actually receives.

10. An apple cake recipe has the following ingredients:
    300 g apples
    250 g flour
    200 g caster sugar
    250 g butter
    4 eggs (240 g)
    10 g ground cinnamon
    a. Calculate the individual percentages of each ingredient.
    b. Draw a pie graph to represent this data.
    c. How much flour would be needed if the total weight of the ingredients was 1.5 kg?

11. For the pie graph shown calculate the number of schools for each sector if, in 2018, there were 9529 schools in Australia.

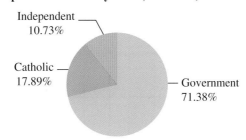

Proportion of school by sector, Australia, 2009

Independent 10.73%

Catholic 17.89%

Government 71.38%

12. The table shown displays a typical family budget. Construct a pie graph to display this data.

Area	Cost ($)
Housing	420
Food	180
Clothing	72
Transport	96
Entertainment	144
Health	60
Savings	120
Miscellaneous	108

7.4 Misleading graphs

7.4.1 Methods of misrepresenting data

Many people have reasons for misrepresenting data: politicians may wish to magnify the progress achieved during their term, or business people may wish to accentuate their reported profits. There are numerous ways of misrepresenting data. In this section, only graphical methods of misrepresentation are considered.

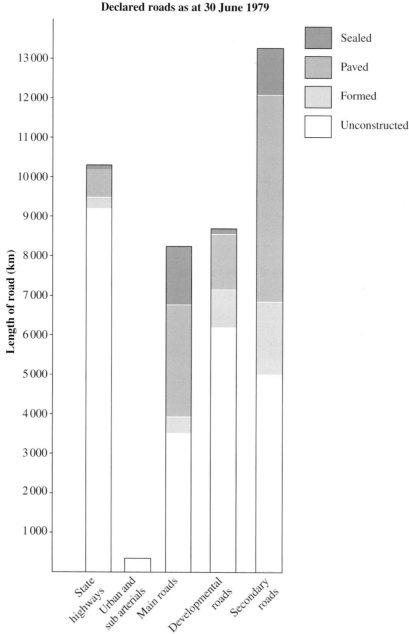

Declared roads as at 30 June 1979

Source: Dept of Mapping & Surveying (1980), *Queensland resources atlas*, 2nd rev. ed. (Courtesy Dept of Lands).

7.4.2 Vertical axis and horizontal axis

It is a truism that the steeper the graph, the better the growth appears. A 'rule of thumb' for statisticians is that for the sake of appearances, the vertical axis should be two-thirds to three-quarters the length of the horizontal axis. This rule was established in order to have some comparability between graphs.

The figure above illustrates how distorted the graph appears when the vertical axis is disproportionately large.

7.4.3 Changing the scale on the vertical axis

The following table gives the holdings of a corporation during a particular year.

Quarter	Holdings in $000 000
J-M	200
A-J	200
J-S	201
O-D	202

Here is one way of representing these data:

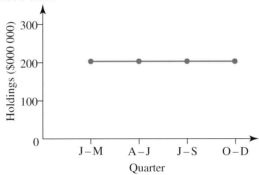

But it is not very spectacular, is it? Now look at the following graph showing the same information. The shareholders would be happier with this one.

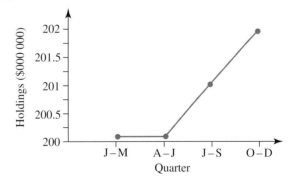

7.4.4 Omitting certain values

If one chose to ignore the second quarter's value, which shows no increase, then the graph would look even better.

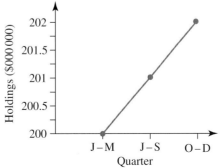

7.4.5 Foreshortening the vertical axis

Look at the figures below. Notice in graph (a) that the numbers from 0 to 4000 have been omitted. In graph (b) these numbers have been inserted. The rate of growth of the company looks far less spectacular in graph (b) than in graph (a). This is known as foreshortening the vertical axis.

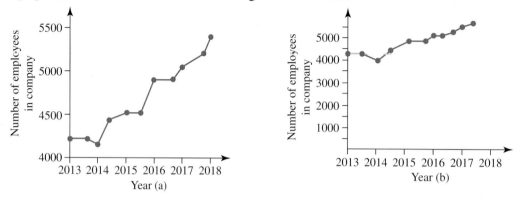

Foreshortening the vertical axis is a very common procedure. It does have the advantage of giving extra detail but it can give the wrong impression about growth rates.

7.4.6 Visual impression

In this graph, height is the property that gives the true relation, yet the impression of a much greater increase is given by the volume of each money bag.

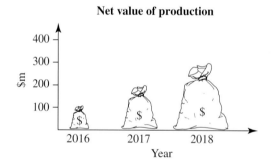

7.4.7 A non-linear scale on an axis or on both axes

Consider the following two graphs.

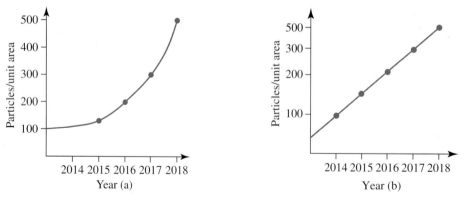

Both of these graphs show the same numerical information. But graph (a) has a linear scale on the vertical axis and graph (b) does not. Graph (a) emphasises the ever-increasing rate of growth of pollutants while graph (b) suggests a slower, linear growth.

The following data give wages and profits for a certain company. All figures are in millions of dollars.

Year	2015	2016	2017	2018
Wages	6	9	13	20
% increase in wages	25	50	44	54
Profits	1	1.5	2.5	5
% increase in profits	20	50	66	100

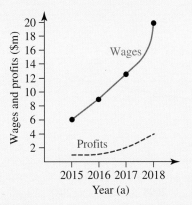
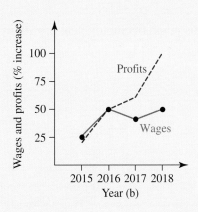

a. Do the graphs accurately reflect the data?
b. Which graph would you rather have published if you were:
   i. an employer dealing with employees requesting pay increases?
   ii. an employee negotiating with an employer for a pay increase?

THINK

a. 1. Look at the scales on both axes. Both scales are linear.
   2. Look at the units on both axes. Graph (a) has y-axis in $ while graph (b) has y-axis in %.

b. i. 1. Compare wage increases with profit increases.
   2. The employer wants high profits.

   ii. 1. Consider again the increases in wages and profits.
   2. The employee doesn't like to see profits increasing at a much greater rate than wages.

WRITE

Graphs do represent data accurately. However, quite a different picture of wage and profit increases is painted by graphing with different units on the y-axis.

The employer would prefer graph (a) because he/she could argue that employees' wages were increasing at a greater rate than profits.

The employee would choose graph (b), arguing that profits were increasing at a great rate while wage increases clearly lagged behind.

Exercise 7.4 Misleading graphs

1. **WE6** This graph shows the dollars spent on research in a company for 2010, 2014 and 2018. Draw another bar graph that minimises the appearance of the fall in research funds.

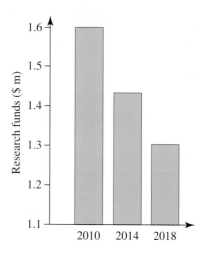

2. Examine this graph of employment growth in a company.

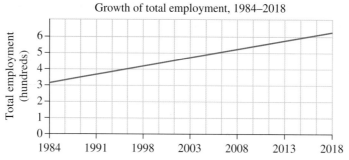

Why is this graph misleading?

3. Examine this graph.

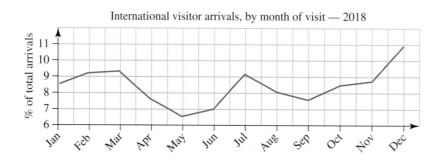

a. Redraw this graph with the vertical axis showing percentage of total arrivals starting at 0.
b. Does the change in visitor arrivals appear to be as significant as the original graph suggests?

4. This graph shows the student-to-teacher ratio in Australia for the years 2008 and 2018.
   Note: Ratio = Number of full-time equivalent students divided by the number of full-time equivalent teaching staff.

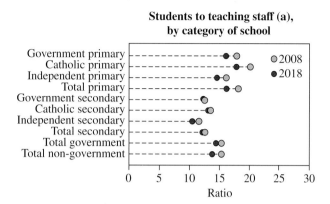

Students to teaching staff (a), by category of school

a. Describe what has generally happened to the ratio of students to teaching staff over the 10-year period.
b. A note says that the graph should not be used as a measure of class size. Explain why.

5. You run a company that is listed on the Stock Exchange. During the past year you have given substantial rises in salary to all your staff. However, profits have not been as spectacular as in the year before. The following table gives the figures for the mean salary and profits for each quarter. Draw two graphs, one showing profits, the other showing salaries, that will show you in the best possible light to your shareholders.

	1st quarter	2nd quarter	3rd quarter	4th quarter
Profits $000 000	6	5.9	6	6.5
Salaries $000 000	4	5	6	7

6. You are a manufacturer and your plant is discharging heavy metals into a waterway. Your own chemists do tests every 3 months and the following table gives the results for a period of 2 years. Draw a graph that will show your company in the best light.

	2017				2018			
Date	Jan.	Apr.	July	Oct.	Jan.	Apr.	July	Oct.
Concentration (parts per million)	7	9	18	25	30	40	49	57

7. This pie graph shows the break-up of national health expenditure in 2017–18 from three sources: Australian Government, state and local government, and non-government. (This expenditure relates to private health insurance, injury compensation insurers and individuals.)

Expenditure source ($ m) %	($ m)	%
Australian Government	37 229	45
State and local government	21 646	25
Non-government	28 004	30

Break-up of national expenditure

- Australian Government
- Non-government
- State and local governments

a. Comment on the claim that $87 000 m was spent on health from these three sources.
b. Which area contributes least to national health expenditure? Comment on its quoted percentage.

c. Which area contributes the next greatest amount to national health expenditure? Comment on its quoted percentage.

d. The Australian Government contributes the greatest amount. Comment on its quoted percentage.

e. Consider the pie chart.

    i. Based on the percentages shown in the table, what should the angles be?

    ii. Based on the actual expenditures, what should the angles be?

    iii. Measure the angles in the pie chart and comment on their values.

8. This graph shows how the $27 that a buyer pays for a new album by their favourite singer-songwriter is distributed among the departments of a major recording company involved in its production and marketing.

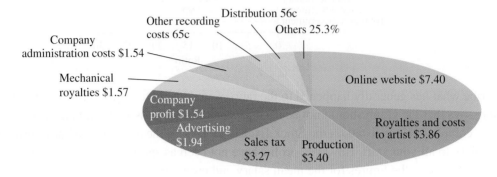

You are required to find out whether or not the graph is misleading, to explain fully your reasoning, and to support any statements that you make.

a. Comment on the shape of the graph and how it could be obtained.

b. Does your visual impression of the graph support the figures?

7.5 Graphing with technology

7.5.1 Drawing lines in Excel worksheets

Excel worksheets allow us to draw accurately and easily various charts from the datasets entered, as well as providing a convenient way of constructing line graphs.

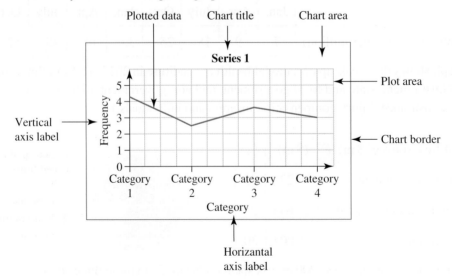

7.5.2 Chart Tools

The Chart Tools allow us to easily format the fonts of the text and the colours of various parts of the chart.

WORKED EXAMPLE 7

Using an Excel worksheet draw a line graph for the data recorded in the table of values shown.

Time, hours	0	1	2	3	4	5	6	7	8
Temperature, °C	21	25	29	31	33	30	31	32	33

THINK

1. Open a new Excel worksheet.

2. Label the two columns for the two variables. The independent variable will be written in column A.

   The dependent variable will be written in column B.

3. Insert the data.

WRITE/DRAW

	A	B	C	D	E	F	G	H	I	J
1										

Click in cell A1 and type 'Time, min'.

	A	B	C	D	E
1	Time, min				
2					

Click in cell B1 and type 'Temperature, °C'.

	A	B	C	D	E
1	Time, min	Temperature, °C			
2					

Click in cell A2 and start typing in the data given for the independent variable.

	A	B	C
1	Time, min	Temperature, °C	
2	0		
3	1		
4	2		
5	3		
6	4		
7	5		
8	6		
9	7		
10	8		
11			

▶

Click in cell B2 and start typing in the data given for the dependent variable.

	A	B	C
1	Time, min	Temperature, °C	
2	0	21	
3	1	25	
4	2	29	
5	3	31	
6	4	33	
7	5	30	
8	6	31	
9	7	32	
10	8	33	
11			

4. Draw the line graph.

Select the two columns starting from cell A1 to cell B10. To select these cells, click in cell A1, hold, move to cell B10 and release.

	A	B	C
1	Time, min	Temperature °C	
2	0	21	
3	1	25	
4	2	29	
5	3	31	
6	4	33	
7	5	30	
8	6	31	
9	7	32	
10	8	33	
11			

Go to Insert and then click on the 'Lines' icon on the 'Charts' menu.

Click any of the templates for a line. The chart shown is the default chart Excel draws. It draws the two datasets as two separate sets of data rather than as related to each other.

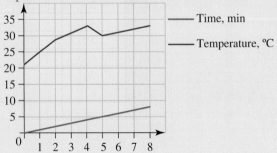

5. Format the line graph.

This chart can be formatted the way we want.
Click 'Select data' on the menu bar. The screen below will open.

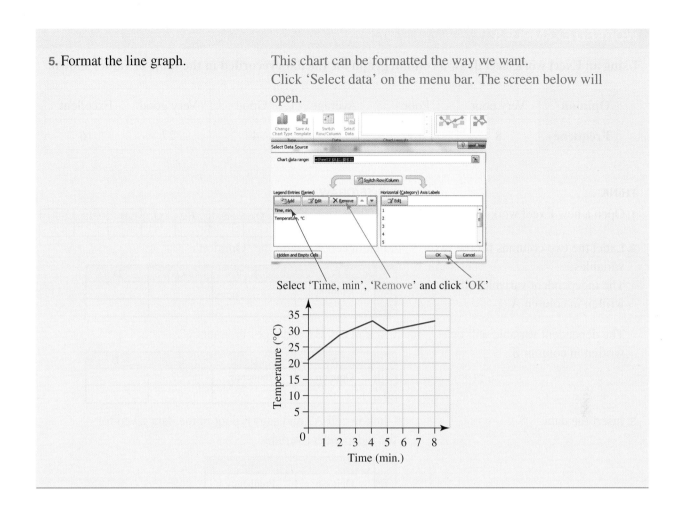

Select 'Time, min', 'Remove' and click 'OK'

7.5.3 Drawing a column graph in an Excel worksheet

Excel worksheets provide a convenient way of constructing column graphs.

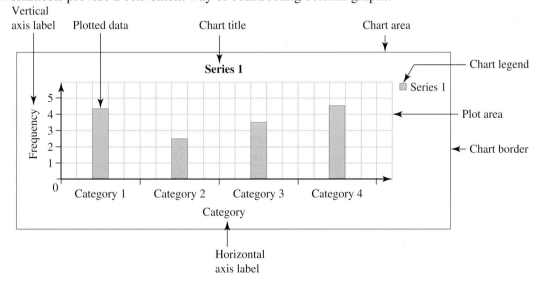

Using an Excel worksheet draw a column graph for the data recorded in the table of values shown.

Opinion	Very poor	Poor	Average	Good	Very good	Excellent
Frequency	8	15	52	44	27	16

THINK

1. Open a new Excel worksheet.

2. Label the two columns for the two variables.
   The independent variable will be written in column A.

   The dependent variable will be written in column B.

3. Insert the data.

WRITE/DRAW

	A	B	C	D	E	F	G	H	I	J
1										

Click in cell A1 and type 'Opinion'.

	A	B	C	D
1	Opinion			
2				

Click in cell B1 and type 'Frequency'.

	A	B	C	D
1	Opinion	Frequency		
2				

Click in cell A2 and start typing in the data given for the independent variable.

	A	B
1	Opinion	Frequency
2	Very poor	
3	Poor	
4	Average	
5	Good	
6	Very good	
7	Excellent	

Click in cell B2 and start typing in the data given for the dependent variable.

	A	B
1	Opinion	Frequency
2	Very poor	8
3	Poor	15
4	Average	52
5	Good	44
6	Very good	27
7	Excellent	16

4. Draw the column graph.

Select the two columns starting from cell A1 to cell B7. To select these cells, click in cell A1, hold, move to cell B7 and release.

	A	B
1	Opinion	Frequency
2	Very poor	8
3	Poor	15
4	Average	52
5	Good	44
6	Very good	27
7	Excellent	16

Go to Insert and then click on the 'Column' icon on the 'Charts' menu.

Click any of the templates for a column. The chart shown is the default chart Excel draws.

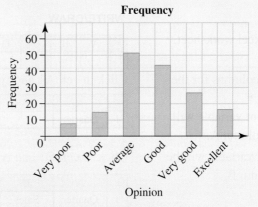

5. Format the column graph.

This chart can be formatted the way we want. Click 'Chart Tools' on the menu bar and format the chart as desired.

7.5.4 Drawing a pie graph in an Excel worksheet

Pie charts are easy to draw in Excel worksheets because the program calculates the percentages and angles required automatically.

Sales

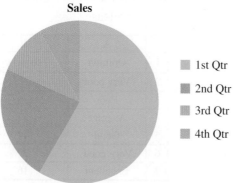

- 1st Qtr
- 2nd Qtr
- 3rd Qtr
- 4th Qtr

WORKED EXAMPLE 9

Using an Excel worksheet draw a pie graph for the data recorded in the table of values shown.

Quarter	1st	2nd	3rd	4th
Sales	16	8	32	40

THINK

1. Open a new Excel worksheet.

2. Label the two columns for the two variables.
   The independent variable will be written in column A.
   The dependent variable will be written in column B.

3. Insert the data.

WRITE/DRAW

	A	B	C	D	E	F
1						

Click in cell A1 and type 'Quarter'.

	A	B	C	D
1	Quarter			

Click in cell B1 and type 'Sales'.

	A	B	C	D
1	Quarter	Sales		

Click in cell A2 and start typing in the data given for the independent variable.

	A	B
1	Quarter	Sales
2	1st Qtr	
3	2nd Qtr	
4	3rd Qtr	
5	4th Qtr	

Click in cell B2 and start typing in the data given for the dependent variable.

	A	B
1	Quarter	Sales
2	1st Qtr	8.2
3	2nd Qtr	3.2
4	3rd Qtr	1.4
5	4th Qtr	1.2

4. Draw the pie graph.

Select the two columns starting from cell A1 to cell B5. To select these cells, click in cell A1, hold, move to cell B5 and release.

	A	B
1	Quarter	Sales
2	1st Qtr	8.2
3	2nd Qtr	3.2
4	3rd Qtr	1.4
5	4th Qtr	1.2

Go to Insert and then click on the 'Pie' icon on the 'Charts' menu.

Click any of the templates for a pie graph. The chart shown is the default chart Excel draws.

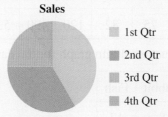

5. Format the pie graph.

This chart can be formatted the way we want.
Click 'Chart Tools' on the menu bar and format the chart as desired.

Exercise 7.5 Graphing with technology

1. **WE7** Using an Excel worksheet draw a line graph for the data recorded in the table of values shown.

Day of the week	Mon	Tue	Wed	Thu	Fri	Sat	Sun
Temperature, °C	32	35	34	29	21	26	34

2. Using an Excel worksheet draw a line graph for the data recorded in the table of values shown.

Time, days	0	1	2	3	4	5	6	7	8	9	10
Height, cm	3.2	3.5	4.1	4.7	5.9	6.0	6.2	6.5	7.0	7.2	7.3

3. a. What symbols are used to label rows and columns in Excel worksheets?
   b. State the row and the column of cell A4.
   c. Name the cell shown.

	A	B
1		
2		

4. Using an Excel worksheet construct a line graph for the data recorded in the table shown.

Day	Mon	Tue	Wed	Thu	Fri	Sat	Sun
Temperature (°C)	25	21	24	29	18	21	22

5. **WE8** Using an Excel worksheet draw a column graph for the data recorded in the table of values shown.

Month	Jan	Feb	Mar	Apr	May	Jun	Jul
Electricity, kWh	4.1	3.7	3.9	3.8	4.3	5.6	5.2

6. Using an Excel worksheet draw a line graph for the data recorded in the table of values shown.

Month	Jan	Feb	Mar	Apr	May	Jun	Jul	Aug	Sep	Oct	Nov	Dec
Gas, mJ	70	60	65	75	82	120	490	530	380	420	120	45

7. Using an Excel worksheet, construct a column graph for the data recorded in the table shown.

Pet	Cat	Dog	Rabbit	Fish
Frequency	15	36	8	17

8. The table shown displays the total road length in Australia by state and territory. Draw a column graph for the dataset given using an Excel worksheet.

NSW km	VIC km	QLD km	SA km	WA km	TAS km	NT km	ACT km
184 794	153 000	183 041	97 319	154 263	25 599	22 239	2 963

9. The table shown displays the total road length in Australia for the period from 2014 to 2018. Draw a line graph for the dataset given using an Excel worksheet.

Year	Total km
2014	815 588
2015	816 949
2016	822 649
2017	825 592
2018	823 217

10. **WE9** Using an Excel worksheet draw a pie graph for the data recorded in the table of values shown. Label the sectors of the pie graph with the names of the ingredients and their quantities, in grams.

Ingredients	Flour	Sugar	Butter	Cinnamon	Cocoa
Quantity, g	500	250	100	20	60

11. Using an Excel worksheet draw a pie graph for the data recorded in the table of values shown. Label the sectors of the pie graph with the names of the categories and their percentages.

Type of payment	Net wage	Tax	Superannuation	Medicare levy
Amount, $	57 000	10 230	1200	855

12. The total crash costs in 1993 were $6.1 billion. These costs are estimated for the following categories:
Lost earnings of victims $829.1 million
Family and community losses $587.8 million
Vehicle damage $1868.2 million
Pain and suffering $1463.3 million
Insurance administration $571.1 million
Other $816.4 million
a. Calculate the percentages for each category.
b. Construct a pie graph for this dataset. Label each sector with the category and the corresponding percentage.

13. The data recorded in the table shown represents the number of international passengers carried by major airlines for the year ended December 2018 to/from Australia. Construct a pie graph for the data given using an Excel worksheet.

Airline	Percentage
Qantas Airways	17.7%
Singapore Airlines	9.2%
Emirates	8.4%
Virgin Australia	8.3%
Jetstar	8.3%
Air New Zealand	8.0%
Cathay Pacific Airways	4.9%
Malaysia Airlines	3.7%
Thai Airways International	3.5%
AirAsia X	2.8%
Others	25.3%

14. The data recorded in the table shown represents the number of international passengers carried (in thousands) to/from Australia for the year ended December 2018. Construct a line graph for the two datasets given using an Excel worksheet.

	Inbound	Outbound
Dec-17	1220	1451
Jan-18	1523	1281
Feb-18	1202	1019
Mar-18	1141	1184
Apr-18	1177	1203
May-18	1001	1098
Jun-18	1094	1267
Jul-18	1411	1186
Aug-18	1191	1229
Sep-18	1241	1294
Oct-18	1410	1164
Nov-18	1207	1247
Dec-18	1314	1539

7.6 Review: exam practice

7.6.1 Drawing and using graphs: summary

Drawing and using line graphs

- Line graphs are the most popular representations of datasets because they are easy to draw and easy to read.
- Line graphs clearly show trends in the data.
- Line graphs show the relationship between two numerical continuous variables or a discrete numerical variable and a categorical variable displayed in a chronological **order.**
- Line graphs can be used to compare two sets of information related to the same situation.

Drawing and using column and pie graphs

- Column graphs consist of vertical or horizontal bars of equal width. The frequency is measured by the height of the column.

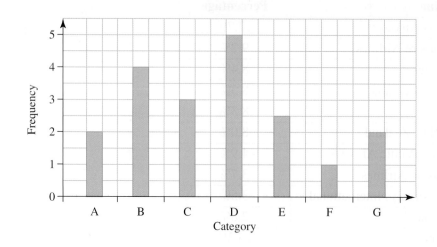

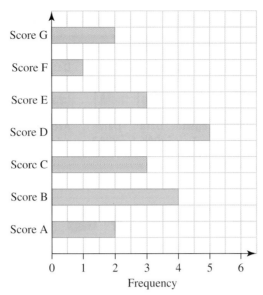

- Pie graphs are circular graphical representations of data. Each sector represents one item of the data.
- Formula to calculate a percentage: $x\% = \dfrac{x}{100} \times$ sum of all data
- Formula to calculate the angle of the sector: angle $= x\% \times 360°$
- Each sector has to be coloured differently. Other patterns can be used instead of a plain colour.
- All labels and other writing have to be horizontal and equally distanced from the circle.

Wage, tax, Medicare levy and superannuation

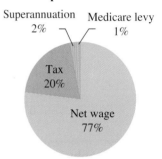

Misleading graphs

When reading graphs to obtain clear and correct information, look for the following:
- the scale on both axes is linear
- scales on vertical and horizontal axes have not been lengthened or shortened to give a biased impression
- certain values have not been omitted in the graph
- picture graphs are drawn to represent a height and not a volume.

Graphing with technology

- **Line graphs** can be drawn using Excel worksheets.

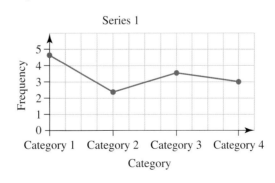

- **Chart Tools** allow us to easily format the fonts of the text and the colours of various parts of the chart.

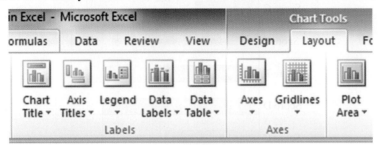

- **Chart types** in Excel worksheets.

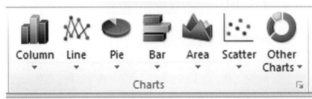

- **Column graphs** in Excel worksheets.

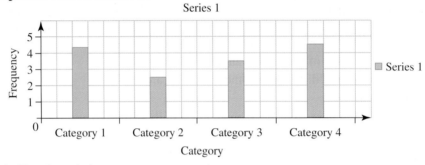

- **Pie graphs** in Excel worksheets.

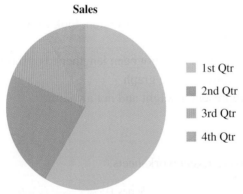

Exercise 7.6 Review: exam practice

Simple familiar

1. Every driver's licence has a number ID. A police officer recorded the licence numbers of 100 drivers who have been tested for drink driving. The data collected is:
   A. categorical nominal.
   B. numerical continuous.
   C. categorical ordinal.
   D. numerical discrete.

2. **MC** Which of the following statements is not correct?
   A. Line graphs display numerical continuous data.
   B. Line graphs display categorical data only.
   C. A line graph is made of independent points joined by straight lines.
   D. The dependent variable is displayed on the vertical axis.

3. **MC** Which of the following graphs displays the data recorded in the table shown?

Time (days) / Height (cm)	0	1	2	3	4	5
Plant A	10	15.6	21.5	29.3	34.7	35.1
Plant B	10	13.2	17.4	20.8	25.6	29.2

A.

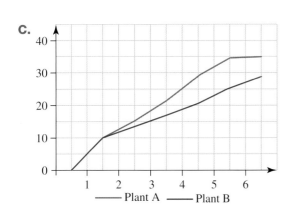

B.

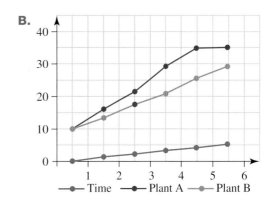

C.

D.

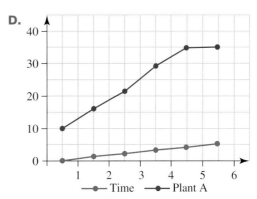

4. **MC** Which of the following statements is not correct?
  A. Column graphs consist of columns of equal width.
  B. Each score is represented by one rectangle.
  C. There are no spaces between columns.
  D. There is a half space before the first column.

5. **MC** Which of the following column graphs correctly displays the data recorded in the table shown?

Opinion	People
Excellent	29
Very good	24
Good	20
Poor	5
Very poor	2
Total	**80**

A.

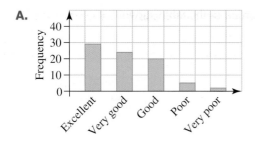

B.

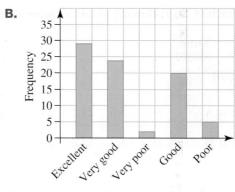

C.

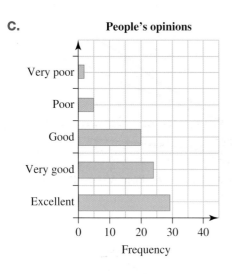

D.

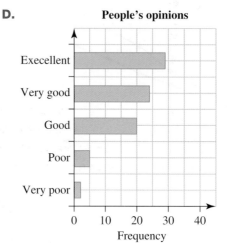

6. The column graph shown displays the monthly mean number of cloudy days recorded at the Canberra airport meteorological station for the period 1939−2010.

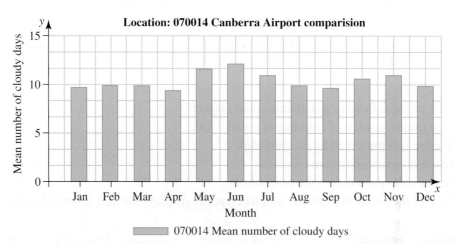

The months with exactly 10 cloudy days are:

A. February and December.

B. February and August.

C. August and December.

D. February, March and August.

7. The line graph shown displays the relationship between inches and centimetres.

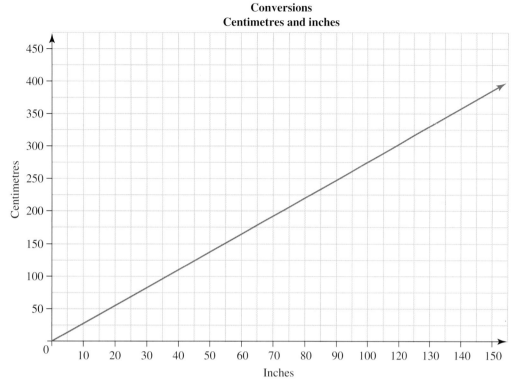

Conversions
Centimetres and inches

The closest measure of 39 inches in centimetres is:

A. 100 cm. **B.** 30 cm. **C.** 200 cm. **D.** 80 cm.

8. The graph shown displays the superannuation contributions, in $billion, from June 2009 to June 2018.

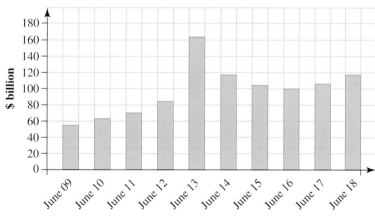

Super contributions bounce back

The highest increase in the superannuation contributions was:

A. from June 2009 to June 2010. **B.** from June 2013 to June 2014.
C. from June 2017 to June 2018. **D.** from June 2012 to June 2013.

9. Select the incorrect statement:
   A. A pie chart displays categorical data.
   B. A pie chart displays the data as percentages of the whole.
   C. In a pie chart every sector represents a category.
   D. A pie chart displays numerical data.

10. The Bureau of Meteorology collects data on rainfall over catchment areas. The rainfall is measured in millimetres. What type of data is being collected?

   A. Nominal   **B.** Ordinal   **C.** Discrete   **D.** Continuous

11. To draw a line graph in an Excel worksheet click on:

   A.   **B.**   **C.**   **D.**

12. To construct a vertical column graph in an Excel worksheet use:

   A.   **B.**   **C.**   **D.**

Complex familiar

13. The table shown displays the number of words that users from Australia typed in the search engines during March 2018.

   a. What type of data is displayed in this table?
   b. Construct a column graph to represent this dataset.
   c. From a total of 5000 searches, how many would be two word searches?

Words:	🇦🇺 Au
1	47.14%
2	21.98%
3	18.06%
4	6.51%
5	3.37%
6	1.43%
7	0.80%

14. The graph below shows monthly car sales for a local car yard over the past year.

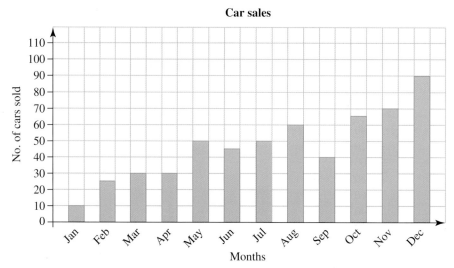

Use the graph to answer the following questions.

   a. During what month were the lowest sales figures recorded?
   b. In which month were the number of cars sold equal?
   c. What was the highest number of cars sold and in which month did this occur?
   d. By how much did the December sales exceed the January sales?
   e. What was the difference between the highest and the second highest sales figures recorded over the last year?
   f. What were the total sales for the year?
   g. Is there any pattern in sales that you can see?

15. The firing temperature in a kiln was collected every 5 minutes and recorded in the table shown. Construct a line graph to display this dataset using an Excel worksheet.

Time, min	5	10	15	20	25	30	35	40	45
Temperature, °C	204	288	338	382	410	438	450	466	471

16. A vegetable shop surveyed 60 of its customers on their favourite vegetable. The data was collected and recorded in the table shown.

Favourite vegetable	Frequency
Carrots	11
Tomatoes	27
Cabbage	8
Broccoli	5
Cauliflower	9

a. Choose five pictures to represent each vegetable.
Construct a picture graph to represent this dataset using
b. the picture to represent three items.
c. the picture to represent five items.
d. the picture to represent ten items.
e. Which graph is clearest to read, b, c, or d?

17. Shirly recorded the approximate times she spends on various task during a 24-hour period in the table shown.

Activity	Time, hours
School	7.0
Going to and from school	0.5
Homework	3.5
Sleep	8.5
Exercise	1.0
Relaxing	1.5
Other	2.0

a. Display this data in a column graph.
b. Calculate the percentages that each activity represents out of the 24 hour period.
c. Display this data as a pie chart.
d. Calculate the number of hours Shirly spends on her homework over five days.

18. The daily maximum temperatures in May 2018 in Perth are displayed in the table shown.

Day	Max temperature, °C	Day	Max temperature, °C
1st	25.7	17th	20.3
2nd	22.9	18th	20.5
3rd	22.6	19th	21.7
4th	23.6	20th	19.2
5th	25.4	21st	21.2
6th	22.3	22nd	21.6
7th	24.1	23rd	20.8
8th	20.8	24th	22.0
9th	20.8	25th	23.2
10th	20.1	26th	24.9
11th	19.7	27th	23.0
12th	20.7	28th	20.0
13th	21.1	29th	19.6
14th	21.8	30th	14.6
15th	22.6	31st	16.2
16th	23.4		

a. Construct a line graph for this dataset.
b. What day in May 2018 recorded the highest maximum temperature for the month?
c. What day in May 2018 recorded the lowest maximum temperature for the month?
d. What was the temperature difference between 7 and 8 May? Was this a decrease or an increase in temperature?
e. What was the temperature difference between 25 and 26 May? Was this a decrease or an increase in temperature?

Answers

Chapter 7 Drawing and using graphs

Exercise 7.2 Drawing and using line graphs

1.

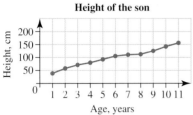

Height of the son

2.

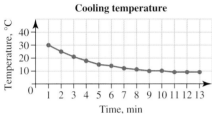

Cooling temperature

3.

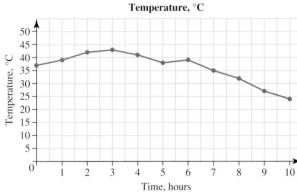

Temperature, °C

4. a.

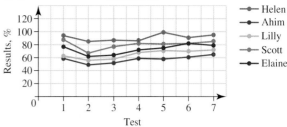

Test results

b. Test 2

c. Helen – Test 5, Ahim – Test 7, Lilly – Test 7, Scott – Test 1, Elaine – Test 6

5. See bottom of the page*

6. See bottom of the page**

7. a. $400   **b.** $800   **c.** $1200   **d.** $1200

e. Do-it-today would be cheaper after 5 hours of hire.

8.

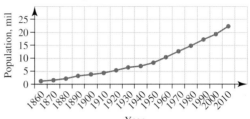

Population of Australia (mil)

9. 0.4 mil, 0.6 mil, 1.0 mil, 0.6 mil, 0.6 mil, 1.0 mil, 1.1 mil, 0.5 mil, 1.3 mil, 2.1 mil, 2.3 mil, 2.1 mil, 2.4 mil, 2.1 mil, 3.0 mil

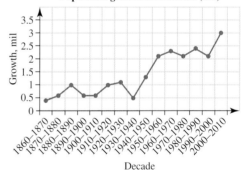

Population growth in Australia (mil)

*

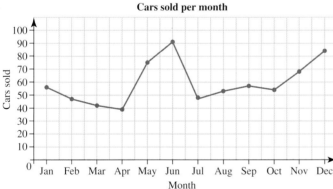

Cars sold per month

**

GST paid per month, $1000

10.

Rainfall (mm)

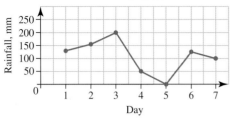

11.

Share prices of Stock A and Stock B

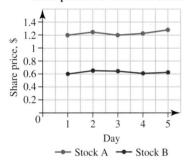

12.

Number of customers in a day

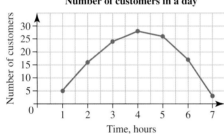

13.

Water temperature, °C

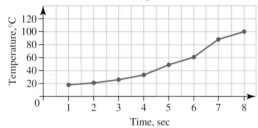

14.

Mean maximum temperature in Perth Metro, December 2000–2017

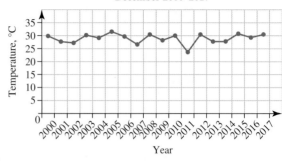

15.

Mean maximum temperature, December, Perth and Adelaide

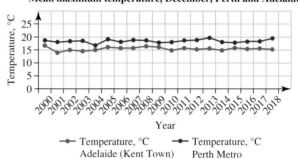

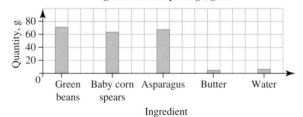

16. a. 20 000 L

   b. 14 000 L

   c. Tank A 500 mins, Tank B 700 mins

   d. After 300 mins

Exercise 7.3 Drawing and using column and pie graphs

1.

Ingredients on a package, grams

2. See bottom of the page*

*

Victoria ML (million litres)

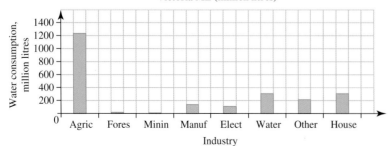

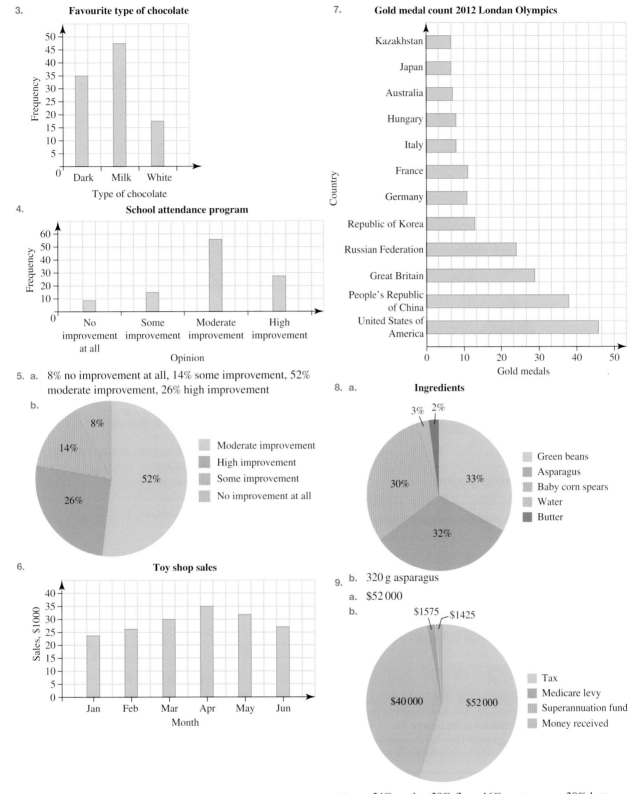

3. Favourite type of chocolate

4. School attendance program

5. a. 8% no improvement at all, 14% some improvement, 52% moderate improvement, 26% high improvement

b.

- Moderate improvement
- High improvement
- Some improvement
- No improvement at all

6. Toy shop sales

7. Gold medal count 2012 Londan Olympics

8. a. Ingredients

- Green beans
- Asparagus
- Baby corn spears
- Water
- Butter

b. 320 g asparagus

9. a. $52 000

b.

- Tax
- Medicare levy
- Superannuation fund
- Money received

10. a. 24% apples, 20% flour, 16% caster sugar, 20% butter, 19% eggs, 1% ground cinnamon

b.

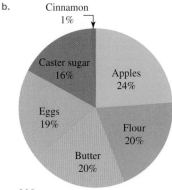

c. 300 g

11. 6802 government schools, 1705 Catholic schools, 1022 independent schools

12.

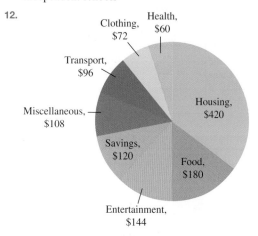

Exercise 7.4 Misleading graphs

1. See the worked solutions online for a sample response.

2. Horizontal axis uses same division for 5 and 7 year periods

3. **a.** See the worked solutions online for a sample response.
   b. No

4. **a.** Student to teacher ratios have improved slightly.
   b. Country schools have smaller class sizes.

5. See the worked solutions online for a sample response.

6. See the worked solutions online for a sample response.

7. **a.** The claim is accurate enough in the context ($86 879 m actually).
   b. State and local governments. The stated 25% is correct (rounded up from 24.9%).
   c. Non-government organisations. The stated 30% is rounded down from 32.2%. The percentages being quoted seem to be rounded to the nearest 5%.
   d. The quoted percentage (45%) has been rounded up from 42.9%. This could be considered misleading in some contexts.
   e. i. 162°, 90°, 108°
      ii. 154°, 90°, 116°
      iii. 154°, 78°, 128°. Even though the pie chart gives a rough picture of the relative contributions of the three sectors, it has not been carefully drawn.

8. **a.** It is a circle viewed on an angle to produce an ellipse.
   b. No, because it causes some angles to be larger and others to be smaller.

Exercise 7.5 Graphing with technology

1.

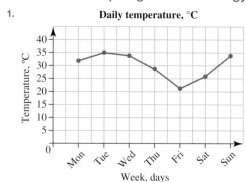

2.

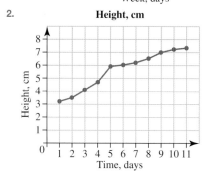

3. **a.** Rows are labelled with numbers and columns are labelled with capital letters
   b. Row 4, column A
   c. Cell B2

4.

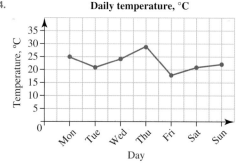

5.

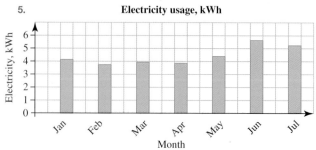

6. See bottom of the page*

7.

Family pets

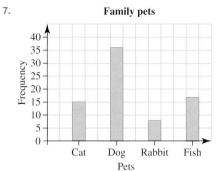

8.

Total road length per state and territory

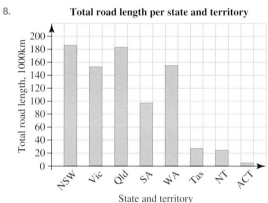

9.

Total road length in Australia, 1000 km

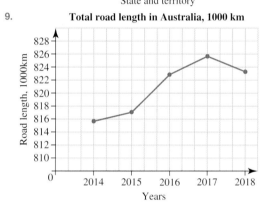

10.

Ingredients, g

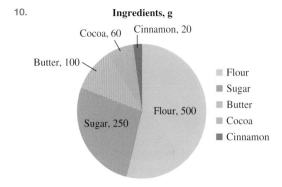

11.

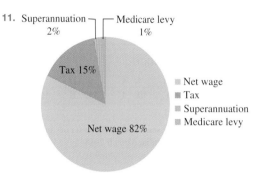

12. a. Lost earnings of victims 14%
Family and community losses 10%
Vehicle damage 30%
Pain and suffering 24%
Insurance administration 9%
Other 13%

b.

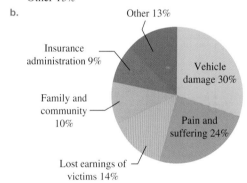

*

Gas usage, mJ

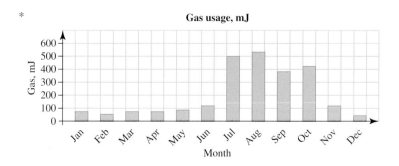

13.

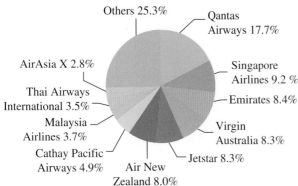

Others 25.3%
Qantas Airways 17.7%
AirAsia X 2.8%
Singapore Airlines 9.2 %
Thai Airways International 3.5%
Emirates 8.4%
Malaysia Airlines 3.7%
Virgin Australia 8.3%
Cathay Pacific Airways 4.9%
Jetstar 8.3%
Air New Zealand 8.0%

14. See table at bottom of the page*

7.6 Review: exam practice

1. **A**   2. **B**   3. **A**   4. **C**   5. **D**   6. **B**
7. **A**   8. **D**   9. **D**   10. **C**   11. **B**   12. **E**

13. a. Categorical nominal.

b.

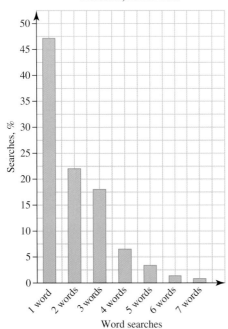

Type of word searches on the Internet in Australia, March 2018

Word searches

c. 1099 two-word searches

14. a. January
b. March and April; May and July
c. December (90 cars)
d. By 80 cars
e. 20 cars
f. 565 cars
g. Although sales figures fluctuate from month to month, overall a definite increasing trend is observed.

15.

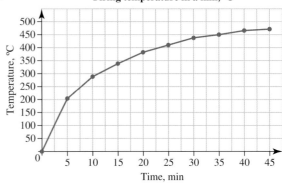

Firing temperature in a kiln, °C

16. b.

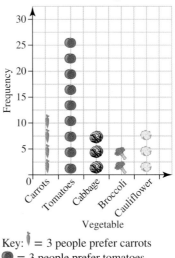

Favourite vegetable

Key:
= 3 people prefer carrots
= 3 people prefer tomatoes
= 3 people prefer cabbage
= 3 people prefer broccoli
= 3 people prefer cauliflower

*

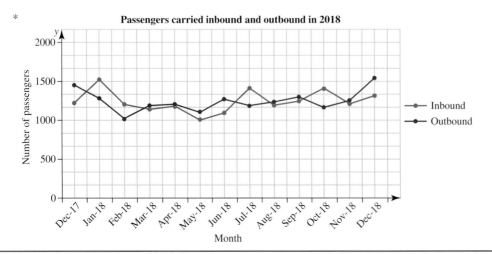

Passengers carried inbound and outbound in 2018

- Inbound
- Outbound

Month

c.

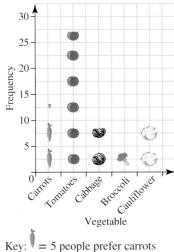

Favourite vegetable

Key: = 5 people prefer carrots
= 5 people prefer tomatoes
= 5 people prefer cabbage

= 5 people prefer broccoli
= 5 people prefer cauliflower

d.

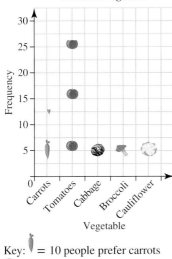

Favourite vegetable

Key: = 10 people prefer carrots
= 10 people prefer tomatoes
= 10 people prefer cabbage

= 10 people prefer broccoli
= 10 people prefer cauliflower

e. Graph **b** because the fewer items a picture represents, the easier it is to read the information required.

17. a.

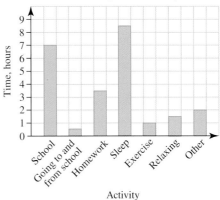

Activities per 24-hour day

b. School 29.2%, Going to and from school 2%, Homework 14.6%, Sleep 35.4%, Exercise 4.2%, Relaxing 6.3%, Other 8.3%.

c.

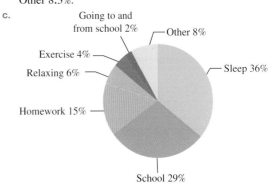

d. 17.5 hours.

18. a. see table at bottom of the page**
   b. 1 May
   c. 30 May
   d. A decrease of 3.3°C
   e. An increase of 1.7°C

**

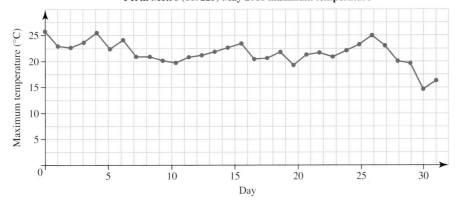

Perth Metro (009225) May 2018 maximum temperature

PRACTICE ASSESSMENT 2

Essential Mathematics: Unit 1 examination

Unit
Unit 1: Number, data and graphs

Topic
Topic 1: Number
Topic 2: Representing data
Topic 3: Graphs

Conditions

Technique	Response type	Duration	Reading
Paper 1: Simple (24 marks) Paper 2: Complex (16 marks)	Short response	60 minutes	5 minutes

Resources	Instructions
• QCAA formula sheet: • Notes not permitted • Scientific calculator permitted	• Show all working. • Write responses using a black or blue pen. • Unless otherwise instructed, give answers to **two decimal places**.

Criterion	Marks allocated	Result
Foundational knowledge and problem solving	40	

For the most up to date assessment information, please see www.qcaa.qld.edu.au/senior.

A detailed breakdown of examination marks summary can be found in the PDF version of this assessment instrument in your eBookPLUS.

Question 1 (3 marks)

During a stocktake sale there was 15% off all televisions. Peter liked a television with an original price of $1500.

a. Determine how much Peter would save if he bought it during the sale.

b. Determine how much he would pay for the television during the sale.

Question 2 (3 marks)

A cyclist covers 1000 metres in 1 minute and 15 seconds.

a. Calculate the cyclist's average speed in km/h.

b. If the cyclist could maintain this average speed, calculate how long it would take them to complete 2.5 kilometres, in minutes.

Question 3 (4 marks)

A smoothie contains 300 grams of milk, 100 grams of bananas, 150 grams of strawberries and 25 grams of honey. Calculate the percentage of:

a. milk that was used

b. bananas and honey that were used.

Question 4 (4 marks)

To make a mixture of concrete, the ratio of cement to sand to stones is $1 : 2 : 3$. To make 48 kilograms of concrete, calculate how many kilograms of the following you need.

a. Sand

b. Stones

Question 5 (6 marks)

The following data represents the daily maximum temperature for a two-week period:
18, 21, 23, 19, 27, 29, 23, 22, 29, 31, 24, 18, 25, 29

Determine the following from the data.

a. The mean

b. The median

c. The mode

_____

_____

_____

_____

_____

_____

Question 6 (4 marks)

A class of 30 students were surveyed on their favourite sport. The results are graphed below. From the graph, answer the following.

a. What was the favourite sport?

b. How many students chose golf as their favourite sport?

c. What sport received 6 votes?

d. What two sports received the same number of votes?

Favourite sport

_____

_____

_____

_____

_____

_____

_____

Part B: Complex – total marks: 16

Question 7 (6 marks)

Contestants in a house renovation competition want to budget for a number of jobs they need done to complete work. Calculate the total cost of the following jobs.

- Carpenter at $50 per hour for 24 hours
- Electrician at $95 per hour for 12.5 hours
- Tiler at $70 per hour for 6 hours and 45 minutes
- Plasterer at $28 per half hour for 7 hours and 20 minutes
- Plumber with a call-out charge of $100 and $90 per hour for 8.5 hours

Question 8 (6 marks)

The following pie graph represents where the money goes in a family budget.

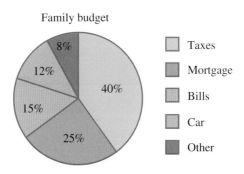

Family budget

8%
12%
15%
25%
40%

- Taxes
- Mortgage
- Bills
- Car
- Other

Determine the following.

a. What percentage is spent on taxes?

b. What percentage is spent on the car?

c. If the family earned $120 000 per year, how much would they spend on bills per year?

d. If the family earned $120 000 per year, how much would they spend on the mortgage and bills per year?

_____

_____

_____

_____

_____

_____

_____

_____

_____

_____

_____

_____

_____

Question 9 (4 marks)

Draw a step graph to represent the hire cost of a stand-up paddle board.

Paddle board hire	
Hours	**Rate ($)**
> 0–1	$15
> 1–2	$25
> 2–4	$45
Daily charge	$60

CHAPTER 8
Earning money

8.1 Overview

Most people would like to earn more money, but just as important is learning how to manage that money once you've earned it. The first step on that path is knowing where the money you've earned goes – how much of it goes towards tax and superannuation, and how much you're left with afterwards.

LEARNING SEQUENCE

8.1 Overview
8.2 Wages and salaries
8.3 Commission, piecework and royalties
8.4 Taxation and deductions
8.5 Performing financial calculations
8.6 Review: exam practice

CONTENT

In this chapter, students will learn to:
- find earnings, including salary, wages, overtime, piecework and commission
- convert between annual, monthly, fortnightly, weekly and hourly rates of earning [complex]
- understand the purpose of superannuation
- interpret entries on a selection of wage or salary pay slips and timesheets
- understand the purpose of taxation and the use of tax file numbers
- use tax tables to determine PAYG tax for periodic (weekly/fortnightly/monthly) earnings [complex]
- interpret entries on a simple PAYG summary
- apply the concepts of taxable income, gross income, allowable deductions and levies in simple contexts [complex]
- calculate a simple income tax return and net income using current income tax rates [complex].

Fully worked solutions for this chapter are available in the Resources section of your eBookPLUS at www.jacplus.com.au.

8.2 Wages and salaries

8.2.1 Salaries, wages and penalty rates

Salaries

A person on a **salary** gets a fixed amount of money for a year, but is usually paid a portion of this each fortnight (2 weeks) or month. Each payment is an equal amount and is calculated by dividing the annual amount by the number of payment periods per year.

Hourly wages

A person on an **hourly wage** gets paid a set amount for each hour worked.

If a person works more than a certain number of hours per week, any additional hours are called **overtime**, and are paid at a higher rate. The overtime hourly rate is usually a multiple of the regular hourly rate:

- $1.5 \times$ regular hourly wage (time-and-a-half)
- $2 \times$ regular hourly wage (double-time), and so on.

Penalty rates

Workers may receive a higher pay rate when they are required to work on weekends and public holidays, do shift work (early mornings or late nights), and work overtime. The pay rate paid is called a **penalty rate**. Penalty rates will vary according to the working agreements and will be a certain percentage more than the normal pay rate.

WORKED EXAMPLE 1

a. A bank employee earns a salary of **$61 000** per year. How much does he get paid each fortnight?

b. A fast-food employee is paid **$15.50** per hour. If she worked 72 hours last fortnight, how much did she get paid?

THINK	WRITE
a. 1. Determine the number of fortnights in a year by dividing the number of days in a year by the number of days in a fortnight. Note that the result will be different in a leap year (366 days).	$\dfrac{365}{14} \approx 26.07$ (correct to 2 decimal places)
2. Divide the annual salary by 26.07 and round to the nearest cent.	$61\,000 \div 26.07 \approx 2339.85$
3. Answer the question.	The bank employee's fortnightly pay is $2339.85
b. 1. The worker is paid $15.50 for each hour they work, so multiply this amount by the number of hours worked.	$15.50 \times 72 = 1116.00$
2. Answer the question.	Last fortnight the fast-food employee was paid $1116.00.

Note: We will use the approximations of 52.14 weeks and 26.07 fortnights in a year in this text.

A hospitality worker is paid time-and-a-half for working on Saturday nights and double-time for working on Sundays or public holidays. The normal hourly pay rate is $15.50. If the worker worked 5 hours on Saturday and 6 hours on Sunday, calculate the amount of money she earned over the weekend.

THINK	WRITE
1. Calculate the hourly rate of pay for Saturday. Time-and-a-half means × 1.5.	$15.50 \times 1.5 = \$23.25$ per hour
2. Calculate the amount paid for Saturday.	$23.25 \times 5 = \$116.25$
3. Calculate the hourly rate of pay for Sunday. Double-time means × 2.	$15.50 \times 2 = \$31$
4. Calculate the amount paid for Sunday.	$6 \times 31 = \$186$
5. Answer the question. *Note*: This is the amount she earned before tax was taken out.	The amount earned for the weekend is $\$116.25 + 186 = \$302.25$.

8.2.2 Other allowances

Employees may receive an **allowance** to complete certain tasks, such as to use their own tools at work, or to work in certain conditions. Common allowances include uniforms, tools and equipment, travel costs, car and phone costs, and first-aid training.

Allowances are included in wages and are documented separately on pay sheets.

8.2.3 Superannuation

All workers in Australia get an additional sum of money for their retirement. This sum is called **superannuation**. From 2017 the law requires employers to pay an additional 9.5% of annual salary into a recognised superannuation fund. The amount is based on usual earnings before tax and is calculated each pay cycle.

Some workers choose to contribute additional funds from their wages to increase their superannuation. There are tax incentives, such as paying lower tax rates on superannuation lump sums, to encourage workers to save for their retirement.

Additional government financial support, known as the **pension**, is also available for some people who have retired. The amount a retired person receives in the pension depends on their personal financial security and wealth.

WORKED EXAMPLE 3

A worker's hourly rate is $28.75 and he works 38 hours each week. The worker is paid fortnightly. Calculate the amount of superannuation the employer pays on his behalf each fortnight.

THINK	WRITE
1. Calculate the fortnightly wage. Note that if the employee works 38 hours per week, then he works 76 (38 x 2) hours per fortnight.	$76 \times 28.75 = \$2185.00$
2. Calculate 9.5% of the fortnightly wage. Remember to convert 9.5% to a decimal by dividing by 100.	$9.5\% \times 2185 = 0.095 \times 2185$ $\approx \$207.58$
3. Answer the question.	The employer pays $207.58 into the superannuation fund each fortnight.

8.2.4 Annual leave loading

Some workers will receive an extra payment on top of the 4-week annual leave pay. It is usually 17.5% of their normal pay for 4 weeks. It will depend on the working agreement workers have with their employers if they receive an **annual leave loading**.

The purpose of annual leave loading is to compensate workers who are unable to earn additional money through overtime while on leave, and to help cover the costs associated with taking holidays.

WORKED EXAMPLE 4

A worker's annual salary is $58 056. Calculate the amount, in dollars, paid for annual leave loading before tax.

THINK	WRITE
1. Calculate the weekly salary by dividing the annual salary by 52.14.	$\dfrac{58\,056}{52.14} \approx \$1113.46$
2. Calculate the wage for 4 weeks.	$\$1113.46 \times 4 = \$4453.84$
3. Calculate 17.5% of the 4-week wage. Remember to convert 17.5% to a decimal by dividing by 100.	$17.5\% \times 4453.84 = 0.175 \times 4453.84$ ≈ 779.42
4. Answer the question.	The amount paid for annual leave loading, before tax, is $779.42.

 **Resources**

Interactivity: Special rates (int-6068)

Exercise 8.2 Wages and salaries

Note: Assume 52.14 weeks and 26.07 fortnights in a year.

1. A lawyer is offered a job with a salary of $74 000 per year, or $40 per hour. Assuming that she works 80 hours every fortnight, which is the greater pay?

2. **MC** An hourly wage of $32.32 for 77.5 hours per fortnight results in a fortnightly pay of:
   A. $1252.40. **B.** $2504.80. **C.** $842.58. **D.** $1292.80.

3. **WE1** Calculate the wage for the following hourly rates and hours worked.
   a. $19.75 for 74.75 hours
   b. $24.85 for 45.75 hours
   c. $45.30 for 35.25 hours

4. A worker earns $20.40 an hour. He needs to earn a minimum of $700 each week before tax, to buy food, pay rent and bills and have some money for entertainment. Calculate the minimum number of hours he has to work each week. Give your answer correct to 2 decimal places.

5. **WE2** A retail worker is paid time-and-a-half for working on Saturday and double-time for working on Sundays or public holidays. The normal pay rate is $18.75. The worker worked over a long weekend, with the Monday being a public holiday. The hours worked for Saturday, Sunday and Monday (public holiday) were 8, 5 and 6 hours respectively. Calculate the amount of money she earned over the weekend.

6. Over the last four weeks, a woman has worked 35, 36, 34 and 41 hours. If she earns $24.45 per hour, how much did she earn for each of the two fortnights?

7. **WE3** A worker's hourly rate is $29.45 and she works 25.75 hours each week. The worker is paid fortnightly. Calculate the amount of superannuation the employer pays on her behalf each fortnight.

8. A school principal is on an annual salary of $155 750.
   a. How much does she earn each month?
   b. What is her superannuation fund payment if the fund is paid 9.5% of her annual salary?

9. A salary earner makes $62 000 per year.
   a. How much does he earn each month?
   b. What is his superannuation fund payment each month if he receives 9.5% superannuation?
   c. What is the total amount deposited into the fund in a year?

10. **WE4** A worker's annual salary is $85 980. Calculate the amount, in dollars, paid for annual leave loading before tax.

11. Some employers offer a superannuation bonus scheme. They pay you 9.5%, then for every additional 1% of your salary that you contribute, they match it with a further 1%. For example, if you contribute 2% of your salary to superannuation, you will receive a total of 9.5% + 2% + 2% = 13.5% put into your fund. A software saleswoman makes $70 000 per year and decides to contribute 3% of her salary into superannuation.
   a. How much does the saleswoman contribute from her own salary into superannuation each year?
   b. The saleswoman's employer matches her superannuation contributions. In total, how much does she receive into her superannuation fund each year?

12. There is a proposal to have the same penalty rates for both Saturday and Sunday of time-and-a-half. Currently workers receive time-and-a-half to work Saturdays and double-time to work Sundays.

    A waiter receives penalty rates and her hourly rate is $17.75.

    a. By what percentage, to the nearest whole number, will she be worse off if the penalty rates change?

    b. By what percentage, to 2 decimal places, will her hourly rate need to increase to ensure she receives the same amount she currently does for working both Saturday and Sundays, assuming she works the same number of hours each day?

13. A factory worker receives an hourly rate of $28.40 to work a standard 38-hour-week. She receives overtime of time-and-a-half for any hour worked above 38 hours.

    a. On average, she works 42.5 hours each week. Calculate her weekly wage before tax.

    b. The worker is offered a salary of $68 000. Will she be better off remaining on a wage or taking the salary offer? Justify your answer using calculations.

14. An employer is proposing a new working agreement: the removal of the 17.5% annual leave loading, an increase of 5% in annual wages and 10.5% superannuation contributions. Employees are considering whether to accept the agreement. Should they accept the agreement? Justify your answer using calculations.

15. A secretary's current annual salary is $56 890. She is offered two packages:

    Package 1: pay increase of 1.5% each year for 5 years
    Package 2: superannuation contribution of 10.75% for 5 years.

    By calculating her salary each year for 5 years and the additional amount of superannuation contributions that will be deposited into her superannuation fund, which package should she choose?

8.3 Commission, piecework and royalties

8.3.1 Commission

A **commission** is paid to a person when an item is sold. For example, if a real estate agent sells a house, then he or she is paid a commission.

Commissions are calculated as a percentage of the sale price. If no sales are made, no commission is received. A commission table is often used to determine the value of a commission.

WORKED EXAMPLE 5

When selling real estate, an agent is paid according to the table below.

Sale price	Commission	Plus
Between $0 and $80 000	2% of sale price	0
Between $80 001 and $140 000	1.5% of amount over $80 000	$1600
$140 000 and over	1.1% of amount over $140 000	$2500

If a house is sold for $200 000, what is the commission paid to the real estate agent?

THINK	WRITE
1. Since the amount of commission varies according to the sale price, determine which of the 3 ranges the sale falls into. 200 000 > 140 000, so the price falls into range 3 ($140 000 and over).	200 000 −140 000 60 000
The commission for this range is based on the amount of the sale *over* $140 000. Calculate the amount over $140 000.	
2. The commission is 1.1% of 60 000; write 1.1% as a decimal.	$1.1\% = \dfrac{1.1}{100} = 0.011$
3. Multiply 0.011 by the amount over $140 000, that is $60 000.	$1.1\% \text{ of } 60\,000 = 0.011 \times 60\,000$ $= 660$
4. The table also specifies an additional commission of $2500, based on the amount of the sale *under* $140 000. Calculate the total of the commission.	$660 + 2500 = 3160$
5. Answer the question.	The total commission is $3160.

8.3.2 Salary and commissions

In some industries the rate of sales varies widely, so it is not practical to live on commissions only. For example, car salespeople are given a base salary in addition to a commission.

WORKED EXAMPLE 6

A car salesperson earns a salary of $1200 per month plus a commission of 3% of the total sales that she makes that month. In February, she sold cars worth $198 500. What is her total wage (salary + commission) for that month?

THINK	WRITE
1. Her commission is 3% of sales. Write 3% as a decimal.	$3\% = \dfrac{3}{100} = 0.03$
2. Calculate 3% of the total sales in order to determine the commission.	$3\% \text{ of } 198\,500 = 0.03 \times 198\,500$ $= 5955$

3. Her total wage is the commission plus her base salary. Add the commission of $5955 to the base salary of $1200.

$5955 + 1200 = 7155$

4. Answer the question.

The total wage paid in February is $7155.

8.3.3 Piecework

A person paid by **piecework** is paid a fixed amount for each item produced. Often the work is done at home. There may be bonuses for faster workers.

WORKED EXAMPLE 7

A seamstress gets paid $4.75 for every dress she sews. If she sewed 124 dresses last week, how much money did she earn?

THINK

1. The seamstress gets $4.75 for each dress, so multiply this amount by the number of dresses sewn.

2. Answer the question.

WRITE

$4.75 \times 124 = 589$

Last week the seamstress was paid $589.00.

8.3.4 Royalties

A **royalty** is a payment made to an author, composer or creator for each copy of the work or invention sold. For example, if a pop star sold $2 million worth of music last year, she is entitled to some of the profits. Other people, such as managers, also get a share. Royalties are usually calculated as a percentage of the total sales (not total profit).

WORKED EXAMPLE 8

Last year, a new rock star sold music worth $2 156 320. His recording contract specifies that he receives 2.4% royalty on the total sales. How much money did he earn last year?

THINK

1. He earned 2.4% of the sales ($2 156 320). Write 2.4% as a decimal.

2. Calculate 2.4% of the total sales; multiply 0.024 by the total sales.

3. Answer the question.

WRITE

$2.4\% = \dfrac{24}{100}$

$= 0.024$

2.4% of $2 156 320 = 0.024 \times 2 156 320$
$= \$51 751.68$

The rock star earned $51 751.68 in royalties.

Exercise 8.3 Commission, piecework and royalties

1. **WE5** Using the commission table below, calculate the commission on the sale of a house for $945 000.

SOLD for $945 000

Sale price	Commission	Plus
Between $0 and $400 000	0.5% of sale price	0
Between $400 001 and $700 000	0.3% of amount over $400 000	$2000 (0.5% of $400 000)
$700 001 and over	0.2% of amount over $700 000	$2900 (0.5% of $400 000 + 0.3% of $300 000)

2. Using the commission table from question 1, calculate the commission for the following sales.
   a. $330 000   b. $525 000   c. $710 000   d. $1 330 000

3. **MC** Using the commission table from question 1, the commission on the sale of a $500 000 house would be:
   A. $1000.   **B.** $2000.   **C.** $2300.   **D.** $2500.

4. **WE6** A car salesperson earns a salary of $1400 per month plus a commission of 3.5% of the total sales he makes that month. In April, he sold cars worth $155 000. What is his total wage (salary + commission) for that month?

5. **WE7** A shoemaker is paid $5.95 for each pair of running shoes he can make. If the shoemaker made 235 pairs of shoes last week, what was the amount paid?

6. **MC** If a software engineer gets paid $3.40 for every line of computer code she writes, then how much will she make if she writes 865 lines in a fortnight?
   A. $29 410.00   **B.** $2941.00
   C. $294.10   **D.** $2329.00

7. A seamstress gets paid per item she sews. If she received $545 for sewing 45 items, how much did she get paid per item?

8. **WE8** Last year, a pop star sold $5 342 562 worth of music on her website and via CD sales. Her contract calls for a 3.5% royalty on all sales. How much did she earn last year?

9. **MC** If an author earned $25 560 on the basis of sales of $568 000, then the royalty payment is:
   A. 4.5%.   **B.** 45%.
   C. 0.045%.   **D.** 0.45%.

10. An actress is paid a 2.5% royalty plus $300 000 cash to act in the latest blockbuster movie.

    a. Complete the following table of royalties.

Time	Jan–Mar	Apr–Jun	Jul–Aug	Sep–Dec
Sales	$123 400	$2 403 556	$432 442	$84 562
Royalty payment				

    b. What is the total amount the actress received, including the cash?

11. To keep matters simple, the winner of a television talent show receives a royalty of 1% on sales of their first album. If this year's winner sells $4 563 453 worth of music, what is the royalty payment?

12. A new children's book author is offered a choice of $100 000 cash or a 4% royalty on sales. What would the sales need to be to match the cash offer?

13. A hot-shot used car salesman earns $1500 per month plus 6.2% commission on all sales. If he sold $243 540 worth of cars last month, what is his total wage?

14. A songwriter receives royalties on the sales of her songs. She received $11 473.75 in royalties for the sale of $458 950 in songs. Explain how the royalty percentage she receives can be determined, and hence find the royalty percentage she receives.

15. A shoe manufacturer decides to offer his shoemakers two pay options:
    Option 1: $8.25 per pair of shoes made
    Option 2: receive a 3.5% commission on the shoes sold.

    A shoemaker can make 15 pairs of shoes each day (over 5 days) and each pair of shoes sells for $75.95. In one week the manufacturer will sell 250 pairs of shoes. Which option will earn the shoemaker the most money? Justify your answer using calculations.

16. A telecommunications salesperson receives a base wage of $450 per week plus 2.5% commission on all new telephone plans sold during the week. In one week, his earnings (before tax) were $502.25.

    a. What was the total amount (in dollars) of telephone plans he sold during the week?

    b. If each plan was $95, how many plans did he sell?

    c. The salesperson is offered the choice to remain on a weekly base wage of $530 with an increase of 3.5% in commission on all sales, or to receive an annual salary of $37 500. Which option would you recommend he choose? Justify your answers using calculations.

8.4 Taxation and deductions

8.4.1 The purpose of taxation

Taxation is a means by which State and Federal governments raise revenue for public services, welfare and community needs by imposing charges on citizens, organisations and businesses.

Services include education, health, pensions for the elderly, unemployment benefits, public transport and much more.

8.4.2 Tax file numbers

A tax file number (TFN) is a personal reference number for every taxpaying individual, company, funds and trusts. Tax file numbers are for life and are issued by the Australian Taxation Office (ATO).

8.4.3 Income tax

Income tax is a tax levied on the financial income of people. It is deducted from each fortnightly or monthly pay.

The amount of income tax is based upon **total income** and **tax deductions**, which determines a worker's **taxable income**.

$$\text{Taxable income} = \text{total income} - \text{tax deductions}$$

The calculation of income tax is based upon an 'income tax table'. The income tax table at the time of writing is:

Taxable income	Tax on this income
0–$18 200	Nil
$18 201–$37 000	19 c for each $1 over $18 200
$37 001–$87 000	$3572 plus 32.5 c for each $1 over $37 000
$87 001–$180 000	$19 822 plus 37 c for each $1 over $87 000
$180 001 and over	$54 232 plus 45 c for each $1 over $180 000

NOTE: The income tax table is subject to change.

8.4.4 Tax deductions

Workers who spend their own money for work-related expenses are entitled to claim the amount spent as tax deductions. Tax deductions are recorded in the end-of-the-financial-year tax return. The deductions are subtracted from the taxable income, which lowers the amount of money earned and hence reduces the amount of tax to be paid.

What can be claimed as tax deductions is determined by the Australian Taxation Office. Some examples of deductions that can be claimed include using your car to travel to work-related events, purchasing materials, union fees, donations made to charities and using a home office. Deductions must be work-related expenses and evidence, such as receipts, must be provided.

WORKED EXAMPLE 9

Calculate the income tax payable by the teacher who earns a salary of $67 400. Tax deductions for the teacher are $4240 and the teacher earned $1240 in bank interest.

THINK	WRITE
1. Determine the total income by adding the salary and bank interest together.	$67 400 + 1240 = \$68 640$
2. Subtract the tax deductions from the total income.	Taxable income = $\$68 640 - 4240$ $= \$64 400$
3. Determine the tax bracket based on the taxable income of $64 400.	Taxable income is between $37 001 and $87 000.
4. Determine the amount of tax to be paid.	$3572 plus 32.5 c for every $1 over $37 000
5. Determine the amount of taxable income over $37 000 by subtracting $37 000 from the teacher's taxable income.	$64 400 - 37 000 = \$27 400$

▶

6.	Determine the tax rate amount as 32.5% of $27 400. Remember to convert 32.5% to a decimal.	$0.325 \times 27\,400 = \$8905$
7.	Add the 'plus amount' to find the payable tax.	$\$3572 + 8905 = \$12\,477$
8.	Answer the question.	The total tax payable by the teacher is $12 477.

8.4.5 The Medicare levy

Australian residents have access to healthcare through Medicare, which is partly funded by taxpayers through the payment of a **Medicare levy**. The current Medicare levy is 2% of taxable income, and may be reduced if the taxable income is below a certain amount.

In addition to the Medicare levy, residents who do not have adequate private healthcare may also be required to pay a Medicare levy surcharge.

WORKED EXAMPLE 10

A plumber's taxable income for the financial year is $65 850. He claims $5680 in work-related expenses (deductions) and he also has private health insurance. Determine the amount he has to pay for the Medicare levy.

THINK	WRITE
1. Calculate the taxable income.	Taxable income $= \$65\,850 - \$5680$ $\qquad\qquad\quad = \$60\,170$
2. Calculate 2% of taxable income. Remember to convert 2% to a decimal.	$2\% \times 60\,170 = 0.02 \times 60\,170$ $\qquad\qquad\quad = \$1203.40$
3. Answer the question.	He will pay $1203.40 in Medicare levy.

8.4.6 Pay As You Go (PAYG) tax

The Australian Taxation Office administers a **Pay As You Go (PAYG) tax**, which is a withholding tax system. It requires employers to calculate the amount of income tax to withhold from employees. The amount withheld is determined by taxation tables provided by the Tax Office.

This amount is withheld from the employee's regular earnings (gross pay) and contributes towards the employee's tax to be paid at the end of the financial year.

$$\text{Net pay} = \text{gross pay} - \text{tax withheld}$$

At the end of the financial year, workers submit a tax return which lists their deductions and all money earned.

- If the total amount of tax paid is less than is required, the worker will have to pay the difference.
- If the amount of tax paid over the year is more than is required, the worker receives a refund.

Australian workers who provide their tax file number and do not work another job claim the tax-free threshold. This means that they do not pay tax on the first $18 200 earned in the financial year.

A factory worker is paid $19.94 per hour. Each week she works 38 hours, with an additional 4 hours overtime paid at time-and-a-half. Using the taxation table, calculate her net pay (the amount she receives).

Weekly earnings ($)	Tax withheld ($)	Weekly earnings ($)	Tax withheld ($)
856.00	133.00	881.00	141.00
857.00	133.00	882.00	142.00
858.00	133.00	883.00	142.00
859.00	134.00	884.00	142.00
860.00	134.00	885.00	143.00
861.00	134.00	886.00	143.00
862.00	135.00	887.00	143.00
863.00	135.00	888.00	144.00
864.00	135.00	889.00	144.00
865.00	136.00	890.00	144.00
866.00	136.00	891.00	145.00
867.00	136.00	892.00	145.00
868.00	137.00	893.00	145.00
869.00	137.00	894.00	146.00
870.00	137.00	895.00	146.00
871.00	138.00	896.00	146.00
872.00	138.00	897.00	147.00
873.00	138.00	898.00	147.00
874.00	139.00	899.00	147.00
875.00	139.00	900.00	148.00
876.00	139.00		
877.00	140.00		
878.00	140.00		
879.00	141.00		
880.00	141.00		

THINK	WRITE
1. Calculate the weekly wage for normal hours.	$38 \times 19.94 = \$757.72$
2. Calculate the overtime.	$19.94 \times 1.5 \times 4 = \$119.64$
3. Calculate the total weekly wage.	$\$757.72 + 119.64 = \$877.36$
4. Using the table, locate the weekly wage and read the amount of tax withheld.	Weekly wage $877, tax withheld $140
5. Calculate the net pay by subtracting the tax withheld from the total wages.	Weekly income $= 877.36 - 140$ $= \$737.36$

Exercise 8.4 Taxation and deductions

1. **WE9** Calculate the income tax payable by a baker whose salary is $51 260. Tax deductions of the baker are $2120, and the baker earned $1850 in bank interest. Refer to the tax table provided on page 317.

2. **WE10** Calculate the Medicare levy for the following taxable incomes.
   a. $60 400   b. $77 300   c. $89 400   d. $108 423

3. **MC** A shop worker earns $18.95 per hour. She works 52 hours in a fortnight. Which of the amounts is closest to the amount of Medicare levy she will be expected to pay at the end of the financial year? *Note:* Assume there are 26.07 fortnights in a year.
     A. $20   **B.** $51
     C. $200   **D.** $514

4. **WE11** A garage mechanic is paid by the hour.
   a. If she works 84 hours in a fortnight and is paid $22.50 an hour, what is her gross pay (amount before tax)?
   b. Using the taxation table shown, calculate her net pay.

Fortnightly earnings ($)	Tax withheld ($)
1832.00	306.00
1834.00	308.00
1836.00	308.00
1838.00	308.00
1840.00	310.00
1842.00	310.00
1844.00	310.00
1846.00	312.00
1848.00	312.00
1850.00	314.00
1852.00	314.00
1854.00	314.00
1856.00	316.00
1858.00	316.00
1860.00	316.00
1862.00	318.00
1864.00	318.00
1866.00	318.00
1868.00	320.00
1870.00	320.00
1872.00	320.00
1874.00	322.00
1876.00	322.00

Fortnightly earnings ($)	Tax withheld ($)
1878.00	322.00
1880.00	324.00
1882.00	324.00
1884.00	324.00
1886.00	326.00
1888.00	326.00
1890.00	326.00
1892.00	328.00
1894.00	328.00
1896.00	330.00
1898.00	330.00
1900.00	330.00
1902.00	332.00
1904.00	332.00
1906.00	332.00
1908.00	334.00
1910.00	334.00
1912.00	334.00
1914.00	336.00
1916.00	336.00
1918.00	336.00
1920.00	338.00

5. For the worker in question **4**, any hours worked over 76 hours per fortnight are paid overtime at a rate of $45 an hour. Adjust her gross and net pay accordingly.

6. Using the taxation table provided:
   i. write down the percentage tax payable for the following annual salaries.
   ii. hence, calculate the amount of tax payable.
   a. $37 500   b. $15 879   c. $85 670   d. $112 500

Taxable income	Tax on this income
0–$18 200	Nil
$18 201–$37 000	19 c for each $1 over $18 200
$37 001–$87 000	$3572 plus 32.5 c for each $1 over $37 000
$87 001–$180 000	$19 822 plus 37 c for each $1 over $87 000
$180 001 and over	$54 232 plus 45 c for each $1 over $180 000

7. The table shows the tax payable for fortnightly wages. Using the table, calculate the tax payable on the following fortnightly wages by workers who claim the tax-free threshold.

Fortnightly income ($)	Tax to be withheld		Fortnightly income ($)	Tax to be withheld	
	With tax-free threshold ($)	No tax-free threshold ($)		With tax-free threshold ($)	No tax-free threshold ($)
1000	74	262	2000	366	608
1100	96	296	2200	434	676
1200	118	330	2500	540	780
1500	192	434	3000	712	968
1800	296	538	4000	1086	1358

   a. $1500   b. $2000   c. $2500   d. $4000

8. Calculate the percentage of tax payable for the incomes from question **7** by workers who don't claim the tax-free threshold. Give your answers to 2 decimal places where necessary.

9. An apprentice electrician is paid $17.14 per hour. She is paid the normal rate for the first 38 hours worked in any week and then the overtime rate of $25 for hours worked over 38 hours. In one week, she worked 45 hours. Using the taxation table, calculate her net wage for the week (the amount she receives).

Weekly earnings ($)	Amount to be withheld ($)	Weekly earnings ($)	Amount to be withheld ($)
816.00	119.00	821.00	120.00
817.00	119.00	822.00	121.00
818.00	119.00	823.00	121.00
819.00	120.00	824.00	121.00
820.00	120.00	825.00	122.00

Weekly earnings ($)	Amount to be withheld ($)	Weekly earnings ($)	Amount to be withheld ($)
826.00	122.00	836.00	126.00
827.00	122.00	837.00	126.00
828.00	123.00	838.00	126.00
829.00	123.00	839.00	127.00
830.00	123.00	840.00	127.00
831.00	124.00	841.00	127.00
832.00	124.00	842.00	128.00
833.00	125.00	843.00	128.00
834.00	125.00	844.00	128.00
835.00	125.00	845.00	129.00

10. A nurse's gross annual salary is $58 284 and he is paid monthly.
    a. Calculate his gross monthly salary.
    b. Using the tax table shown, determine the amount of tax withheld from his monthly salary.

Monthly earnings ($)	Tax withheld ($)	Monthly earnings ($)	Tax withheld ($)
4792.67	949.00	4870.67	979.00
4797.00	953.00	4875.00	979.00
4801.33	953.00	4879.33	979.00
4805.67	958.00	4883.67	984.00
4810.00	958.00	4888.00	984.00
4814.33	958.00	4892.33	984.00
4818.67	962.00	4896.67	988.00
4823.00	962.00	4901.00	988.00
4827.33	962.00	4905.33	988.00
4831.67	966.00	4909.67	992.00
4836.00	966.00	4914.00	992.00
4840.33	966.00	4918.33	997.00
4844.67	971.00	4922.67	997.00
4849.00	971.00	4927.00	997.00
4853.33	971.00	4931.33	1001.00
4857.67	975.00	4935.67	1001.00
4862.00	975.00	4940.00	1001.00
4866.33	975.00		

c. Determine the percentage of tax, to 2 decimal places, he pays each month.

d. Calculate the Medicare levy he will pay.

11. An accountant pays $1719.40 in Medicare levy.

a. If she has private health insurance, calculate her taxable income for the year.

b. Calculate the tax payable for the financial year. Use the tax table provided on page 317.

12. A truck driver earns $24.07 per hour for working a 76-hour fortnight. An overtime rate of time-and-a-half is paid for additional hours worked over the 76 hours during Monday to Friday, and double-time is paid for hours worked on weekends (Saturday and Sunday). Assume 52 weeks in 1 year.

Over a fortnight, the truck driver worked 80 hours Monday–Friday and 15 hours over the weekend.

a. Calculate the gross fortnightly wage for the truck driver.

Fortnightly earnings ($)	Tax withheld ($)	Fortnightly earnings ($)	Tax withheld ($)
2682.00	602.00	2712.00	612.00
2684.00	602.00	2714.00	614.00
2686.00	604.00	2716.00	614.00
2688.00	604.00	2718.00	614.00
2690.00	604.00	2720.00	616.00
2692.00	606.00	2722.00	616.00
2694.00	606.00	2724.00	616.00
2696.00	606.00	2726.00	618.00
2698.00	608.00	2728.00	618.00
2700.00	608.00	2730.00	618.00
2702.00	608.00	2732.00	620.00
2704.00	610.00	2734.00	620.00
2706.00	610.00	2736.00	620.00
2708.00	610.00	2738.00	622.00
2710.00	612.00	2740.00	622.00

b. Show that the truck driver's net fortnightly pay was $2089.84 by using the PAYG tax table shown above.

c. If the truck driver claims $2486 in deductions, calculate his taxable income, and hence calculate the amount of payable tax for the year by using the tax table provided on page 317.

d. How much Medicare levy should the driver expect to pay, given that he has private health insurance?

13. A computer technician's annual salary is $67 374. He claims $1580 in deductions and receives $225 in dividends from shares.

   a. The technician is paid monthly. Calculate his gross monthly salary.
   b. His employer withholds $1200 PAYG tax each month. Calculate the total amount of tax withheld for the year.
   c. Determine the tax payable by the computer technician for the financial year. Hence, determine whether he receives a tax refund or is required to pay more tax, and state the amount.

14. A hairdresser receives a weekly wage of $930.60, which includes 38 hours of normal pay plus 4 hours of overtime, paid at time-and-a-half.
   a. Explain how his hourly rate can be determined, and hence find his hourly rate.
   b. Determine his annual salary if he works an average of 40 hours a week (38 hours at normal pay and 2 hours overtime). Assume 52 weeks in 1 year.
   c. Using your answer from part b, determine the amount of tax payable at the end of the financial year.
   d. The hairdresser forgot to claim $1980 in deductions and is looking forward to receiving $1980 in a tax refund. By recalculating the amount of tax payable at the end of the financial year based on his new taxable income, explain why he won't receive $1980 in a tax refund.

15.

Taxable income	Tax on this income
0–$18 200	Nil
$18 201–$37 000	19 c for each $1 over $18 200
$37 001–$87 000	$3572 plus 32.5 c for each $1 over $37 000
$87 001–$180 000	$19 822 plus 37 c for each $1 over $87 000
$180 001 and over	$54 232 plus 45 c for each $1 over $180 000

The Australian tax system is a tiered system, as shown in the table. A dentist's annual salary is $97 065.
   a. Explain why the tax payable is not found by calculating 37% of $97 065.
   b. Explain how the additional 'plus amount' of $19 822 is calculated.

16. A mechanical engineer pays $14 812 in tax.
   a. Explain how her taxable income can be determined.
   b. Determine her taxable income.
   c. After the financial year she realises that she forgot to claim $985 in tax deductions. Determine her taxable income and the tax refund she should expect.

8.5 Performing financial calculations

8.5.1 Calculating percentage change

Percentage increase and decrease are examples of percentage change. It is the extent to which an amount grows or loses.

$$\text{Percentage change} = \frac{\text{difference between two numbers}}{\text{original number}} \times 100$$

A spreadsheet can be used to find the percentage change by inserting a formula that calculates the difference between the cell numbers, and then divides by the original number and multiplies by 100.

WORKED EXAMPLE 12

A shop assistant receives a pay increase. Her original weekly pay was $545 and her new pay is $567. Using a spreadsheet, calculate the percentage change of her wage. Write your answer to the nearest whole per cent.

THINK

1. Set up a spreadsheet with headings 'original pay', 'new pay' and 'percentage change'.

WRITE

	A	B	C
1	original pay	new pay	percentage change
2			
3			
4			
5			
6			

2. Enter her original pay ($545) in cell A2 and her new pay ($567) in cell B2.

	A	B	C
1	original pay	new pay	percentage change
2	545	567	
3			
4			
5			
6			

3. In Cell C2, enter the formula that calculates the difference between cells A2 and B2: (B2-A2) and then divides by the original pay: (B2-A2)/A2, and finally converts to a percentage by multiplying by 100: (B2-A2)/A2 * 100. Be sure to place an = sign before your equation.

	A	B	C
1	original pay	new pay	percentage change
2	545	567	= (B2-A2)/A2*100
3			
4			
5			
6			

4. Answer the question.

The percentage change of her wage is 4% (a 4% pay rise).

8.5.2 Calculating tax payable

Calculating the tax payable on gross incomes can be found using a spreadsheet and inserting formulas that determine the amount of tax payable for each tax tier.

A law clerk's annual salary is \$45 675. She claims \$450 in tax deductions and receives \$250 in interest from investments. Using a spreadsheet, calculate the amount of tax payable for the financial year.

THINK

WRITE

1. Create a spreadsheet with headings 'gross income', 'interest', 'deductions' and 'taxable income'. Enter the values.

	A	B	C	D
1	gross income	interest	deductions	taxable income
2	45 675	250	450	
3				
4				
5				
6				

2. In cell D2 calculate the taxable income: add the interest $(\$250)$ to the annual salary $(\$45\,675)$ and subtract the deductions $(\$450)$ by writing the formula = A2+B2-C2.

D2			fx	= A2 + B2 − C2

	A	B	C	D
1	gross income	interest	deductions	taxable income
2	45 675	250	450	45 475
3				
4				
5				
6				

3. Insert a column to represent the tax bracket that the law clerk falls into.

	A	B	C	D	E
1	gross income	interest	deductions	taxable income	tax tier (37 001–87 000)
2	45 675	250	450	45 475	
3					
4					
5					
6					

4. Insert a formula to calculate the tax payable for this tax tier.

Type the following formula into E2:
= 3575 + (D2-37000) ∗ 0.325

5. Calculate the tax payable.

	A	B	C	D	E
1	gross income	interest	deductions	taxable income	tax tier (37 001–87 000)
2	45 675	250	450	45 475	6329.375
3					
4					
5					
6					

6. Answer the question, giving the answer to the nearest cent.

The tax payable by the law clerk is \$6329.38.

8.5.3 Preparing a wage sheet

A wage sheet shows a list of workers and details of their earnings. These include net wages, pay rates, gross deductions and allowances. A wage sheet may look similar to the one shown.

Employee	Pay rate ($)	Normal hours worked	Overtime 1.5	Penalty rate 1.5	Penalty rate 2	Allowance ($)	Gross pay ($)	Tax withheld ($)	Net pay ($)
Dave	19.50	38	2	5		25.5	945.75	163.00	782.75
Jose	21.80	32	4	3.5	5		1253.50	271.00	982.50
Neeve	25.70	25		4		15	1002.30	183.00	819.30
Niha	22.45	38				10	853.10	131.00	722.10
Yen	29.85	36			6		1074.60	208.00	866.60

A pay sheet (or payslip) shows individual workers' details, including the hours worked, pay rate, net wage, tax deduction, leave days and superannuation paid.

A wage sheet can be set up in a spreadsheet containing formulas that perform the necessary calculations.

WORKED EXAMPLE 14

Using a spreadsheet, complete the following wage sheet.

Employee	Pay rate ($)	Normal hours worked	Overtime 1.5	Penalty rate 1.5	Penalty rate 2	Allowance ($)	Gross pay ($)	Tax withheld ($)	Net pay ($)
Honura	38.95	38	4		6	15		620	
Skye	28.40	38	3	5				289	
Beau	19.15	35		5				234	

THINK

1. Set up a spreadsheet.

2. Calculate pay for hours worked:
   Pay for normal hours formula (e.g. 38*38.95)

   Overtime pay calculation (e.g. 4*1.5*38.95)

   Pay for working on penalty rates (e.g. 0*1.5*38.95 + 6*2*38.95)

WRITE

⊿	A	B	C	D	E	F	G	H	I	J
1			Normal hours worked	Overtime	Penalty rate		Allowance ($)	Gross pay ($)	Tax with held ($)	Net pay ($)
2		Pay rate ($)		1.5	1.5	2				
3	Employee									
4	Honura	38.95	38	4		6	15		620	
5	Skye	28.40	38	3	5				289	
6	Beau	19.15	35		5				234	

B4*C4

D4*D3*B4

E4*E3*B4+F4*F3*B4

3. Calculate gross pay (= hours worked plus allowances).

In cell H4H4 enter the formula
H4=B4*C4+D4*D3*B4+E4*E3*B4+F4*F3*B4+G4

4. Complete the table by filling the formula in column H down for all workers.

	A	B	C	D	E	F	G	H	I	J
1			Normal	Overtime	Penalty rate		Allowance	Gross	Tax with	Net
2		Pay	hours						with	
3	Employee	rate ($)	worked	1.5	1.5	2	($)	pay ($)	held ($)	pay ($)
4	Honura	38.95	38	4		6	15	2196.20	620	
5	Skye	28.40	38	3	5			1420	289	
6	Beau	19.15	35		5			813.88	234	

5. Calculate the net pay by deducting tax withheld.

In cell J4 type
= H4 − I4

6. Answer the question.

	A	B	C	D	E	F	G	H	I	J
1			Normal	Overtime	Penalty rate		Allowance	Gross	Tax with	Net
2		Pay	hours						with	
3	Employee	rate ($)	worked	1.5	1.5	2	($)	pay ($)	held ($)	pay ($)
4	Honura	38.95	38	4		6	15	2196.20	620	1576.20
5	Skye	28.40	38	3	5			1420	289	1131
6	Beau	19.15	35		5			813.88	234	579.88

Exercise 8.5 Performing financial calculations

1. **WE12** Calculate the percentage change for the following using a spreadsheet. Give your answers correct to 2 decimal places.
    a. Original tax paid is $450 and the new tax paid is $435.
    b. A person receives a $45 pay rise. The fortnightly pay was originally $1680.
    c. The original price of an item was $1580 and after two years it is worth $1375.
    d. After a $115 pay increase, a person's new monthly pay is $5890.

2. A bus driver receives a pay increase. Her original weekly pay was $875 and her new pay is $897. Using a spreadsheet, calculate the percentage change of her wage. Give your answer correct to 2 decimal places.

3. A steel worker receives a pay increase of 5.5%. If her original weekly pay (before tax) is $1023.45, calculate her new weekly pay.

4. **WE13** A fast-food manager's annual salary is $48 131. He claims $980 in tax deductions and receives $500 in interest from investments. Using a spreadsheet, calculate the amount of tax payable for the financial year by using the tax table provided on page 317.

5. Using a spreadsheet, calculate the tax payable by the following workers.

Employee	Gross salary ($)	Interest ($)	Deductions ($)	Taxable income	Tax free (0–18 200) 0c	Tax tier (18 201–37 000) 19c	Tax tier (37 001–87 000) 32.5c	Tax tier (87 001–180 000) 37c	Tax tier (180 000+) 45c
Stan	52 895	350							
Bett	65 845		650						
Xiao	36 080	125							
Mohamed	75 040	870	1500						

6. Workers in a textile factory are paid per item. The number of items 4 employees made for the week are shown.

	Rate/item	Monday	Tuesday	Wednesday	Thursday	Friday
Harry	$4.75	25	26	24	30	20
Theo	$6.50	19	18	16	20	15
Maria	$3.50	35	40	32	29	32
Marcia	$7.95	22	18	25	21	15

Using a spreadsheet, calculate each employee's weekly earnings, before tax.

7. **WE14** Using a spreadsheet, complete the following wage sheet.

Employee	Pay rate ($)	Normal hours worked	Overtime 1.5	Penalty rate 1.5	Penalty rate 2	Allowance ($)	Gross pay ($)	Tax withheld ($)	Net pay ($)
Rex	15.85	28		7	7			124	
Tank	22.15	35		5		15		167	
Gert	30.10	32	1.5		8	19		367	

8. A pay sheet for an individual employee is shown. Complete the pay sheet using a spreadsheet, and hence state the employee's net weekly pay and amount of superannuation paid into her superannuation fund.

Entitlements	Unit	Rate	Total
Wages for ordinary hours worked	30 hours	$35.05	$
TOTAL ORDINARY HOURS = 30 hours			
Penalty (double-time)	5 hours	$	$
		Gross payment	$

Deductions	
Taxation	$325
Total deductions	
Net payment	

Employer superannuation contribution	
Contribution	$

9. A worker receives a 4.5% pay increase. Their new weekly pay after the pay rise is $1095.
   a. Calculate their annual salary, assuming 52.14 weeks in a year.
   b. Calculate their annual salary before the pay rise.

10. A small business owner employs a manager who receives a monthly salary, 2 office workers who are paid an hourly rate and 3 other workers who are paid per item constructed.
   The following table shows the work information of the 6 employees for one week.

Employee	Hours worked	Hourly rate	Items	Rate/item	Annual salary
Sara (manager)	40 hours	–	–	–	$55 850
Troy (office worker)	25	$16.85	–	–	
Helga	30	$22.50	–	–	
Nina	20	–	19	$4.25	
Bill	15	–	15	$5.75	
Max	35	–	28	$4.95	

Using a spreadsheet, construct a weekly wage sheet that shows the earnings, before tax, for the 6 employees. Assume 52 weeks in 1 year.

11. The taxation table is shown.

Taxable income	Tax on this income
0–$18 200	Nil
$18 201–$37 000	19 c for each $1 over $18 200
$37 001–$87 000	$3572 plus 32.5 c for each $1 over $37 000
$87 001–$180 000	$19 822 plus 37 c for each $1 over $87 000
$180 001 and over	$54 232 plus 45 c for each $1 over $180 000

Using a spreadsheet, explain how the 'plus' calculations of $3572, $19 822 and $54 232 are found for the different tax levels in the table.

12. A waiter is paid an hourly rate of $21.50. He receives penalty rates of time-and-a-half for working Saturday and double-time for working Sundays and public holidays. Over the Christmas period, he worked 5 hours on Friday night, 8 hours on Saturday, 9 hours on Sunday and 7.5 hours on Boxing Day (public holiday). His employer withholds $185 in tax. Prepare a pay sheet using a spreadsheet to represent his weekly pay.

13. A teacher's annual salary is $94 961. He decides to reduce his time to 0.8 (this means he receives 0.8 of his salary and works 4 days out of 5), claiming that he will be on the lower tax level and pay less in tax overall, so his net fortnightly pay will not be that much less. By calculating the tax payable for both the full-time salary and the 0.8 salary, determine how accurate his claim is.

14. Each year, an IT technician moves up one level on a pay scale. The current pay scale is shown.

Level	Annual salary
1–1	$51 758
1–2	$53 258
1–3	$55 333
1–4	$57 508
1–5	$60 088

The technician is currently on level 1–3. At the start of the following year she moves up the pay scale to level 1–4.

a. Determine the percentage change in her wage correct to 2 decimal places.

A new working agreement is reached which increases the pay level of technicians on levels 1 and 2 by 3.75%.

b. Determine the percentage change in the technician's wage from her original 1–3 wage correct to 2 decimal places.

c. After the pay increase of 3.75%, another technician moves to level 2–3 with an annual salary of $65 880. The overall percentage change in his wage is 8.5%. Determine the salary for level 2–2 before and after the pay increase.

15. A new taxation system is being proposed.

Taxable income	Tax on this income
0–$19 500	Nil
$19 501–$65 000	21 c for each $1 over $19 500
$65 001–$125 000	A plus 35.5 c for each $1 over $65 000
$125 001–$180 000	B plus 37 c for each $1 over $125 000
$180 001 and over	C plus 45 c for each $1 over $180 000

a. Calculate the values for A, B and C using a spreadsheet.

b. Write down the formula that would calculate the tax payable for a taxable income within the tax level $65 001–$125 000.

c. By comparing this proposed taxation table with the current table, would this be a fairer taxation system? Justify your answer using calculations.

8.6 Review: exam practice

8.6.1 Earning money: summary

Wages and salaries

- A salary is a payment for work which is a fixed amount of money for a year. It is usually paid in portions each fortnight (2 weeks) or month.
- Salary payment = $\dfrac{\text{annual pay amount}}{\text{number of pay periods per year}}$
- An hourly wage is a set payment for each hour worked.
- Overtime is any additional hours worked over a set number of hours. Overtime is usually paid at a multiple of the regular hourly rate: $1.5 \times$ regular hourly rate (time-and-a-half) or $2 \times$ regular hourly rate (double-time).
- A penalty rate is a higher rate of pay for working unusual hours, such as weekends, public holidays, shift work (early mornings or late nights) and overtime work.
- Penalty rates are often a certain percentage more than the normal pay rate.
- Allowances are payments to cover out of pocket expenses such as uniforms, tools and equipment, travel costs, car and phone costs, and first-aid training. Allowances are included in wages and are documented separately on pay sheets.
- Superannuation is an additional payment by employers of 9.5% of gross pay. It is paid into a recognised superannuation fund to help workers in Australia save for their retirement.
- The amount of superannuation paid is based on usual earnings before tax and is calculated each pay cycle.
- Workers may contribute additional funds to their superannuation accounts.
- Depending on the working agreement workers have with their employers, some workers will receive an extra payment on top of the 4 week annual leave pay (annual leave loading). It is usually 17.5% of the normal pay for 4 weeks.

Commission, piecework and royalties

- Commission is paid to a person when an item is sold and is calculated as a percentage of the sale price.
- If no sale is made, no commission is paid.
- In some industries such as real estate and car sales, the rate of sales varies widely, so it is not practical to live only on commissions. Some workers are paid a base salary in addition to a commission.
- Piecework is a fixed amount payment for each item produced.
- Royalties are a payment made to an author, composer or creator for each copy of the work or invention sold.
- Royalties are calculated as a percentage of the total sales.

Taxation and deductions

- Income tax is a tax levied on the financial income of people and is deducted from each fortnightly or monthly pay.
- The amount of income tax payable is based on total income and tax deductions.
  Taxable income = total income − deductions
- The calculation of income tax is based upon an income tax table.
- Workers who spend their own money for work-related expenses are entitled to claim the amount spent as tax deductions. Deductions are subtracted from taxable income which lowers the amount of money earned, reducing the amount of tax to be paid.
- Tax deductions are recorded in the end of the financial year tax return.
- The Medicare levy is a payment of 2% of taxable income to partially fund healthcare through Medicare.

- The 2% Medicare levy may be reduced if the taxable income is below a certain amount.
- The Medicare levy surcharge is an additional payment for workers who do not have adequate private health insurance.
- Pay As You Go (PAYG) tax is a withholding tax system which requires employers to calculate and withhold from the employees regular earnings (gross pay) an amount of tax to be paid to the government.
- Net pay = Gross pay – tax withheld
- At the end of the financial year, workers submit a tax return which lists their deductions and all money earned.
- If the total amount of tax paid is less than is required, the worker will have to pay the difference. If the amount of tax paid over the year is more than is required, the worker receives a refund.
- If a worker provides their tax file number and does not work another job, they can claim the tax-free threshold, and will not pay tax on the first $18 200 earned in the financial year.

Performing financial calculations

- Percentage change $= \dfrac{\text{difference between two numbers}}{\text{original number}} \times 100$
- The tax payable on gross incomes can be determined by using a spreadsheet and inserting formulas that determine the amount of tax payable for each tax tier.
- A wage sheet contains details of a worker's earnings, including net wages, pay rates, gross deductions and allowances.
- A pay sheet/pay slip contains individual worker's details on the hours worked, pay rate, net wage, tax deduction, leave days and superannuation paid.

A summary of this topic is available in the Resources section of your eBookPLUS at www.jacplus.com.au.

Exercise 8.6 Review: exam practice

Simple familiar

1. A nurse is paid $28.50 per hour for an 80-hour fortnight. Assuming 26.07 fortnights per year, calculate:
   a. the nurse's annual salary
   b. the nurse's annual superannuation (assuming 9.5% superannuation)
   c. the nurse's weekly pay.

2. **MC** Which term represents getting paid for every item constructed?
   A. Wages   **B.** Piecework
   C. Commission   **D.** Superannuation

3. A novelist gets a royalty payment of 2.9% of gross sales of her book. Complete the following table.

	January	February	March	April
Gross sales	$45 000	$125 000	$320 000	
Royalty				$1508

4. Convert the following annual salaries to fortnightly pay. (Assume 26.07 fortnights per year.)
   a. $67 899   b. $98 765   c. $101 010   d. $123 456

5. Calculate the amount of superannuation, each fortnight, that is paid (assume 9.5% superannuation) for the following annual salaries:
   a. $67 899.
   b. $98 765.
   c. $101 010.
   d. $123 456.

6. An architect makes a salary of $85 500 per year and is given 9.5% extra in the form of superannuation.
   a. How much superannuation does she get each year?
   b. In the following year, the architect is not given a pay increase, but instead her superannuation increases to 14% of her salary. Calculate the amount of the increase in superannuation.

7. A first-year teacher makes $49 200 per year.
   a. How much do they get paid each fortnight (assuming 26.07 fortnights in a year)?
   b. How much additional money is paid into their superannuation fund (assuming 9.5% contribution) each fortnight?

8. **MC** In a leap year, an annual salary of $56 200 results in a fortnightly pay of:
   A. $2161.54.
   B. $2155.73.
   C. $2149.73.
   D. $2150.80.

9. Calculate the Medicare levy for the following annual salaries.
   a. $48 501
   b. $32 570
   c. $65 890
   d. $112 570

10. Employees at a cafe are paid penalty rates for working on weekends. The normal hourly rate is $18.50. Penalty rates for Saturday are time-and-a-half, and double-time for Sundays and public holidays.

    Four employees worked on the long weekend and the hours worked are shown:

Employee	Saturday	Sunday	Monday (public holiday)
Tran	5	8	
Gus		6	
Meg	3.5		7
Warren	4		6

    Calculate the earnings before tax for each employee.

11. A librarian pays $1137.60 in Medicare levy. Determine his annual salary.

12. **MC** A bookkeeper receives a pay rise. Her original weekly pay was $758 and her new weekly pay is $775. The percentage change of her pay is approximately:
    A. 0.0219%.
    B. 0.024%.
    C. 2.19%.
    D. 2.24%.

Complex familiar

13. A factory worker's hourly rate is $21.50. He works 38 hours a week with 4 hours overtime paid at time-and-a-half. The tax withheld each week is $195. He claims $25 a week for the uniform and earns $120 in interest for the year. (Assume 52.14 weeks in the year.)
    a. Calculate the taxable income and Medicare levy.
    b. Explain why the worker should expect to receive a tax refund. Support your explanation with calculations.

14. The fortnightly pay slip for an employee is shown.

Entitlements	Unit	Rate	Total
Wages for ordinary hours worked	76 hours	$20.95	$
TOTAL ORDINARY HOURS = 76 hours			
Overtime	15 hours	$31.40	$
Travel allowance	350 km	$0.66/km	$
		Gross payment	$

Deductions	
Taxation	$430
Total deductions	
Net payment	

a. Using a spreadsheet, calculate the fortnightly pay.

b. Assuming she receives the same fortnightly pay for the year, calculate her annual taxable income. Assume 26.07 fortnights in the year.

c. Determine if the employee will receive a tax refund or have to pay more in tax.

15. A new working agreement is proposed for workers at a local abattoir. Penalty rates for working on weekends and overtime will be standardised to a rate of 1.25 hourly rate with a pay rise of 4.5%. The current penalty rates for working Saturday and Sunday are time-and-a-half and double-time respectively, and working overtime is paid at time-and-a-half.

If the average hourly rate for a worker at the abattoir is $19.78, will workers be better off under the new working agreement? Justify your answer using calculations.

16. A window cleaner has the opportunity to move from casual to part-time employment. Workers employed on a casual basis do not receive annual or sick leave. All employees in the cleaning business receive 9.5% superannuation. The window cleaner's current casual hourly rate is $25.76 and the part-time hourly rate is $23.70. Should he move to part-time employment? Justify your answer by calculating his annual salary based on working 38 hours each week at the normal hourly rate and his annual leave loading.

Complex unfamiliar

17. A worker receives 17.5% annual loading just before his four-week annual leave. He is paid an hourly rate of $20.50 for working a 38-hour week. He does not receive overtime or penalty rates.

a. Calculate the amount of annual loading he will receive for the four weeks he takes his leave.

b. Complete his pay slip for the month he receives his annual leave loading.

Entitlements	Unit	Rate	Total
Wages for ordinary hours worked	38 hours	$20.50	$
Annual leave loading (17.5%)			$
TOTAL ORDINARY HOURS = 38 hours			
		Gross payment	$

Deductions	
Taxation	$295
Total deductions	
Net payment	

The worker complains that he has paid too much in tax in the month he received his annual leave loading, and asks to have his leave loading paid over a 6 month period (26 weeks). His weekly PAYG tax is $106.

c. Calculate his weekly wage if his annual leave loading is paid over 26 weeks.
The PAYG tax if he is paid his annual leave loading over 26 weeks is $113.

d. Determine if his claim of paying too much tax is valid. Justify by using your answers from the previous parts.

18. A baker is paid an hourly rate of $25.27 and overtime rate of $35. In a usual working week she works 38 hours at normal pay and 6 hours overtime.

She receives her annual leave loading of $672.18 and superannuation contributions of $91.22 for the week.

She complains that she has been underpaid for both her annual loading and superannuation contributions. Explain how her annual leave loading and superannuation contributions were calculated. Should she have received more money?

19. Business groups are seeking a change to the taxation system. Their proposal is shown:

Taxable income	Tax on this income
0–$29 000	Nil
$29 001–$85 000	35 c for each $1 over $29 000
$85 001–$175 000	A plus 40 c for each $1 over $85 000
$175 001	B plus 58 c for each $1 over $175 000

a. Calculate the values of A and B.

b. Compare the tax payable for the following annual salaries under the new proposed system and the current taxation table.
  i. $25 890   **ii.** $45 870   **iii.** $67 950   **iv.** $81 940   **v.** $195 870

c. Based on your calculations from part b, how will workers be better or worse off under the proposed new taxation system?

20. A physiotherapist's annual salary is $63 821. Her annual travel claim is 950 km and she is allowed to claim 66c for each km travelled as a tax deduction. She receives a uniform allowance of $15 each week. She is paid monthly and her employer withholds $1120 in tax. Assume 52.14 weeks in the year.

a. Using a spreadsheet, calculate her taxable income, tax payable, Medicare levy and employer superannuation contribution.

b. Explain why she will have to pay tax. Justify your answer using calculations.

Answers

Chapter 8 Earning money

Exercise 8.2 Wages and salaries

1. Wage (hourly rate) is the greater pay ($3200 fortnight compared to $2838.51).

2. B

3. a. $1476.31   b. $1136.89
   c. $1596.83

4. 34.31 hours

5. $637.50

6. First fortnight: $1735.95
   Second fortnight: $1833.75

7. $144.08

8. a. $12 979.17
   b. $14 796.25 annually, $1233.02 monthly

9. a. $5166.67   b. $490.83
   c. $5890

10. $1154.32

11. a. $2100   b. $10 850

12. a. She will be worse off by 25% for every hour she works on Sunday, but no worse off for working on Saturdays.
    b. 16.67%

13. a. $1270.90
    b. Her weekly salary would be $1304.18, which is higher than her hourly rate calculation ($1270.90). She should take the salary offer.

14. The employees should accept the new agreement. The new agreement is 15.5% more than the base salary compared to 10.8% more under the current agreement.

15. She will receive the most amount over 5 years by choosing package 1 ($322 836.55 compared to $315 028.38).

Exercise 8.3 Commission, piecework and royalties

1. $3390

2. a. $1650   b. $2375
   c. $2920   d. $4160

3. C

4. $6825

5. $1398.25

6. B

7. $12.11

8. $186 989.67

9. A

10. a.

Time	Jan–Mar	Apr–Jun	Jul–Aug	Sep–Dec
Sales	$123 400	$2 403 556	$432 442	$84 562
Royalty payment	$3085.00	$60 088.90	$10 811.05	$2114.05

   b. $376 099

11. $45 634.53

12. $2 500 000

13. $16 599.48

14. Royalty percentage can be found by dividing the royalty amount by the total sale amount and then multiplying by 100. The songwriter's royalty percentage is 2.5%.

15. Option 1:
    $15 \times 5 \times 8.25 = \$618.75$
    Option 2:
    $250 \times 75.95 \times 0.035 = \$664.56$
    Option 2

16. a. $2090
    b. 22
    c. If he sells 22 new telephone plans each week, taking the salary option of $37 500 per annum would give him a higher weekly wage.

Exercise 8.4 Taxation and deductions

1. $8118.75

2. a. $1208   b. $1546
   c. $1788   d. $2168.46

3. D

4. a. $1890   b. $1564

5. $1578.50

6. a. i. 32.5% on amounts over $37 000 + $3572
      ii. $3734.50
   b. i. Nil
      ii. $0
   c. i. 32.5% for amounts over $37 000 + $3572
      ii. $19 389.75
   d. i. 37% on amounts over $87 000 + $19 822
      ii. $29 257

7. a. $192   b. $366   c. $540   d. $1086

8. a. 28.93%   b. 30.4%   c. 31.2%   d. 33.95%

9. $704.32

10. a. $4857   b. $975   c. 20.07%   d. $1165.68

11. a. $85 970   b. $19 487.25

12. a. $2695.84
    b. $2695.84 − $606 = $2089.84
    c. $13 518.90
    d. $13 52.12

13. a. $5614.50
    b. $14 400
    c. He will receive a refund of $1396.83

14. a. $21.15
       Working 4 hours overtime equates to the same pay as working 6 hours normal time ($4 \times 1.5 = 6$). Calculate the number of 'normal' hours worked ($38 + 6 = 44$) and divide into wage.
    b. $45091.80
    c. $6201.84
    d. $5558.34
       He won't receive $1980 as a tax refund because the tax payable is a percentage of his taxable income. His taxable income was reduced by $1980 but the tax payable was reduced by 32.5% of the difference ($0.325 \times (8091.80 - 1980)$).

15. a. When calculating tax using this tiered system, the taxable salary of $97 605 is split into the different tax-bracket amounts, which are found by the difference between the lower and upper limits, e.g.
37 000 − 18 200 = $18 800.
$18 200 + $18 800 + $50 000 + $10 605 = $97 605, and each amount is multiplied by different percentages, as shown.

Tax bracket	Amount of salary ($)	Tax %	Tax paid ($)
0–18 200	18 200	0	
18 201–37 000	18 800	19	0.19 × 18 800 = 3572
37 001–87 000	50 000	32.5	0.325 × 50 000 = 16 250
87 001–180 000	10 605	37	0.37 × 10 605 = 3923.85
Total	$97 605		Total = $23 745.85

This is much lower than 37% × $97 605 = $36 113.85.

b. The dentist's taxable salary falls into the fourth tax bracket, 37% for amounts over $87 000 plus $19 822. The $19 822 is the tax amount from the previous tax brackets (19% × $18 800 + 32.5% × $50 000), as shown.

16. a. Determine the tax bracket her taxable income falls within. Since she paid more than $3572 and less than $19 822, her taxable income is within the third tax bracket: $3572 + 32.5% × (taxable income − $37 000).

b. $71 584.62

c. Refund: $320.13

Exercise 8.5 Performing financial calculations

1. a. 3.33%
b. 2.68%
c. 12.97%
d. 1.99%

2. 2.51%

3. $1079.74

4. $7033.58

5. Stan = $8851.63, Bett = $12 735.38, Xiao = $3420.95, Mohamed = $15 730.25

6. Harry = $593.75, Theo = $572.00, Maria = $588.00, Marcia = $802.95

7. See bottom of the page *

8.

Entitlements	Unit	Rate	Total
Wages for ordinary hours worked	30 hours	$35.05	$1051.50
TOTAL ORDINARY HOURS = 30 hours			
Penalty (double-time)	5 hours	$70.10	$350.50
Gross payment			$1402.00

Deductions	
Taxation	$325
Total deductions	$325
Net payment	$1077.00

Employer superannuation contribution	
Contribution	$133.19

9. a. $57 093.30   **b.** $54 634.74

10. Sara = $1074.04, Troy = $421.25, Helga = $675.00, Nina = $80.75, Bill = $86.25, Max = $138.60

11. The 'plus' calculations are found by adding the tax payable in the previous brackets.

12. See bottom of the page **

13. The claim is incorrect. While the teacher will pay $6530.71 less in tax, reducing his salary by 20% has an overall reduction of $18 922.20 in pay, so he will have $12 461.49 less pay.

*

Employee	Pay rate ($)	Normal hours worked	Overtime 1.5	Penalty rate 1.5	Penalty rate 2	Allowance ($)	Gross pay ($)	Tax with held ($)	Net pay ($)
Rex	15.85	28			7	7		832.13	124	708.13
Tank	22.15	35		5		15	956.38	167	789.38
Gert	30.10	32	1.5		8	19	1531.53	367	1164.53

**

Hourly rate	Normal hours	Saturday 1.5	Sunday 2	Public holiday 2	Gross pay	PAYG tax	Net pay
21.5	5	8	9	7.5	$1075.00	$185.00	$890.00

14. a. 3.93%

b. 7.83%

c. Before pay increase = $60 718.89
After pay increase = $62 995.85

15. a. A = $9555, B = $30 855, C = $51 205

b. = 9555 + 0.355× (taxable salary –65 000)

c. Workers who previously earned between $37 000 and $65 000 will pay less tax in the new system; however, workers previously earning between $65 000 and $87 000 will pay 3% more in tax. Workers who previously earned between $18 200 and $37 000 will pay 1% more in tax, while workers earning more than $180 000 will pay less. So on balance, the middle earners will pay less but those who are low-income earners will pay more and those on high wages pay less; hence it does not seem to be a fairer system.

8.6 Review: exam practice

1. a. $59 439.60   **b.** $5646.76   **c.** $1140

2. B

3.

	January	February	March	April
Gross sales	$45 000	$125 000	$320 000	$52 000
Royalty	$1305	$3625	$9280	$1508

4. a. $2604.49   **b.** $3788.45
   c. $3874.57   **d.** $4735.56

5. a. $247.43   **b.** $359.90
   c. $368.08   **d.** $449.88

6. a. $8122.50   **b.** $3847.50
   Superannuation increase is $3847.50.

7. a. $1887.23   **b.** $179.29

8. C

9. a. $970.02   **b.** $651.40
   c. $1317.80   **d.** $2251.40

10.

Employee	Saturday	Sunday	Monday (public holiday)	Gross wage
Rate	$27.75	$37	$37	
Tran	5	8		$434.75
Gus		6		$222
Meg	3.5		7	$356.13
Warren	4		6	$333

11. $56 880

12. D

13. a. Taxable income: $48 140.94
   Medicare levy: $962.82

b. He has paid $2011.60 more in tax than is required, so he will receive a tax refund.

14. a. $1864.20

b. $59 809.79

c. A tax return of $224.92

15. For workers who work only standard hours (38 hours per week), the new agreement will result in more money (an increase of 4.5% more). However, workers who work on weekends or overtime will be worse off under the new agreement.

16. Casual employment will give him more money overall ($52 048.64 compared to $51 599.99 including annual leave loading) based on working 38 hours each week for 52 weeks. He will not have sick or annual leave. Considering taking annual leave, working part-time would be the better deal. He will have sick leave and job security.

17. a. $545.30

b.

Entitlements	Unit	Rate	Total
Wages for ordinary hours worked	38 hours	$20.50	$779.00
Annual leave loading (17.5%)			$545.30
TOTAL ORDINARY HOURS = 78 hours			
		Gross payment	$1324.30

Deductions		
Taxation		$295
Total deductions		$295
Net payment		$1029.30

c. $686.97

d. Taking the annual leave over 26 weeks reduces his PAYG tax for the year by $7.00. So his claim is correct.

18. Average working week: $38 \times \$25.27 = \$960.26$
Annual leave loading is 17.5% of 4 weeks' pay:
$0.175 \times \$960.25 \times 4 = \$672.18$
Superannuation is 9.5% of normal pay:
$0.095 \times \$960.25 = \$91.22$
Both annual leave loading and superannuation are calculated on the average working week paid at normal rates. Additional pay from working overtime is not included in the calculation. Therefore both calculations for the baker are correct and she should not receive more money.

19. a. A: $19 600

B: $55 600

b.

	Taxable income	Tax payable current	Tax payable proposed
i.	$25 890	$1461.10	$0
ii.	$45 870	$6454.75	$5904.50
iii.	$67 950	$13 630.75	$13 632.50
iv.	$81 940	$18 177.50	$18 529.00
v.	$195 870	$61 373.50	$67 704.60

c. Under the proposed taxation system, lower income earners will pay less in tax due to an increase in the upper limit within the tax-free threshold from $18 200 to $25 890. Higher income earners pay more ($67 704.60 compared to $61 373.50) due to a higher percentage tax rate (58% compared to 45% currently). Middle income workers will pay slightly more ($18 529 compared to $18 177.50) under the proposed system due to an increase in the tax rate (35% compared to 32.5% under the current system).

20. a. Taxable income = $63 976.10
Tax payable = $12 339.23
Medicare levy = $1279.52
Superannuation contribution = $6063.00

b. Her PAYG tax is $13 440 and amount-payable tax based on her taxable income ($63 976.10) is $12 339.23, not including the 2% Medicare levy ($1279.52). Including the Medicare levy, the total amount of tax she needs to pay is $13 618.75, which is tax owing of $178.75.

CHAPTER 9
Budgeting

9.1 Overview

CONTENT

In this chapter, students will learn to:
- investigate the costs involved in independent living [complex]
- prepare a personal budget plan [complex].

Fully worked solutions for this chapter are available in the Resources section of your eBookPLUS at www.jacplus.com.au.

9.2 Household expenses

9.2.1 Understanding household electricity and gas bills

Electricity and gas are supplied to our homes to help us keep warm and cool, to allow us to cook, and to provide lighting.

There are three main parts to our electricity and gas supply systems: wholesalers (or generators), distributors and retailers.

- Wholesalers produce electricity and extract gas.
- Distributors own and maintain infrastructure such as power poles, wiring and gas lines. When there are disruptions to the electricity or gas supply, the distributors are contacted to fix the problem.
- Retailers purchase electricity and gas from wholesalers and then sell them to customers. There are many electricity and gas options available to customers, and customers can select whichever deal best suits their needs. Therefore, it is important to know how much you will be charged per unit of energy used and what discounts are available before you sign any contracts.

The standard measurement of electricity consumption is the **kilowatt hour** (kWh). One kWh is the amount of energy produced by an appliance outputting 1 Kw (1000 W) of power for an hour. Gas consumption is measured in **megajoules** (MJ).

To understand electricity and gas bills, it is important to know some of the terms that are used.

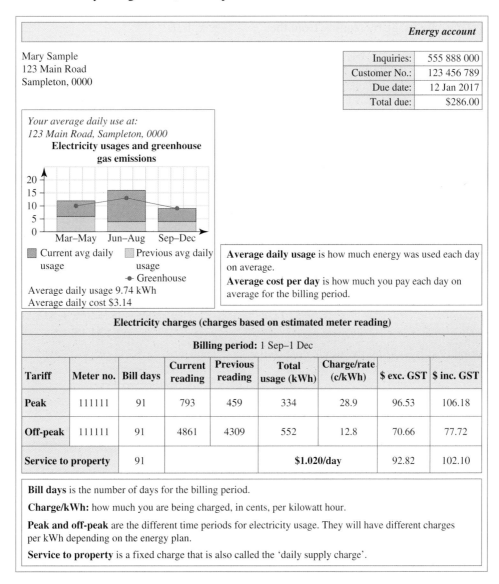

					Energy account

Mary Sample
123 Main Road
Sampleton, 0000

Inquiries:	555 888 000
Customer No.:	123 456 789
Due date:	12 Jan 2017
Total due:	$286.00

Your average daily use at:
123 Main Road, Sampleton, 0000

Electricity usages and greenhouse gas emissions

- Current avg daily usage
- Previous avg daily usage
- Greenhouse

Average daily usage 9.74 kWh
Average daily cost $3.14

Average daily usage is how much energy was used each day on average.
Average cost per day is how much you pay each day on average for the billing period.

Electricity charges (charges based on estimated meter reading)

Billing period: 1 Sep–1 Dec

Tariff	Meter no.	Bill days	Current reading	Previous reading	Total usage (kWh)	Charge/rate (c/kWh)	$ exc. GST	$ inc. GST
Peak	111111	91	793	459	334	28.9	96.53	106.18
Off-peak	111111	91	4861	4309	552	12.8	70.66	77.72
Service to property		91			**$1.020/day**		92.82	102.10

Bill days is the number of days for the billing period.

Charge/kWh: how much you are being charged, in cents, per kilowatt hour.

Peak and off-peak are the different time periods for electricity usage. They will have different charges per kWh depending on the energy plan.

Service to property is a fixed charge that is also called the 'daily supply charge'.

The meter readings and charge/rates for a household are as shown.

| Electricity charges (charges based on meter reading) | | | | | | | |
| Billing period: 1 Jan–31 March | | | | | | | |
Tariff	Meter no.	Bill days	Current reading	Previous reading	Total usage (kWh)	Charge/rate (c/kwh)	$ Exc. GST	$ Inc. GST
Peak		90	12 791	12 353	438	29.85		
Off-peak		90	2 587	1 883	704	13.37		
Service to property		90			**$1.254/day**			

Calculate the total bill amount.

THINK

1. State the electricity usage by reading from the table for both peak and off-peak.

2. Calculate the cost for each time period by multiplying the usage by the charge/rate.

3. Add GST by adding 10% of each amount.

4. Calculate the costs for service to property by multiplying the cost per day by the number of billing days.

5. Add 10% GST.

6. Find the total cost by adding the usage cost to the service cost.

7. Answer the question.

WRITE

Peak: 438
Off-peak: 704

Peak: $438 \times 0.2985 \approx \$130.74$
Off-peak: $704 \times 0.1337 \approx \$94.12$

Peak: $10\% \times \$130.74 = \$13.074$
$\$130.74 + \$13.074 = \$143.81$
Off-peak: $10\% \times \$94.12 = \$9.412$
$\$94.12 + \$9.412 = \$103.53$
Total for usage $= \$143.81 + 103.53$
$\qquad = \$247.34$

Service costs $= 1.254 \times 90$
$\qquad = \$112.86$

$10\% \times \$112.86 = \$11.286$
Sevice costs $= \$112.86 + \$11.286$
$\qquad = \$124.15$

Total cost $= \$247.34 + 124.15$
$\qquad = \$371.49$

The bill total is $371.49.

9.2.2 Understanding household water bills

Water is provided to homes and businesses through a series of water pipes that are managed and serviced by providers.

Home owners are charged a fixed fee to provide water and remove water wastage, known as sewerage. Households are also charged a fee towards the management of stormwater services and some households have to pay a recycled water usage fee, depending on their property.

The amount of water used in the home over a fixed period, usually 3 months, is measured through a water meter fixed to the main water supply to the home. Home owners are charged for their water usage at a rate, in dollars, per kilolitre (kL).

$$1000 \, litres = 1 \, kilolitre$$

Note: GST does not apply to water bills.

WORKED EXAMPLE 2

Calculate the water rates for a home with the following usage and fixed charges.

Bill details as of 31 March		Value	Price
Water service charge	1 Jan to 31 Mar	22.51	
Sewerage service charge	1 Jan to 31 Mar	145.90	
Stormwater	1 Jan to 31 Mar	18.70	
	Volume	**Charge/kL**	
Water usage	55	$2.00	
Recycled water usage	35	$1.79	

THINK

1. Calculate the cost of the water usage by multiplying the volume by the cost.
2. Calculate the cost of the recycled water usage by multiplying the volume by the cost.
3. Calculate the fixed service cost.
4. Add the fixed service cost to the water usage cost.
5. Answer the question.

WRITE

$$\text{Water usage} = 55 \times 2$$
$$= \$110$$

$$\text{Recycled water} = 35 \times 1.79$$
$$= \$62.65$$

$$22.51 + 145.90 + 18.70 = 187.11$$

$$187.11 + 110 + 62.65 = \$359.76$$

The water rates cost $359.76.

9.2.3 Understanding council rates

Councils provide services and infrastructure to people living in communities, such as parklands, libraries, rubbish collection and road maintenance. To support councils in providing these services, home owners pay **council rates**.

Council rates are calculated as a percentage on the capital improved value of the land and buildings within the council municipality. The rate at which rates are charged — the **rate in the dollar** — is determined by finding how much revenue the council plans to raise and dividing this by the amount of capital improvement of land and buildings within the council municipality.

For example, if the council plans to raise $15 million for the annual budget and there is $2 billion in capital improved value within the municipality, then the rate in the dollar is $\dfrac{15\,000\,000}{2\,000\,000\,000} = 0.0075$.

If the capital improvement on a rate payer's house and land is $450\,000$, then the payable council rates for that property are. $0.0075 \times 450\,000 = \$3375$.

Note: GST does not apply to council rates.

WORKED EXAMPLE 3

A council municipality plans to raise $15 million and there is $3.25 billion in capital improved value in the municipality. Calculate the payable council rates for a property with $275\,000 capital improved value. Write your answer correct to the nearest cent.

THINK	WRITE
1. Find the rate in the dollar. Divide the revenue to be raised by the capital improved value.	$\dfrac{15\,000\,000}{3\,250\,000\,000} = \dfrac{15}{3250}$ $\qquad\qquad = 0.004\,62\ldots$
2. Multiply the rate in the dollar by the capital improved value.	$0.004\,62\ldots \times 275\,000 = \$1269.23$
3. Answer the question.	The council rates payable are $1269.23.

---

on Resources

✦ **Interactivity:** Electricity bills (int-6912)

Exercise 9.2 Household expenses

1. **WE1** The meter readings and charge/rates for a household are as shown. Calculate the total bill amount. (Remember to include GST.)

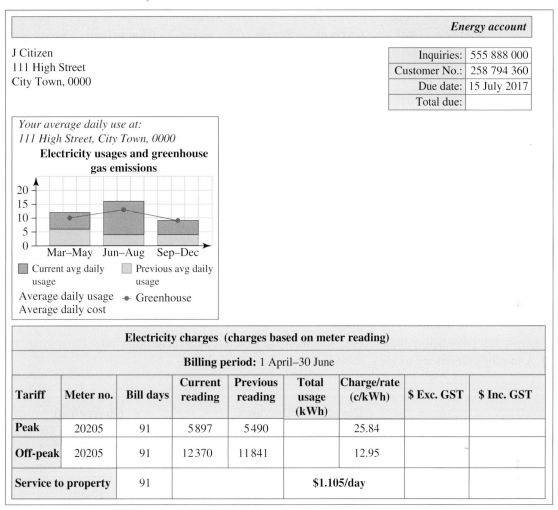

Energy account								

J Citizen
111 High Street
City Town, 0000

Inquiries:	555 888 000
Customer No.:	258 794 360
Due date:	15 July 2017
Total due:	

Your average daily use at:
111 High Street, City Town, 0000

Electricity usages and greenhouse gas emissions

- Current avg daily usage
- Previous avg daily usage
- Average daily usage
- Average daily cost
- Greenhouse

Electricity charges (charges based on meter reading)								
Billing period: 1 April–30 June								
Tariff	**Meter no.**	**Bill days**	**Current reading**	**Previous reading**	**Total usage (kWh)**	**Charge/rate (c/kWh)**	**$ Exc. GST**	**$ Inc. GST**
Peak	20205	91	5897	5490		25.84		
Off-peak	20205	91	12370	11841		12.95		
Service to property		91			**$1.105/day**			

2. Calculate the energy charges (excluding GST) for the following, giving your answers correct to the nearest cent.

	Energy usage	**Charge/rate**
a.	563 kWh	10.56 c/kWh
b.	1070 MJ	2.629 c/MJ
c.	895 kWh	14.64 c/kWh
d.	6208 MJ	1.827 c/MJ

3. The usage component of a family's energy bill for a quarter is $545 excluding GST. The charge/rate is 11.43 c/kWh. Calculate their energy usage for the quarter. Give your answer to the nearest whole number.

4. Part of an electricity bill is shown below.

Electricity charges (charges based on meter reading)								
Billing period: 1 Sep–31 Dec								
Tariff	Meter no.	Bill days	Current reading	Previous reading	Total usage (kWh)	Charge/rate (c/kWh)	$ exc. GST	$ inc. GST
All day	28 710	91	47 890	46 995		30.54		
Service to property		91			$1.017/day			

a. Write down:
 i. the charge/rate
 ii. the daily service to property fee
 iii. the usage amount in kWh.
b. Calculate:
 i. the amount charged for electricity usage excluding GST
 ii. the total amount for the bill, including GST.

5. There is a discount for customers who pay their energy bill before the due date. The energy usage of a customer is:

 • electricity: 456 kWh at 12.46 c/kWh
 • gas: 967 MJ at 1.956 c/MJ
 • $95 fixed fee for service to the property.

 If the customer pays before the due date, calculate the amount paid including GST.

6. **MC** A family's electricity usage is shown in the following table.

Tariff	Bill days	Current reading	Previous reading	Total usage	Charge/rate c/kWh
Peak	90	58 973	58 615		29.05
Off-peak	90	8 569	8 028		15.59
Service to property	90			$1.08/day	

 The total cost for their electricity usage, including GST and correct to the nearest cents, is:
 A. $188.35.   B. $269.00.   C. $285.55.   D. $314.10.

7. **WE2** Calculate the water rates for a home with the following usage and fixed charges.

Bill details as of 31 March		Value	Price
Water service charge	1 Jan to 31 Mar	22.51	
Sewerage service charge	1 Jan to 31 Mar	145.90	
Stormwater	1 Jan to 31 Mar	18.70	
	Volume (L)	Charge/kL	
Water usage	58 000	$2.00/kL	
Recycled water usage	39 000	$1.79/kL	

8. A customer pays $245.85 for their water bill for 3 months. They are charged a fixed service fee of $180.30. The charge/kL is $1.95. The customer does not have recycled water service. Calculate the amount of water used by the customer for the 3 months. Give your answer correct to 2 decimal places.

9. **WE3** A council municipality plans to raise $12 million, and there is $1.8 billion in capital improved value in the municipality. Calculate the payable council rates for a property with $275 000 capital improved value.

10. The council rates for a home that has a capital improved value of $325 000 are $1950.
    a. Calculate the rate in the dollar charged by the council.
    b. Calculate the amount of revenue the council intends to collect if the total amount of capital improved value for the municipality is $2.67 billion.

11. A customer has a choice between two electricity plans, A and B. The two plans are shown below.

Plan A		Plan B	
Charge/rate		**Charge/rate**	
All day	12.03 c/kWh	Off-peak (11 pm–7 am)	9.08 c/kWh
		Peak	13.01 c/KWh
Service to property	$1.01/day	Service to property	$0.99/day
Pay-on-time discount	15%	Pay-on-time discount	12%

The customer's energy usage over 90 billing days is 534 (peak time) and 757 (off-peak time).

Which plan would be the more cost-effective for the customer? Justify your answer by calculating the electricity costs for each plan for if the customer pays on time and if he doesn't pay on time.

12. Here is an incomplete water bill received by a household.

Bill details as of 30 June		Value	Price
Water service charge	1 Apr to 30 Jun	23.80	
Sewerage service charge	1 Apr to 30 Jun	150.20	
Stormwater	1 Apr to 30 Jun	17.95	
	Volume (L)	**Charge/kL**	
Water usage	64 000		130.56
Recycled water usage	38 000		
		Total:	$393.57

a. Complete the water bill by filling in the missing values.
   As of 1 July, the charges per kL will increase by 4%, the water service charge will increase by $2.50 and the sewerage service charge will increase by $1.50.
b. Calculate the new charges for water usage.
c. Determine the percentage increase for water and sewerage service charges. Give your answers correct to 2 decimal places.
d. By what percentage will the average water bill increase (assuming water usage remains the same)? Give your answer correct to 2 decimal places.

13. A family has both electricity and gas connected to their house. Their average electricity usage for 90 billing days is 987 kWh and their average gas usage is 1650 MJ per month (30 days). The family decides to change providers. They have the following choices:

    Choice 1: Separate electricity and gas providers
    - Electricity: all-day rate 10.03 c/kWh
    - Pay-on-time discount (electricity only): 12.5%
    - Gas: all-day rate 1.89 c/MJ

    Choice 2: Bundle both electricity and gas with one provider; receive the following discounts
    - 17% discount on gas
    - 10% discount on electricity
    - Electricity daily rate:12.01 c/kWh
    - Gas daily rate:

    Service to property charges are the same for both providers.

    Which choice will be the most cost-effective for the family, assuming they always pay on time?

14. A council plans to raise $7.5 million in revenue. The capital improved value of home owners in the council area is $2.5 billion. To calculate the rate in the dollar, the council performs the following calculation: $\dfrac{7.5}{2.5} = 3$.

    The capital improved value on a particular home is $175\,000$.

    To calculate the council rates owing for that home, a council worker performs the following calculation: $\dfrac{3}{100} \times 175\,000 = \$5250$.

    The home owner complains that the rates are far too high. Is there an error in the calculation performed by the council and/or by the council worker? Justify your answer by calculating the council rates for the rate payer's property.

15. A family household's bills for 3 months (1 quarter) are as shown.

Electricity bill

Tariff	Bill days	Current reading	Previous reading	Total usage	Charge/rate (c/kWh)
Peak	90	12 589	11 902		31.08
Off-peak	90	2 058	1 511		14.05
Service to property	90				$1.075/day

Water bill

Bill details as of 31 March		Value	Price
Water service charge	1 Jan to 31 Mar	22.51	
Sewerage service charge	1 Jan to 31 Mar	145.90	
Stormwater	1 Jan to 31 Mar	18.70	
	Volume (kL)	**Charge/kL**	
Water usage	55	$2.00/kL	
Recycled water usage	35	$1.79/kL	

Council rates

Capital improved value $280 000, rate in the dollar 0.008

a. Show that the quarterly council rate instalment is $560.
b. Calculate the amount that the family spends on electricity, water and council rates for the quarter. (Remember to include GST where relevant.)
c. The family's monthly net income is $6019.35 (after tax). Calculate the percentage of their income that goes towards paying the household bills. Give your answer correct to 2 decimal places.
d. If the family wishes to reduce their household bills, what advice would you give?

9.3 Purchasing, running and maintaining a vehicle

9.3.1 Planning for the purchase of a vehicle

Buying a vehicle for the first time can be very exciting. However, it pays to do research and know exactly which type of vehicle will best suit your needs. For example, do you need a small, medium-sized or large vehicle? How many seats and how big an engine do you need?

Many costs are associated with buying a vehicle, but the primary cost is the purchase price or sale price. The advertised price for the vehicle is not always the sale price. Many vehicle dealerships are willing to negotiate on price or match other competitive prices for a similar or same-model vehicle. It is important to know how much the vehicle is worth and then shop around for the best price.

Paying for the vehicle

Many financial institutions offer loans to purchase vehicles. It is also possible to obtain finance through vehicle dealerships. As with all loans, it is important to know exactly how much interest is charged and what additional costs apply, such as administration fees and loan servicing fees. It is worth noting that if the customer does not meet the loan repayments, then the vehicle may be repossessed by the dealership or financial institution.

- A **comparison rate** gives a more accurate indication of the interest rate for the loan. It includes all the additional fees in the calculation.
- At times, 0% finance is advertised for the purchase of a vehicle. This means no interest is charged on the loan. However, this type of finance may apply only to certain models and the sale price may not be negotiable.

WORKED EXAMPLE 4

A customer takes out a loan to purchase a car for $15 090. The monthly repayments are $294.26 and the loan is for 5 years. The customer pays a 10% deposit. Calculate:
a. the deposit paid.
b. the total amount paid for the car.
c. the interest rate charged for the loan.

▶

THINK	WRITE
a. 1. Find 10% of $15 090.	$\dfrac{10}{100} \times 15\,090 = 1509$
2. Answer the question.	The deposit paid is $1509.
b. 1. Find the total amount paid in repayments.	Monthly repayments = $294.26 for 5 years $294.26 \times 12 \times 5 = \$17\,655.60$
2. Calculate the amount paid for the car by adding the deposit paid to the total amount in monthly repayments.	Amount paid for the car = $1509 + $17 655.60 $\qquad\qquad\qquad\quad = \$19\,164.60$
3. Answer the question.	The total amount paid for the car is $19 164.60.
c. 1. Find the principal by subtracting the deposit from the cost of the car.	$\$15\,090 - \$1509 = \$13\,581$
2. Find the interest paid for the car by subtracting the principal from the total amount in monthly repayments.	Interest = $17 655.60 − $13 581 $\qquad\;\; = \$4074.60$
3. Use the formula $I = \dfrac{PRT}{100}$ to find the rate. $P = \$13581,\ I = \$4074.60,\ T = 5$	$4074.60 = \dfrac{13\,581 \times R \times 5}{100}$ $407\,460 = 13\,581 \times R \times 5$ $\dfrac{407\,460}{13\,581 \times 5} = R$ $R = 6\%$
4. Answer the question.	The interest rate charged is 6% per year.

9.3.2 On-road costs for new and used vehicles

A **registration fee** is a combination of administration fees, taxes and charges paid to legally drive a vehicle on the roads. Registered vehicles have number plates that help emergency services identify to whom the vehicle belongs and help transport authorities keep track of vehicle ownership. Registration costs depend on the type of vehicle and the location.

Vehicle insurance covers the costs associated with vehicle accidents, such as repairs, replacement vehicles and medical expenses. The different types of vehicle insurance available are:

- **compulsory third-party (CTP)** — mandatory insurance to cover personal injuries in motor vehicle accidents. This is required in all states and territories of Australia. It covers any person whom you might injure while you are driving. In New South Wales, CTP is required before a vehicle can be registered.
- **comprehensive** — covers damage to your own vehicle and other people's property, as well as theft and some other risks, plus legal costs
- **third-party property** — covers damage to other people's property and legal costs, but not damage to your own vehicle
- **third-party fire and theft** — third-party property with some add-on features that cover your vehicle in the event of fire or theft or both.

The cost of vehicle insurance depends on the driver's experience, the type of vehicle, the age of the car, the address where the vehicle is kept and the type of insurance required. For comprehensive insurance, vehicles can be insured at the market value (the vehicle's current value) or the agreed value (a value that is agreed on by the owner and insurer).

There is an excess cost for insurance claims. This is the cost the insured driver needs to pay before the claim is processed. Drivers under a certain age usually have to pay higher excess costs for insurance claims.

Stamp duty is a tax levied by all Australian states and territories on purchases of property such as homes, land and vehicles. The tax rate for stamp duty differs between states. In Queensland, the stamp duty depends on whether the vehicle is new or used and on the type of vehicle.

Type of vehicle	Rate
• hybrid–any number of cylinders • electric	Up to $100 000: $2 for each $100, or part of $100 More than $100 000: $4 for each $100, or part of $100
• 1 to 4 cylinders • 2 rotors • steam powered	Up to $100 000: $3 for each $100, or part of $100 More than $100 000: $5 for each $100, or part of $100
• 5 or 6 cylinders • 3 rotors	Up to $100 000: $3.50 for each $100, or part of $100 More than $100 000: $5.50 for each $100, or part of $100
• 7 or more cylinders	Up to $100 000: $4 for each $100, or part of $100 More than $100 000: $6 for each $100, or part of $100

WORKED EXAMPLE 5

Calculate the amount of stamp duty to be paid for the following vehicle purchases.
a. a 4-cylinder car purchased for $14 560
b. a 4-cylinder car purchased for $104 900

THINK

a. 1. Determine the percentage to be paid by reading the table.

2. Calculate 3% of the vehicle purchase price.

3. Answer the question.

b. 1. Determine the percentage to be paid.

2. Calculate 5% of the vehicle purchase price.

3. Answer the question.

WRITE

The vehicle costs less than $100 000 and has 4 cylinders.
Therefore, the percentage is 3%.

$\dfrac{3}{100} \times 14\,560 = \$436.80$

The stamp duty to be paid on a 4-cylinder vehicle purchased for $14 560 is $436.80.

The vehicle costs more than $100 000 and has 4 cylinders. Therefore, the percentage is 5%.

$\dfrac{5}{100} \times 104\,900 = \$5245$

The stamp duty to be paid on a 4-cylinder vehicle purchased for $104 900 is $5245.

9.3.3 Sustainability, fuel consumption rates, servicing and tyres

The running costs of a vehicle include the fuel consumption and maintenance. The latter includes the cost to service the vehicle and replace the tyres.

Fuel consumption rates show the amount of fuel (petrol or diesel) used per 100 km. For example, 7.8 L/100 km means that the vehicle uses an average of 7.8 litres of fuel for every 100 km travelled.

Fuel consumption rates will vary depending on where you drive. Driving in city areas usually uses more fuel because of the constant stopping and starting. Driving on freeways, where the speed travelled is more constant, generally has a lower fuel consumption rate.

WORKED EXAMPLE 6

A vehicle travels 580 km and uses 62 litres of fuel. Calculate the fuel consumption rate in L/100 km. Give your answer correct to 2 decimal places.

THINK	WRITE
1. Divide the total distance travelled by 100 km (this gives the distance as 100 km lots).	$\dfrac{580}{100} = 5.8$
2. Divide the amount of fuel in litres by the answer in step 1. Round to decimal places.	$\dfrac{62}{5.8} = 10.6896$ ≈ 10.69
3. Answer the question.	The fuel consumption rate is 10.69 L/100 km.

Vehicles require servicing by experienced mechanics to maintain performance. Serviced vehicles generally perform more efficiently and therefore cost less to run.

The cost of servicing a vehicle varies according to the distance travelled and general wear on moving parts such as brakes. Services are generally required every 10 000–15 000 km, depending on the manufacturer's guidelines. The service cost will include labour plus the cost to replace parts.

WORKED EXAMPLE 7

A driver travels on average 450 km every week.

a. If the vehicle requires servicing every 15 000 km, how often will this vehicle need servicing over 3 years?

b. At the 30 000 km service, the vehicle needs the following parts. (The prices for the parts include GST.)
- **Oil filter $30**
- **Fuel filter $25**
- **Oil $35**
- **Windscreen wiper blades $15**

The service will take 3 hours of labour at $105 per hour excluding GST.

Calculate the cost for the 30 000 km service.

THINK	WRITE
a. 1. Calculate the number of weeks the driver takes to travel 15 000 km.	$\dfrac{15\,000}{450} = 33.33$

2. Calculate the number of weeks in 3 years.

$52 \times 3 = 156$

3. Divide the number in step 2 by the number in step 1.

$\dfrac{156}{33.33} = 4.68$

4. Answer the question.

The number of 15 000 km services in 3 years is 4.

b. 1. Calculate the labour costs.

$3 \times \$105 = \$315$
Add 10%
GST: 10% of $315 = \$31.50$
Labour costs $= \$315 + \$31.50$
$= \$346.50$

2. Calculate the cost of parts.

$\$30 + 25 + 35 + 15 = \$105$

3. Add the labour costs to the cost of the parts.

$\$105 + \$346.50 = \$451.50$

4. Answer the question.

The total cost for the 30 000 km service is $451.50.

Tyres help keep a vehicle on the road and are subject to roadworthy inspection by police. Vehicles without the regulated amount of tyre tread are deemed unsafe and the owner is required to have the tyres replaced. The wear on tyres depends on the driving conditions, such as the road surface, vehicle performance and style of driving.

The tyres are usually checked during routine vehicle servicing to ensure that they are safe to drive on and wearing evenly. If there is an issue with the vehicle's steering alignment, the tyres may wear more on one side.

Many different qualities and styles of tyres are available on the market at varying costs.

WORKED EXAMPLE 8

For a particular brand of tyres, the manufacturer recommends replacement after 40 000 km. A driver using these tyres travels an average of 550 km each week. How many times should she expect to replace her tyres in 5 years?

THINK

1. Find the number of weeks it takes to travel 40 000 km. Divide 40 000 km by the average weekly distance.

2. Find the number of years by dividing the answer by 52.

WRITE

$\dfrac{40\,000}{550} = 72.73$

$\dfrac{72.73}{52} = 1.3986$

3. Find the number of times in 5 years by dividing 5 by the answer from step 2.

$$\frac{5}{1.398} = 3.575$$

4. Answer the question.

In 5 years she should expect to replace her tyres 3 times.

9.3.4 Calculate and compare the cost of purchasing different vehicles using a spreadsheet

A spreadsheet can be used to calculate and compare the cost of purchasing different vehicles. Many resources that compare vehicle performance and costs are available.

To compare vehicles it is important to compare similar vehicles and common elements, such as fuel consumption and servicing costs. Registration and insurance costs will vary due to other factors, such as location of the vehicle and state of registration.

WORKED EXAMPLE 9

Using a spreadsheet, which one of the following three cars is the most cost effective to buy? Assume that the average distance travelled each year is 15 000 km, the average price of fuel per litre is $1.20, and all of the cars have the same safety rating.

		Four-door sedan 1.6 litre engine		
Model	Sale price	Fuel consumption (L/100 km)	Servicing (km)	Average service costs
Car A	$15 480	8.2	15 000	$450
Car B	$16 250	7.5	20 000	$350
Car C	$14 999	9.8	10 000	$475

THINK

1. Create a spreadsheet.

2. Calculate the number of litres used (on average) for one year of driving.

WRITE

◢	A	B	C	D	E
1	Vehicle	Sales price	Fuel consumption (L/100 km)	Servicing (km)	Average service costs
2	Car A	$15 480.00	8.2	15 000	$450.00
3	Car B	$16 250.00	7.5	20 000	$350.00
4	Car C	$14 999.00	9.8	10 000	$475.00

Average distance travelled = 15 000 km

Number of 100 km units: $\frac{15\,000}{100} = 150$

Insert the formula '= 150 ∗ C2' into the cell F2.
Copy the formula into cells F3 and F4.

3. Calculate the cost to fuel the car for one year at $1.20/L.

Enter the formula '= F2 ∗ 1.2' into cell G2.
Copy the formula into cells G3 and G4.

	A	B	C	D	E	F	G
1	Vehicle	Sales price	Fuel consumption (L/100 km)	Servicing (km)	Average service costs	Litres/ years	Fuel cost
2	Car A	$15 480.00	8.2	15 000	$450.00	1230	$1476.00
3	Car B	$16 250.00	7.5	20 000	$350.00	1125	$1350.00
4	Car C	$14 999.00	9.8	10 000	$475.00	1470	$1764.00

4. Calculate the servicing costs. To compare the costs, we need to find the proportional service costs by dividing the average distance travelled (15 000 km) by the service distance for each car, then multiply by the average service costs.

Enter the formula '= 15 000/D2 ∗ *E2*' into cell H2.
Copy the formula into cells H3 and H4.

	A	B	C	D	E	F	G	H
1	Vehicle	Sales price	Fuel consumption (L/100 km)	Servicing (km)	Average service costs	Litres/ years	Fuel cost	Service cost
2	Car A	$15 480.00	8.2	15 000	$450.00	1230	$1476.00	$450.00
3	Car B	$16 250.00	7.5	20 000	$350.00	1125	$1350.00	$262.50
4	Car C	$14 999.00	9.8	10 000	$475.00	1470	$1764.00	$712.50

5. Calculate the total cost for each car over the year.

Insert the formula '= B2 + G2 + H2' into cell I2.
Copy the formula into cells I3 and I4.

	A	B	C	D	E	F	G	H	I
1	Vehicle	Sales price	Fuel consumption (L/100 km)	Servicing (km)	Average service costs	Litres/ years	Fuel cost	Service cost	Total cost
2	Car A	$15 480.00	8.2	15 000	$450.00	1230	$1476.00	$450.00	$17 406.00
3	Car B	$16 250.00	7.5	20 000	$350.00	1125	$1350.00	$262.00	$17 862.50
4	Car C	$14 999.00	9.8	10 000	$475.00	1470	$1764.00	$712.00	$17 475.50

6. Answer the question.

The most cost-effective car to buy is Car A, with an overall annual cost in the first year of $17 406.

---

on Resources

Interactivity: Car loans (int-6913)

Exercise 9.3 Purchasing, running and maintaining a vehicle

1. **WE4** A customer takes out a loan to purchase a car for $17 995. The monthly repayments are $313.38 and the loan is for 5 years. The customer pays a 10% deposit. Calculate:
   a. the deposit paid.
   b. the total amount paid for the car.
   c. the interest rate charged for the loan, correct to 2 decimal places.

2. A $24 995 car is purchased through finance of 5.99% per annum. A 10% deposit is paid plus monthly repayments for 5 years. Calculate:
   a. the amount borrowed for the car (principal).
   b. the monthly repayments for the 5 years.
   c. the total amount paid for the car.

3. **WE5** Stamp duty payable on the purchase of 4-cylinder vehicles is:
   $100 000 or less — 3 per cent ($3 per $100 or part thereof)
   More than $100 000 — 5 per cent ($5 per $100 or part thereof).
   Calculate the amount of stamp duty to be paid for the following vehicle purchases.
   a. $22 995
   b. $125 000

4. The stamp duty paid for a 4-cylinder vehicle was $250. Determine the price paid for the vehicle.

5. **WE6** A vehicle travels 640 km and uses 58 litres of fuel. Calculate the fuel consumption rate in L/100 km. Give your answer correct to 2 decimal places.

6. A vehicle's fuel consumption rate for a journey was calculated at 8.9 L/100 km. If the vehicle used a total of 71 L of fuel for the journey, how far in km did it travel? Give your answer to the nearest whole number.

7. A family is on a road trip. The vehicle they are travelling in averages 765 km for 99 L of fuel.
   a. Calculate the fuel consumption rate correct to 2 decimal places.
   b. If the average fuel price is $1.34/L, calculate the cost per 100 km.

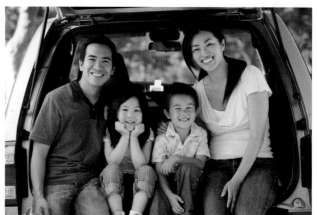

8. **WE7** A driver travels on average 350 km every week.
   a. If the vehicle requires servicing every 10 000 km, how often will his vehicle need servicing over 2 years?
   b. At the 30 000 km service, the vehicle needs the following parts and labour. (The prices for the parts include GST.)

   - Oil filter $20
   - Fuel filter $35
   - Oil $40
   - Windscreen wiper blades $10
   - 2.5 hours labour at $85 per hour excluding GST.

   Calculate the cost for the 30 000 km service.

9. **WE8** For a particular brand of tyres, the manufacturer recommends replacement after 35 000 km. A driver using these tyres travels an average of 2050 km each month. How many times should she expect to replace her tyres in 10 years?

10. **MC** A vehicle's average fuel consumption is 9.8 L/100 km. The driver fills the tank with 65 litres of fuel. The average number of kilometres she can travel before the tank is empty is:
    A. 6.63.
    B. 65.
    C. 151.
    D. 663.

11. **WE9** Using a spreadsheet, work out which one of the following three cars is the most cost-effective to buy. Assume that the average distance travelled each year is 15 000 km and the cost of fuel is $1.15.

Four-door sedan (medium size car)				
Model	Sale price	Fuel consumption (L/100 km)	Servicing (km)	Average service costs
Car A	$37 990	12.7	15 000	$255
Car B	$36 748	8.6	20 000	$595
Car C	$38 490	10.5	10 000	$235

12. A female 19-year-old driver has three options to purchase comprehensive car insurance for a small car that will be parked in a locked garage in a suburban property. All options include replacing her car, providing transportation if the car is unable to be driven, and costs for injuries for all passengers.
    Option A: $1143, $850 excess, 10% online discount
    Option B: $1149, $850 excess
    Option C: $1155, $800 excess, 15% online discount
    The excess fee is the amount that is paid to the insurer if a claim is made.
    Which option is the most cost-effective?

13. The fuel consumption for a vehicle travelling in city traffic is recorded as 9.8 L/100 km. When travelling along highways, the fuel consumption is recorded as 7.5 L/100 km.
    a. Show that the average fuel consumption is 8.65 L/100 km.
    b. The vehicle travels 126 km along city roads and 458 km along the highway. Calculate the total number of litres of fuel used, correct to 1 decimal place.
    c. The cost of fuel is $1.25 per litre. Calculate the cost of the fuel for this trip, correct to the nearest cent.
    d. When calculating the fuel costs using the average fuel consumption, the value is different to the actual cost. Explain why.

14. A vehicle is purchased for $24 980 through finance that attracts 6% interest for 5 years. The owner pays a 10% deposit and then pays $496.28 monthly. This value for the monthly repayments includes a $250 one-off administration fee to set up the loan and a monthly $5 loan-servicing fee. Show that the comparison rate for the loan is 6.489% per annum.

15. A person is looking to buy a new car. The person drives to work in the city, a return distance of 45 km, Monday to Friday. On the weekend, the person drives along a major highway to visit a relative who lives in a regional town 112.5 km from the city.

Car	Fuel consumption for city driving (L/100km)	Fuel consumption for highway driving (L/100 km)	Average fuel consumption (L/100 km)
A	12.5	7.8	
B	11.2	8.5	

a. Calculate the average fuel consumption for both cars.
b. Based on the weekly distances driven by the person, which car would be most cost-effective to purchase? Justify your answer using calculations.

9.4 Budgeting

9.4.1 Monthly income

A **budget** is a list of all planned income and costs. A budget can be prepared for long periods of time, but for individuals it is usually a monthly plan.

In order to create a monthly budget, monthly income and expenses need to be known. **Monthly income** is $\frac{1}{12}$ of the total annual income.

WORKED EXAMPLE 10

Calculate the monthly income of a worker who has an annual salary of **$71 400** and earns **$1590** each year in bank interest.

THINK	WRITE
1. Find the total annual income as the sum of salary and bank interest.	Total annual income after tax $= 71\,400 + 1590$ $= 72\,990$
2. Find the total monthly income after tax by dividing the answer in step 1 by 12.	Total monthly income after tax $= \dfrac{72\,990}{12}$ $= 6082.50$
3. Answer the question.	The worker's monthly income after tax is $\$6082.50$.

9.4.2 Fixed and discretionary spending

Fixed spending is spending that occurs every month (or week, fortnight, year or other period) of the same amount. Examples of fixed spending are rent, school fees, health insurance and vehicle insurance. **Discretionary spending** is spending that occurs from time to time and is not a fixed amount. Examples of discretionary spending are food, clothing, entertainment and medical bills.

WORKED EXAMPLE 11

Determine the total monthly spending from the following list of fixed and discretionary spending. The discretionary spending represents the amounts spent in a month.

	Fixed spending		Discretionary spending	
Item	**Frequency**	**Amount**	**Item**	**Amount**
Rent	Weekly	$230	Food	$423
Health insurance	Monthly	$78	Clothing	$107
Vehicle registration	Yearly	$620	Entertainment	$85
Vehicle insurance	Yearly	$389	Car repairs	$325

THINK	WRITE
1. Convert any weekly spending to monthly spending by multiplying by $\frac{52}{12}$. The only weekly cost is rent of $230.	$\text{Monthly rent} = 230 \times \frac{52}{12}$ $= 996.67 \text{ (correct to 2 decimal palces)}$
2. • Add the annual spending. The annual spending is vehicle registration and insurance $(620 + 389)$. • Convert the annual spending to monthly spending by dividing by 12.	$\text{Annual spending} = 620 + 389$ $= 1009$ $\text{Monthly cost} = \frac{1009}{12}$ $= 84.08 \text{ (correct to 2 decimal places)}$
3. Find the total fixed spending by adding the numbers from steps **1** and **2** to the monthly fixed cost of health insurance $(\$78)$.	$\text{Total monthly fixed spending} = 996.67 + 84.08 + 78$ $= 1158.75$
4. Find the total discretionary spending by adding up the numbers in the last column of the table.	$\text{Total monthly discretionary spending} = 423 + 107 + 85 + 325$ $= 940$
5. To find the total monthly spending, add the total fixed spending to the total discretionary spending obtained in step **4**.	$\text{Total monthly spending} = \text{fixed spending} + \text{discretionary spending}$ $= 940 + 1158.75$ $= 2098.75$
6. Answer the question.	Total spending is $2098.75 per month.

9.4.3 Preparing a personal budget

Personal budgets can be prepared using spreadsheets to calculate the total fixed and discretionary spending for each month and subtract this from the net monthly income. A profit is made when the net monthly income is greater (more) than the total monthly spending. A loss is made when the net monthly income is less than the total monthly spending.

A nurse's net annual salary is $59 500. Her spending is shown in the following table.

Spending description	Amount
Rent (monthly)	$658
Electricity (quarterly)	$295
Water usage (quarterly)	$150
Health insurance (monthly)	$205
Car finance (fortnightly)	$172.50
Vehicle registration (yearly)	$595
Vehicle insurance (half-yearly)	$680
Food (weekly)	$75
Entertainment (monthly)	$200
Fuel (weekly)	$68

Using a spreadsheet, determine whether the nurse has a profit or loss at the end of each month.

THINK

1. Calculate the monthly costs. For yearly, divide by 12. For half-yearly, divide by 6. For weekly, multiply by 52, then divide by 12.
For quarterly, divide by 3. For fortnightly, multiply by 26, then divide by 12.

WRITE

Insert the formulas into the relevant cells in column C:
'= (cell in column B)/12'
'= (cell in column B)/6'
'= (cell in column B)*52/12'
'= (cell in column B)/3'
'= (cell in column B)*26/12'.

	A	B	C
1	Spending description	Amount	Monthly expenses
2	Rent (monthly)	$658.00	$658.00
3	Electricity (quarterly)	$295.00	$98.33
4	Water usage (quarterly)	$150.00	$50.00
5	Health insurance (monthly)	$205.00	$205.00
6	Car finance (fortnightly)	$172.50	$373.75
7	Vehicle registration (yearly)	$595.00	$49.58
8	Vehicle insurance (half yearly)	$680.00	$113.33
9	Food (weekly)	$75.00	$325.00
10	Entertainment (monthly)	$200.00	$200.00
11	Fuel (weekly)	$68.00	$294.67

2. Calculate the total monthly spending.

Create new row: Monthly expenses
Insert '= sum (C 2: C 11)' into cell C12.

3. Calculate the net annual income.

Create new row: Net monthly income
Insert '59500' into B13.
Insert '=B13/12' into C13.

4. Calculate the profit/loss for the month.

Create new row: Profit/loss
Insert '=C13 - C12' into cell C14

	A	B	C
1	Spending description	Amount	Monthly expenses
2	Rent (monthly)	$658.00	$658.00
3	Electricity (quarterly)	$295.00	$98.33
4	Water usage (quarterly)	$150.00	$50.00
5	Health insurance (monthly)	$205.00	$205.00
6	Car finance (fortnightly)	$172.00	$373.75
7	Vehicle registration (yearly)	$595.00	$49.58
8	Vehicle insurance (half yearly)	$680.00	$113.33
9	Food (weekly)	$75.00	$325.00
10	Entertainment (monthly)	$200.00	$200.00
11	Fuel (weekly)	$68.00	$294.67
12	**Monthly expenses**		**$2 367.67**
13	**Net monthly income**		**$4 958.33**
14	**Profit/loss**		**$2 590.67**

5. Answer the question.

She makes a profit of $2590.67.

 Resources

 Interactivity: Budgeting (int-6914)

Exercise 9.4 Budgeting

1. **WE10** Calculate the net monthly income of a worker who has a net annual salary of $63 200 and earns $3745 each year in bank interest.
2. **MC** A person earns $74 600 net annual salary and also receives a bonus payment from their employer of $3500. They have a fortnightly salary of:
   A. $1501.92. **B.** $3003.85. **C.** $3124.00. **D.** $6369.23.
3. A bank teller earns $1100 per week (after tax) plus $2200 per year in bank interest (after tax). Find her net monthly income.
4. Complete the following table.

	Net annual salary	Total net annual investment income	Total net monthly income
a.	$70 500	$4 812	
b.	$81 234	$10 387	
c.	$90 349	$9 885	
d.	$24 500	$42 500	

5. **WE11** Determine the total monthly spending from the following list of fixed and discretionary spending. The discretionary spending represents the amounts spent in a month.

	Fixed spending		Discretionary spending	
Item	**Frequency**	**Amount**	**Item**	**Amount**
Rent	Weekly	$205	Food	$494
Health insurance	Monthly	$89	Clothing	$205
Vehicle registration	Yearly	$540	Entertainment	$123
Vehicle insurance	Yearly	$499	Car repairs	$72

6. A baker earns $18.50 an hour after tax. She works an average of 38 hours a week plus 7 hours overtime, paid at time and a half. Her monthly spending is 47% of her net monthly income. Calculate how much she spends.

7. Determine the total monthly cost from the following list of fixed and discretionary spending. Discretionary spending represents the amounts spent in a month.

	Fixed spending		Discretionary spending	
Item	**Frequency**	**Amount**	**Item**	**Amount**
Home mortgage	Monthly	$1254	Food	$563
Health insurance	Monthly	$124	Clothing	$321
Vehicle registration	Yearly	$702	Entertainment	$389
Vehicle insurance	Yearly	$899	Car repairs	$396
School fees	Half-yearly	$5300	Travel	$1379

8. **MC** Which of the following can be considered a fixed cost?
   A. Daughter's speeding tickets
   B. Son's soccer club membership
   C. Dad's shoe purchases
   D. Winter holiday cost

9. Complete this table of income and costs for a family of five over 4 months.

	Net annual salary	Net monthly bank interest	Monthly fixed spending	Monthly discretionary spending	Monthly profit/loss
a.	$73 700	$784	$3623	$2563	
b.	$73 700	$792	$3623	$639	
c.	$73 700	$804	$3623	$4456	
d.	$73 700	$823	$3623	$1065	

10. **WE12** The net annual salary for a childcare worker is $43 470. Her spending is shown in the following table.

Spending description	Amount
Rent (weekly)	$225
Electricity (quarterly)	$150
Water usage (quarterly)	$75
Travel to work (weekly)	$60
Food (weekly)	$45
Entertainment (monthly)	$250

Using a spreadsheet, determine how much profit or loss the childcare worker has at the end of each month.

11. The following information shows the spending and income for a couple.

Spending	Mortgage	$672 per fortnight
	Food	$123 per week
	Clothing	$78 per week
	Transport	$35 per day
	Entertainment	$120 per week
	Health insurance	$200 per month
	Car (insurance, registration, fuel, servicing)	$600 per month
Net income	Partner A	$1680 per fortnight
	Partner B	$1100 per week

a. Using a spreadsheet, calculate the total monthly spending.
b. Show that the couple make a monthly profit.
   The couple would like to save to buy a house for sale for $350 000. To buy the house, they will need to make a 10% deposit.
c. If the couple's spending does not change, how many months will it take to save the 10% deposit? Give your answer correct to 1 decimal place.
d. The couple's financial adviser recommends they reduce their spending to be less than 50% of their net monthly income. Where should they adjust their spending? Justify your answer by using calculations.

12. The following shows a personal budget prepared by a graduate IT technician.

Item	Frequency	Amount	Monthly
Rent	Weekly	$275	$1100
Bills	Monthly	$145	$145
Travel	Monthly	$350	$350
Entertainment	Monthly	$330	$330
Clothes	Fortnightly	$220	$440
Insurance	Yearly	$1150	$95.83
Monthly spending			$2460.83
Monthly net income			
Profit/loss			**$2355.92**

a. Determine his monthly net income and hence calculate his annual net income.

b. Find the percentage amount he spends of his net income. Give your answer correct to 2 decimal places.

c. He would like to save up for an overseas holiday and has budgeted $5500 for his trip. If he can save 50% of his monthly net income, how many months will it take for him to have saved enough money for his trip?

13. A family's net annual income is $69 480. Their spending is shown in the following table.

Item	Frequency	Amount
Holiday/entertainment	Yearly	$4500
Mortgage	Weekly	$245
Insurance	Yearly	$1980
Car costs	Monthly	$950
Food	Weekly	$145
Electricity, water	Quarterly	$580
School fees	Yearly	$8500
Health insurance	Monthly	$368

a. Prepare a monthly budget and show whether the family is living within their means (i.e. making a profit).

b. The family needs to adjust their spending. By identifying the fixed and discretionary spending items, explain which items they should consider reducing.

14. A person's monthly spending is $1850. His current spending is 55% of his net monthly income.

a. Determine his net monthly income and net annual salary.

b. If he gets paid an hourly rate of $22.50, how many hours a week does he work? Give your answer correct to 1 decimal place.

c. He gets a job that pays 4% more than he is currently earning. If his spending does not change, what percentage will his spending be of his net monthly income? Give your answer correct to 2 decimal places.

15. A family of four are going on an overseas holiday for 6 weeks. They have saved $20 000 for the trip and recorded the costs for the essentials but don't know how much to spend on food.

Item	Cost
Passports	$832
Flights	$5668
Travel insurance	$658
Accommodation	$245 per night
Food	

a. They estimate food costs to be about $45 per day. How much money do they have remaining from the $20 000 to spend on their trip?

b. The family reassess and decide to adjust their spending to save more money for their trip. Their monthly spending is $2585 and net monthly income is $4580. If they can save half of their monthly profit for 4 months, how much more money will they have for their overseas trip?

9.5 Review: exam practice

9.5.1 Budgeting: summary

Household expenses

- The standard measurement of electricity consumption is kilowatt-hours, kWh. It is the average amount of electricity (in kilowatts) used per hour (1000 watts make up a kilowatt).
- Gas consumption is measured in megajoules, MJ.
- Water is provided and managed by providers. Home owners are charged a fixed fee to provide water and remove water wastage, known as sewerage.
- Water usage is measured through a water meter fixed to the home and is charged at a rate, in dollars, per kilolitre (kL). 1000 L = 1 kL.
- Council rates are payments by home owners to local councils to pay for parks, libraries, rubbish collection, road maintenance, etc.
- Council rates are calculated as a percentage on the capital improved value of the land and buildings within the council municipality. The rate in the dollar is determined by how much revenue the council needs to raise and dividing by the amount of capital improvement of land and buildings within the council municipality.

$$\text{Rate in the dollar} = \frac{\text{revenue}}{\text{capital}}$$

Purchasing, running and maintaining a vehicle

- The comparison rate for a loan gives a more accurate indication of the interest rate for the loan. It includes all the additional fees and includes this in the calculation.
- A car registration fee is a combination of administration fees, taxes and charges paid to legally drive a vehicle on the roads. Registration costs depend on the type of vehicle and the location.
- Vehicle insurance can be either Compulsory Third Party (CTP), Comprehensive, Third Party Property and Third Party Fire and Theft.
- Compulsory Third Party (CTP) insurance is mandatory motor vehicle accident personal injury insurance; required by each state and territory. It covers any person that you might injure while you are driving. In NSW CTP is required before a vehicle can be registered.
- Comprehensive insurance covers damage to your own vehicle and other people's property, as well as theft and some other risks, plus legal costs. Vehicles can be insured at a market value or an agreed value.
- Third Party Property insurance covers damage to other people's property and legal costs, but not damage to your own vehicle.
- Third Party Fire and Theft insurance is Third Party Property insurance with some add-on features that cover your vehicle
- Excess costs are cost the insured driver needs to pay before the claim is processed. Drivers under a certain age usually have to pay additional excess costs for insurance claims.
- Stamp Duty is a tax levied by all Australian territories and states on property purchases, such as homes, land, and vehicles.
- A fuel consumption rate is the amount of fuel (petrol/diesel) used per 100 km.
- Vehicles need to be serviced every 10 000–15 000 km depending on the manufacturer's guidelines, and costs will include the cost to replace parts and labour.
- Tyres are usually checked during routine vehicle servicing to ensure that they safe to drive on and wearing evenly. Tyres are subject to road worthy inspection by police.

Budgeting

- A budget is a list of all planned income and costs. Often a monthly plan, all income and expenses need to be known.
- Fixed spending is regular payments of the same amount, such as rent, school fees, health insurance and vehicle insurance.
- Discretionary spending is spending that occurs from time to time and is not a fixed amount, such as food, clothing, entertainment and medical bills.
- A profit is made when net monthly income is greater (more) than the total monthly spending.
- A loss is made when net monthly income is less than the total monthly spending.

Exercise 9.5 Review: exam practice

Simple familiar

1. **MC** The energy usage and rates for a household are as shown. There is a 25% discount for paying the bill before the due date. If the bill was paid before the due date, what was the amount paid? (Remember to include GST.)

Electricity

Tariff	Bill days	Current reading	Previous reading	Total usage (kWh)	Charge/rate(c/kWh)
All day	90	78 250	77 682		25.08
Service to property	90			$1.105/day	

Gas

Tariff	Bill days	Current reading	Previous reading	Total usage (kWh)	Charge/rate (c/MJ)
All day	90	8920	8138		2.452
Service to property	90			$0.985/day	

    A. $107.82   **B.** $241.90   **C.** $262.29   **D.** $288.53

2. A household's daily water usage is 1700 litres. The total water bill for 60 billing days is $392.12, which includes $187.10 for water services. Calculate the charge per kilolitre.

3. A worker's net annual income is $34 105. The worker makes a monthly loss of $145. Calculate the monthly spending.

4. A vehicle is purchased for $18 570 on finance of 5.5% per annum for 3 years. A 10% deposit is paid and there are no other administration costs. Stamp duty payable on the purchase of vehicles is:
$100 000 or less — 3 per cent ($3 per $100 or part thereof)
more than $100 000 — 5% ($5 of $100 or part thereof).
Calculate, correct to the nearest cent:
    a. the stamp duty   b. the monthly instalments cost.

5. **MC** A car can travel 485 km on 68 litres of fuel. The fuel consumption in L/100 km for the car is:
    A. 0.14   **B.** 7.01   **C.** 7.13   **D.** 14.02

6. Complete the following table.

	Energy usage	Charge/rate	Cost
a.	987 kWh	19.02 c/kWh	
b.	2058 MJ		$53.28
c.		15.48 c/kWh	$98.61
d.	5439 MJ	1.534 c/MJ	

7. According to the manufacturer's guidelines, a vehicle has its four tyres replaced 3 times in 9 years. If the vehicle travels an average of 17 500 km each year, how many kilometres (on average) does the driver get out of a set of tyres?

8. The quarterly council rates for a property with capital improved value of $385 000 are $721.88.
   a. Calculate the amount in the dollar charged annually by the council.
   b. If the total capital improved value for all properties within the council municipality is $1.5 billion, calculate the amount of revenue the council plans to raise this year.

9. **MC** In determining the budget for a small business, which of the following can be considered a discretionary cost?
   A. Rent
   B. Employee's wages
   C. Public liability insurance
   D. An electricity bill

10. **MC** In its annual budget, the local council plans to raise $25 million in revenue to repair local infrastructure. The capital improvement for the area is valued at $1.5 billion. The council rate quarterly instalment for a rate payer whose house has a capital improved value of $475 000 is:
    A. $712.50
    B. $791.67
    C. $1979.17
    D. $2850.00

11. The monthly repayments for a car are $396.25 for 48 months.
    a. Calculate the total repayment amount.
    b. The interest charged was $3170 and the owner paid $2000 deposit. How much was the car purchased for? Answer to the nearest dollar.

12. Calculate the monthly net income for each of the following.
    a. A net annual income of $63 480
    b. A weekly net income of $945.35
    c. A fortnightly net income of $1450

Complex familiar

13. The following table shows the net annual incomes and monthly spending of 4 people. Calculate the profit or loss for each person.

Net annual income	Monthly spending	Profit/loss
$51 890	$3950	
$84 755	$8640	
$75 120	$6580	
$62 308	$4980	

14. There is a 12% discount for customers who pay their energy bill before the due date. The energy usage of a customer is:
    • electricity: 875 kWh at 19.02 c/kWh
    • gas: 720 MJ at 2.034 c/MJ.
    The customer is also charged a $112 fixed fee for service to the property.
    If the customer pays before the due date, calculate the amount they paid, including GST.

15. Calculate the monthly running costs for the following vehicles. (Assume fuel costs $1.00/L.)

Vehicle	Fuel consumption (L/100 km)	Tyres (every 30 000 km)	Servicing (every 15 000 km)	Insurance (yearly)	Registration (yearly)	Average distance travelled in 1 year
Van	13.5	$550	$220	$558	$666.40	28 500
SUV	11.8	$870	$390	$680	$735.70	19 400

16. A council raises its council rates by 5% to meet the growing costs associated with improving infrastructure. Last year the council raised $19 million in revenue with total capital improved value of $1.6 billion.
    a. Calculate the rate in the dollar for last year's revenue.
    b. Determine the amount of revenue the council plans to raise this year based on the 5% increase, assuming the total capital improvement value is the same as last year.
    c. A rate payer's council rates last year were $4135.25. Calculate this year's council rates for this rate payer after the 5% increase.
    Over the year, the capital improved value for properties in the municipality increased by 2.5% due to the property market.
    d. Show that the rate in the dollar charged by the council for this year is approximately 0.012.
    e. Explain why the overall increase in the rate in the dollar is closer to 2.5% than the 5$% stated by the council. Use calculations to support your explanation.

17. The monthly credit card spending of a family of four is $8650, which includes all bills and household expenses. Partner A's annual net salary is $74 486 and Partner B's fortnightly wage is $1685.
    a. Calculate the partners' net monthly salaries.
    b. Show that the family spends approximately 88% of their net monthly income.

18. A person is saving for a new car priced at $31 910 (on-road costs). Their net annual income is $52 680 and their monthly spending is shown in the table.
    a. Show that the person has a monthly profit.
    The person pays $2000 deposit for the car and will borrow the remainder at 6.5% per annum for 4 years with monthly repayments.
    b. Calculate the principal borrowed.
    c. Show that the monthly instalment is $785.14.
    d. The annual registration and insurance costs will be $1625. Fuel costs will replace the person's current spending on travel to work. Will they be able to afford to pay for the car? Justify your answer using calculations.

Item	Amount
Mortgage	$880
Electricity bills	$158
Council rates	$185
Travel to work	$250
Food	$120
Health insurance	$225
Entertainment	$250

19. A family of four are going on a holiday. They have a choice of driving or flying. They can average 650 km on 55 litres of fuel and the average cost of fuel is $1.41 per litre. The distance to their destination is 1716.1 km. The cheapest one-way flight to their destination costs $219 per person.
    a. Which option (driving or flying) is the more cost-effective for the family?
    b. Which option would you advise them to choose? Support your advice by providing a positive and negative for each option.

20. A family of 6 lives in a 5-bedroom house in the outer suburbs of a major city. They own two cars. Partner A uses one car to drive to the local train station; Partner B uses the other car to drive their children to school, sports training and holidays.

They spend $165 on food each week. Their health insurance costs $368 per month, their house insurance costs $1580 a year, and their annual council rates are $3580. The children's school fees total $2850 a year.

Partner A's net annual salary is $74 805.

Partner B works part-time Monday–Friday as an administrator at a local law firm. Her hourly rate is $27.50 and her hours are 9 am–3 pm with a 45-minute lunch break. Her employer withholds $85 in PAYG tax each week.

Their bills and other expenses are shown below.

Electricity bill

Billing period: 1 Jul–30 Sep					
Tariff	Bill days	Current reading	Previous reading	Total usage	Charge/rate (c/kWh)
Peak	91	23 890	23 150		33.58
Off-peak	91	48 901	48 334		15.60
Service to property	91			**$1.148/day**	

Water bill

Bill details as of 30 September		Value	Price
Water service charge	1 Jul to 30 Sep	25.75	
Sewerage service charge	1 Jul to 30 Sep	149.80	
Stormwater	1 Jul to 30 Sep	22.05	
	Volume (kL)	Charge/kL	
Water usage	65	$2.05/kL	
Recycled water usage	45	$1.82/kL	

Cars

Vehicle	Fuel consumption (L/100 km)	Registration (annually)	Insurance (6-monthly)	Servicing (every 15 000 km)	Average distance travelled each year
		Average fuel price: $1.31/litre			
Partner A	8.8	$568	$302	$250	12 500
Partner B	10.2	$568	$257	$320	17 500

a. Calculate their joint net monthly income.

b. Prepare a budget for the family.

c. To be solvent means to be able to pay one's debts (expenses, bills). Are the family solvent? Justify your answer using your calculations.

Answers

Chapter 9 Budgeting

Exercise 9.2 Household expenses

1. $301.65
2. a. $59.45   b. $28.13
   c. $131.03   d. $113.42
3. 4768 kWh
4. a. i. 30.54 c/kWh   ii. $1.017   iii. 895 kWh
   b. i. $273.33   ii. $402.47
5. $159.63
6. D
7. $372.92
8. 33.62 kL
9. $1833.33
10. a. 0.006   b. $16 020 000
11. Plan B is the better option for both when he pays on time ($220.04 compared to $230.21) and when he does not ($250.04 compared to $270.83).
12. a.

Bill details as of 30 June		Value	Price
Water service charge	1 Apr to 30 Jun	23.80	23.80
Sewerage service charge	1 Apr to 30 Jun	150.20	150.20
Storm-water	1 Apr to 30 Jun	17.95	17.95
	Volume (L)	**Charge/kL**	
Water usage	64 000	$2.04	130.56
Recycled water usage	38 000	$1.87	71.06
		Total:	$393.57

    b. Water usage: $2.12; recycled water usage: $1.94
    c. Water service charge increase: 10.50%; sewerage service charge increase: 1.00%
    d. Average percentage increase: 2.99%

13. Choice 1
14. The council rate is incorrect. The council made an error in the first calculation for the rate in the dollar: $\frac{7.5}{2.5} = 3$ should be $\frac{7.5}{2500} = 0.003$. The rates for this property should be $525.

15. a. Council rate: $280\,000 \times 0.008 = \$2240$
      Quarterly instalment: $\frac{2240}{4} = \$560$
    b. $1345.60
    c. 7.45%
    d. Answers will vary but could include the following. The family could reduce their electricity bill by installing energy-saving light bulbs and turning off lights and power appliances at the power point. They could reduce their water usage by ensuring there are no leaking taps, reducing time spent in the shower, washing full loads of washing, and ensuring the dishwasher is full before turning it on.
      Council rates are set and will not be reduced by changing household behaviours.

Exercise 9.3 Purchasing, running and maintaining a vehicle

1. a. $1799.50   b. $20 602.30
   c. 3.22% pa
2. a. $22 495.50   b. $487.22   c. $31 732.40
3. a. $689.85   b. $6250
4. $8333.33
5. 9.06 L/100 km
6. 798 km
7. a. 12.94 L/100 km
   b. $17.34
8. a. 3 times
   b. $338.75
9. 7 times
10. D
11. Car B
12. Option C
13. a. $\frac{9.8 + 7.5}{2} = 8.65$ L/100 km
    b. 46.7 L
    c. $58.38
    d. The average fuel consumption is a value that assumes equal distance (1 : 1) of driving under the two conditions (city and highway). In this case, the ratio of city to highway driving was 126 : 458 (1 : 3.6). This means the overall fuel consumption is lower than the average value would predict, because 3.6 times more driving took place at the lower fuel consumption.

14. Total monthly payments: $496.28 \times 60 = \$29\,776.80$
    Deposit: $10\% \times 24\,980 = \$2498$
    Principal borrowed: $24\,980 - 2498 = \$22\,482$
    Interest paid: $29\,776.80 - 22\,482 = \$7294.80$

$$I = \frac{PRT}{100}$$

$$7294.80 = \frac{22\,482 \times R \times 5}{100}$$

$$\frac{7294.80}{22\,482 \times 5} = R$$

$$R \approx 6.489\%$$

15. a. Car A: 10.15 L/100 km
Car B: 9.85 L/100 km

b. Car B has a lower average fuel consumption (9.85 L/100 km) and will therefore be more cost-effective. (The average is valid here because the distances of city and highway driving come to the same value, 225 km.)

Exercise 9.4 Budgeting

1. $5578.75

2. B

3. $4950.00

4.

Net annual salary	Total net annual investment income	Total net monthly income
$70 500	$4812	$6276.00
$81 234	$10 387	$7635.08
$90 349	$9 885	$8352.83
$24 500	$42 500	$5583.33

5. $1957.92

6. $1827.40

7. $5442.75

8. B

9. see table at the foot of the page*.

10. A profit of $1867.50

11. a. $4711.58

b. The net monthly income of $8406.67 exceeds the monthly spending, for a monthly profit of $3695.09.

c. 9.5 months

d. Areas in which spending can be reduced (adjusted) include clothing, entertainment and food. Mortgage, health insurance and car registration/insurance are fixed spending.

12. a. Monthly net income = $4816.75; annual net income = $57 801

b. 51.09%

c. 23 months

13. a.

Item	Frequency	Amount	Monthly cost
Holiday/ entertainment	Yearly	$4500	$375.00
Mortgage	Weekly	$245	$1061.67
Insurance	Yearly	$1980	$165.00
Car costs	Monthly	$950	$950.00
Food	Weekly	$145	$628.33
Electricity, water	Quarterly	$580	$193.33
School fees	Yearly	$8500	$708.33
Health insurance	Monthly	$368	$368.00
		Total	$4449.67
Income	Yearly	$69 480	$5790.00
		Profit	$1340.33

b.

Fixed	Discretionary
Mortgage	Holiday/entertainment
Insurance	Car costs
Health insurance	Food
School fees	Electricity, water

The items they should consider reducing are the discretionary items, particularly the non-essential items: holidays and entertainment. These items do not have fixed costs, which means that the costs can be adjusted based on usage or need.

14. a. Net monthly income: $3363.64; net annual salary: $40 363.64

b. 34.5 hours

c. 52.88%

15. a. $662   **b.** $3990

*9.

	Net annual salary	Net monthly bank interest	Monthly fixed spending	Monthly discretionary spending	Monthly profit
a.	$73 700	$784	$3623	$2563	$739.67
b.	$73 700	$792	$3623	$639	$2671.67
c.	$73 700	$804	$3623	$4456	−$1133.33
d.	$73 700	$823	$3623	$1065	$2276.67

9.5 Review: exam practice

1. D

2. 2.01 $/kL

3. $2987.08

4. a. $557.10   b. $557.08

5. D

6.

	Energy usage	Charge/rate	Cost
a.	987 kWh	19.02 c/kWh	$187.73
b.	2058 MJ	2.589 c/MJ	$53.28
c.	637 kWh	15.48 c/kWh	$98.61
d.	5439 MJ	1.534 c/MJ	$83.43

7. 52 500 km

8. a. 0.0075   b. $11 250 077

9. D

10. C

11. a. $19 020   b. $17 850

12. a. $5290   b. $4096.52   c. $3141.67

13.

Net annual income	Monthly spending	Profit/loss
$51 890	$3950	$374.17 profit
$84 755	$8640	$1577.08 loss
$75 120	$6580	$320 loss
$62 308	$4980	$212.33 profit

14. $283.69

15. Van: $501.03 ; SUV: $397.66

16. a. 0.011875
    b. $19 950 000
    c. $4342.01
    d. Rate in the dollar: $\frac{19\,950\,000}{1\,640\,000\,000} = 0.012$
    e. Property values increased by 2.5%, which was not included in the council's initial calculations; $5 - 2.5 = 2.5\%$.

17. a. Partner A: $6207.17; Partner B: $3650.83
    b. $\frac{8650}{9858} \times 100 \approx 88\%$

18. a. Monthly income − monthly spending = 4390 − 2068 = $2322
    b. $29 910
    c. Monthly repayments: $\frac{7776.60 + 29\,910}{48} = \$785.14$
    d. Yes, they will be able to pay for the car.
       Net monthly profit = 2322 − (785.14 + 135.42)
       = $1401.44

19. a. Driving
    b. Answers may vary. Driving is more cost-effective and will allow the family quality time together and to see the country. Some possible positives and negatives are shown below.

Driving		Flying	
Positive	Negative	Positive	Negative
Very cost-effective	At least 17 hours of driving (at 100 km/h)	Less travelling time than driving	More costly than driving
Quality family time	Passenger comfort	Passenger comfort	Time spent waiting at the airport for luggage, flight delays
Get to see the country	Accommodation required for overnight stay (in both directions)		

20. a. Net monthly income: $8993.54
    b.

Item	Cost	Frequency	Monthly cost
Food	$165.00	Weekly	$715.00
Health insurance	$368.00	Monthly	$368.00
House insurance	$1580.00	Yearly	$131.67
School fees	$2850.00	Yearly	$237 50
Council rates	$895.00	Quarterly	$298.33
Electricity	$485.55	Quarterly	$161.85
Water	$412.75	Quarterly	$137.58
Car insurance	$559.00	Half-yearly	$93.17
Car registration	$1136.00	Yearly	$94.67
Fuel	$3779.35	Yearly	$314.95
Servicing	$581.67	Yearly	$48.47
		Total	$2601.19
		Net monthly income	$8993.56
		Profit	$6392.37

    c. Yes, the family are solvent. Their net monthly income is greater than their spending.

CHAPTER 10
Time, distance and speed

10.1 Overview

LEARNING SEQUENCE

10.1 Overview
10.2 Time
10.3 Interpret timetables
10.4 Time travel
10.5 Distance: scales and maps
10.6 Speed
10.7 Distance–time graphs
10.8 Calculations with motion
10.9 Review: exam practice

CONTENT

In this chapter, students will learn to:
- use units of time and convert between fractional, decimal and digital representations
- represent time using 12-hour and 24-hour clocks
- calculate time intervals, including time between, time ahead, time behind
- interpret timetables for buses, trains and/or ferries
- use several timetables and/or electronic technologies to plan the most time-efficient routes
- interpret complex timetables, such as tide charts, sunrise charts and moon phases [complex]
- compare the time taken to travel a specific distance with various modes of transport
- use scales to find distances, e.g. on maps
- investigate distances through trial and error or systematic methods [complex]
- apply directions to distances calculated on maps including the eight compass points in relation to the rising and setting of the sun: N, NE, E, SE, S, SW, W, NW [complex]
- identify the appropriate units for different activities, e.g. walking, running, swimming, driving and flying
- use units of energy used for foods, including calories
- use of units of energy to describe the amount of energy in activity, including kilojoules.
- calculate speed, distance or time using the formula, speed = distance/time
- calculate the time and costs for a journey from distances estimated from maps, given a travelling speed [complex]
- calculate average speed
- interpret distance-versus-time graphs, including reference to the steepness of the slope (or average speed) [complex].

Fully worked solutions for this chapter are available in the Resources section of your eBookPLUS at www.jacplus.com.au

10.2 Time

10.2.1 Facts about time

With many people having busy schedules, time is an important factor of everyday life. Whether you need to get to class on time, catch your flight or make your doctor's appointment, our lives are organised around time.

Time is divided into units including seconds, minutes, hours, days, weeks, months and years.

A day can be divided into two 12-hour periods:

Midnight to midday	am	ante meridiem (before noon)
Midday to midnight	pm	post meridiem (after noon)

Time conversions

60 seconds	1 minute
60 minutes	1 hour
24 hours	1 day
7 days	1 week
14 days	1 fortnight
12 months	1 year
365 days	1 year
366 days	1 leap year
10 years	1 decade
100 years	1 century
1000 years	1 millennium

 **Resources**

 Interactivity: Converting between units of time (int-6910)

Displaying time

Time can be shown in 12-hour time or 24-hour time.
- Most analogue clocks show 12-hour time, however they do not specify whether it is am or pm.
- A digital clock can show either 12-hour or 24-hour time. The clock might show a dot to indicate that the 12-hour time is in the pm.

Time shown in writing	Time shown on analogue clock	Time shown on digital clock	
		12-hour time	**24-hour time**
Three o'clock in the morning		 3:00 3:00 am	 03:00 3:00 am
Twenty-five minutes past 1 in the afternoon		 1:25 1:25 pm	 13:25 13:25
Ten minutes to 7 in the evening		 6:50 6:50 pm	 18:50 18:50

WORKED EXAMPLE 1

Write the following times in hours, minutes and seconds.

a. **7.3 hours**

b. **$4\frac{3}{8}$ hours**

THINK	WRITE
a. 1. There are 7 hours and 0.3 of an hour in 7.3 hours.	$7.3 = 7$ hours $+ 0.3$ of an hour
2. There are 60 minutes in an hour so there are 0.3 of 60 minutes.	$7.3 = 7$ hours $+ 0.3 \times 60$ minutes $= 7$ hours $+ 18$ minutes
3. Write the answer.	7.3 hours is the same as 7 hours and 18 minutes.
b. 1. There are 4 hours and $\frac{3}{8}$ of an hour in $4\frac{3}{8}$ hours.	$4\frac{3}{8} = 4$ hours $+ \frac{3}{8}$ of an hour

2. There are 60 minutes in an hour so there are $\frac{3}{8}$ of 60 minutes.

$$4\frac{3}{8} = 4 \text{ hours} + \frac{3}{8} \times 60 \text{ minutes}$$
$$= 4 \text{ hours} + 22.5 \text{ minutes}$$

3. There are 60 seconds in a minute so there are 0.5 of 60 seconds.

$$4\frac{3}{8} = 4 \text{ hours} + 22 \text{ minutes} + 0.5 \times 60 \text{ seconds}$$
$$= 4 \text{ hours} + 22 \text{ minutes} + 30 \text{ seconds}$$

4. Write the answer.

$4\frac{3}{8}$ hours is equal to 4 hours, 22 minutes and 30 seconds.

WORKED EXAMPLE 2

A movie runs for 132 minutes. How long is the movie in hours and minutes?

THINK	WRITE
1. There are 60 minutes in one hour, so divide 132 by 60 to get the number of hours.	$132 \div 60 = 2.2$ hours
2. Two hours is 120 minutes ($2 \times 60 = 120$). Subtract 120 from 132 to find how many minutes over two hours the movie is.	$132 - 120 = 12$ minutes
An alternative way of finding the number of minutes over two hours is to convert 0.2 into minutes by multiplying by 60.	$0.2 \times 60 = 12$ minutes
3. Write the answer.	The movie goes for 2 hours and 12 minutes

 Resources

Interactivity: Analogue clock (int-3797)

10.2.2 Time intervals

Often we want to know the time interval or difference between times. This may be to either find out how many hours we worked between 9 am and 1:30 pm, or to know how long our flight between 13.25 and 15.55 will take.

A flight on which an air steward was rostered landed in Sydney at 8:07 pm and left Adelaide at 6:53 pm Sydney time. How long was the flight?

THINK	WRITE
1. Find the number of minutes from the departure to the next hour (7:00 pm).	6:53 pm to 7:00 pm = 7 minutes
2. Find the time from 7:00 pm to 8:00 pm.	7:00 pm to 8:00 pm = 1 hour
3. Find the number of minutes from 8:00 pm to 8:07 pm.	8:00 pm to 8:07 pm = 7 minutes
4. Add the times together.	7 minutes + 1 hour + 7 minutes = 1 hour and 14 minutes.
5. Write the answer.	The flight took 1 hour and 14 minutes

How many days are there between 15 April and 21 June in the same year?

THINK	WRITE
1. You are counting the days between the dates. Count the remaining days in April; there are 30 days in April.	There are 15 days remaining in April.
2. Count the days in May. There are 31 days in May.	There are 31 days in May.
3. Count the days in June up to, but not including, 21 June.	There are 20 days counted in June.
4. Calculate the total number of days.	Total days = 15 + 31 + 20 = 66 days.

Exercise 10.2 Time

1. **WE1** Write the following times in hours, minutes and seconds.

    a. 5.8 hours
    b. $9\frac{7}{8}$ hours

2. Write the following times in hours, minutes and seconds.

    a. 12.156 hours
    b. $3\frac{2}{9}$ hours

3. Write in words the time displayed on each of the following clocks:

a.

b.

c.

d.

e.

f.

g.

h.

4. Write the following times in 24-hour times:
   a. 10:25 am
   b. 7:33 am
   c. 1:45 pm
   d. 8:12 pm
   e. 10:06 pm
   f. 11:45 pm

5. WE2 A horse trail ride went for 156 minutes.

How long was the trail ride in hours and minutes?

6. A game of Rugby is played over two 40 minute halves. If the game also played 2 minutes injury time in the first half and 3 minutes injury time in the second half, how long was the entire Rugby game in hours and minutes?

7. Write the following 24-hour times as digital am or pm times.
   a. 1551
   b. 2022
   c. 0315
   d. 1131
   e. 0902
   f. 2215
8. **MC** Which of the following is **not** used to represent twelve o'clock at night?
   A. 12:00 am
   B. 1200 hours
   C. 0000 hours
   D. Midnight
9. Convert each of the following time periods to minutes.
   a. 3 hours
   b. $5\frac{1}{4}$ hours
   c. 1 day
   d. $\frac{3}{4}$ hour
   e. 3 hours and 18 minutes
   f. 6 hours and 34 minutes
10. Do '2 hours and 25 minutes' and '2.25 hours' represent the same length of time? Explain your answer.
11. Write the following time periods in hours, minutes and seconds.
    a. 210 minutes
    b. 305 minutes
    c. 77 minutes
    d. 2.45 hours
    e. 9.55 hours
    f. 140.45 hours
    g. 8.815 hours
    h. 153.865 minutes
12. **WE3** A flight left Perth at 1:12 pm and landed in Brisbane at 5:27 pm Perth time. How long was the flight?
13. Harper started playing cricket at 0823 and finished at 1320. How long did she play cricket for?
14. Determine the time difference between:
    a. 4:25 pm and 5:50 pm.
    b. 6:30 pm and 2:45 am.
    c. 7:20 am on Monday and 6:30 pm the following day (Tuesday).
    d. 1:20 pm on Wednesday and 9:09 am the following Friday.
    e. 0125 hours and 2345 hours.
    f. 0715 hours and 1550 hours.

15. Calculate the following times.
    a. 1 hour after 12 noon
    b. 3 hours before 7:15 am
    c. 1 hour and 20 minutes after 8:30 am
    d. 2 hours and 30 minutes before 7:45 pm
    e. 4 hours and 14 minutes after 1:08 pm
    f. 3 hours and 52 minutes before 3:25 pm
16. **WE4** How many days are there between 12 July and 27 October in the same year?
17. How many days are there between 9 January and 5 May in 2018?
18. How many days are there until Christmas from the AFL grand final on 28 September?

19. The following table displays the time that a student spent travelling to school. The student left home at 7:25 am.

Activity	Time
Walking from home to the train station	27 minutes
Waiting for the train	12 minutes
Train journey	33 minutes
Walking from the train station to school	8 minutes

    a. How much time did the student spend travelling to school?
    b. At what time did the student arrive at school?
    c. Were they on time for registration at 0845?
20. Use the time units to help you answer each of the following. (*Note:* Assume that there are 365 days in 1 year.) How many:
    a. seconds are in 1 hour?
    b. seconds are in 1 day?
    c. minutes are in 1 day?
    d. hours are in 1 non-leap year?

10.3 Interpret timetables

10.3.1 Timelines

A **timeline** can be used to record a series of events in the order that they occur. A scale must be used when drawing a timeline so that the time elapsed between events can be determined.

The following timeline shows Australian and international events since 1978.

Australian and international events since 1978

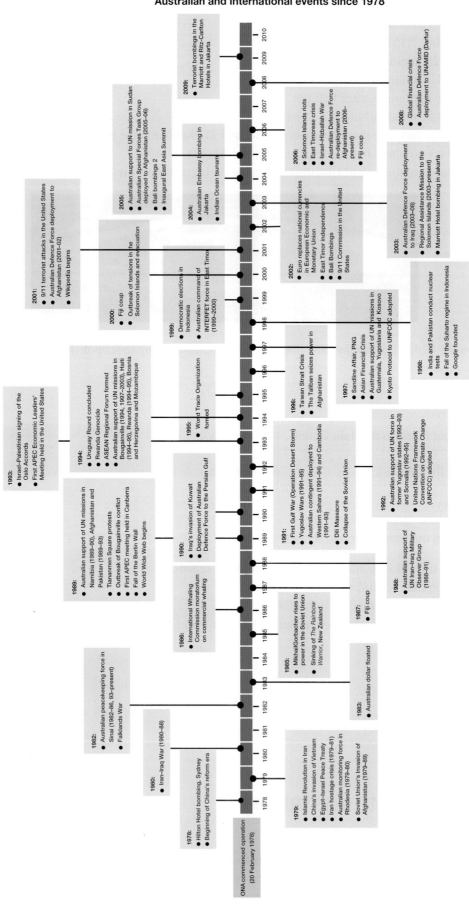

Use the Australian and international events since 1978 timeline to answer the following questions.
a. In what year was the Australian dollar first floated?
b. When did Wikipedia first begin?
c. How long was it between the World Trade Organization being formed and the global financial crisis?

THINK	WRITE
a. Locate the event on the timeline.	The Australian dollar was floated in 1983.
b. Estimate a year and search the timeline around that time.	Wikipedia first began in 2001.
c. 1. Search for the year each of these events occurred.	The World Trade Organization was formed in 1995. The global financial crisis occurred in 2008.
2. Calculate the difference in years and write the answer.	There were 13 years between the World Trade Organization being formed and the global financial crisis.

10.3.2 Timetables

Timelines are useful for plotting events occurring over a particular period of time, however, these can become complex if a number of events are occurring close to each other. This is where timetables are often used.

A **timetable** is a list of events that are scheduled to happen and the timetable usually makes these events easier to read. This is because they allow you to quickly see a lot of information.

Timetables are commonly used for trains and buses.

The timetable below is for a train from Brisbane to Sydney.

	Brisbane to Sydney daily		
	Train A	Train B	Train C
Brisbane Roma Station		05:55	
Kyogle		07:53	
Casino		08:20	19:30
Grafton City	05:15	09:39	20:47
Coffs Harbour	06:26	11:05	22:10
Sawtell	06:35	11:14	22:18
Urunga	06:52	11:31	22:36
Nambucca Heads	07:08	11:47	22:51
Macksville	07:21	12:00	23:05
Eungai	07:37		

	Brisbane to Sydney daily		
	Train A	Train B	Train C
Kempsey	08:05	12:43	23:47
Wauchope	08:44	13:22	00:24
Kendall	09:05	13:54	00:43
Taree	09:52	14:41	01:31
Wingham	10:06	14:55	01:44
Gloucester	10:56	15:44	02:33
Dungog	12:06	16:38	03:26
Maitland	12:53	17:30	04:12
Broadmeadow	13:19	17:52	04:34
Fassifern	13:37	18:12	04:52
Wyong	14:04	18:39	05:23
Gosford	14:19	18:55	05:38
Hornsby	15:03	19:36	06:22
Strathfield	15:28	19:55	06:45
Central (Sydney)	15:43	20:10	07:01

WORKED EXAMPLE 6

Use the Brisbane to Sydney train timetable shown to answer the following questions.

a. If you caught the morning train from Brisbane, what time would you expect to get to Central (Sydney) by?

b. If you wanted to get to Gosford by 6 pm, what time would you need to catch the train from Taree?

THINK	WRITE
a. There is only one morning train from Brisbane at 5:55 am. Follow it down until you get to the end at Central (Sydney) at 20:10.	It arrives at Central (Sydney) at 20:10, which is 8:10 pm.
b. 1. Look down the first column until you find Gosford and see which train gets you there by 6:00 pm.	Train A gets you there at 14:19, which is the same as 2:19 pm.
2. Follow Train A for the previous stations until you find Taree.	The train leaves Taree at 9:52 am.
3. Write the answer.	You need to catch the train at Taree at 9:52 am.

Exercise 10.3 Interpret timetables

1. **WE5** Using the Australian and international events since 1978 timeline on page 384 answer the following:
   a. When did the Solomon islands riots occur?
   b. When was Google founded?
2. Using the Australian and international events since 1978 timeline on page 384 answer the following:
   a. How many years apart were China's invasion of Vietnam and the fall of the Berlin Wall?
   b. How many years separated the start of the World Wide Web and the Bali bombings?

   Questions **3** to **5** refer to the following space travel timeline.

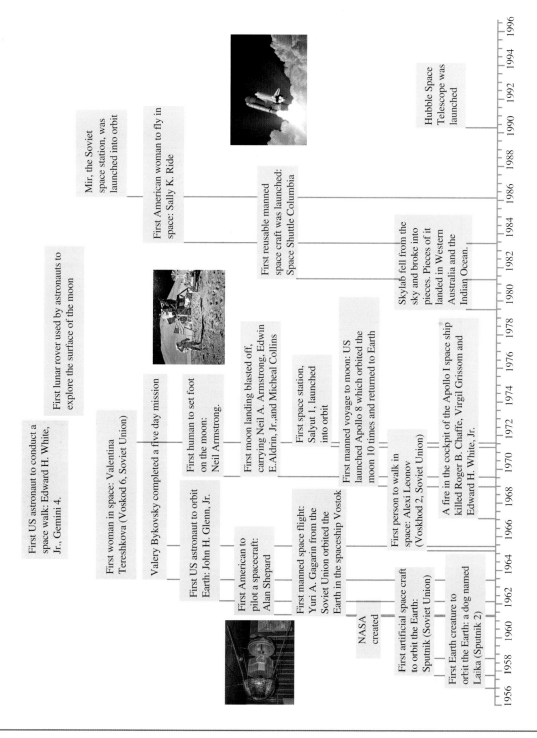

3. a. In what year did Neil Armstrong first step foot on the moon?
   b. In what year was the first woman in space?
4. a. In what year was NASA created?
   b. How many years were between NASA being created and Neil Armstrong setting foot on the moon?
5. a. How many years were between the first woman to fly into space and the first American woman to fly into space?
   b. How many years were between the Mir Soviet space station launch and the Hubble Telescope launch?
6. **WE6** Use the Brisbane to Sydney timetable on pages 385 and 386 to answer the following questions.
   a. If you caught the earliest train from Coffs Harbour, what time would you get into Gosford?
   b. If you wanted to get to Gloucester by 2 pm, what time would you need to catch the train from Kempsey?
7. Use the Brisbane to Sydney timetable on pages 385 and 386 to answer the following questions.
   a. If you caught the evening train from Coff's Harbour, what time would you get into Sydney?
   b. If you caught the evening train from Maitland, how long would it take to travel to Sydney?

Questions 8 and 9 refer to the City Hopper timetable shown (only part of the daily timetable is shown).

Towards North Quay   Towards Sydney Street

Stop	Destination				
	North Quay 2	North Quay 2	North Quay 2	North Quay 2	North Quay 2
Sydney Street ferry terminal, New Farm	8:00 am	8:30 am	9:00 am	9:30 am	10:00 am
Dockside ferry terminal, Kangaroo Point	8:04 am	8:34 am	9:04 am	9:34 am	10:04 am
Holman Street ferry terminal, Kangaroo Point	8:15 am	8:45 am	9:15 am	9:45 am	10:15 am
Eagle Street Pier ferry terminal, Brisbane City	8:20 am	8:50 am	9:20 am	9:50 am	10:20 am
Thornton Street ferry terminal, Kangaroo Point	8:25 am	8:55 am	9:25 am	9:55 am	10:25 am
Maritime Museum ferry terminal, South Brisbane	8:33 am	9:03 am	9:33 am	10:03 am	10:33 am
South Bank 3 ferry terminal, South Brisbane	8:37 am	9:07 am	9:37 am	10:07 am	10:37 am
North Quay 2 ferry terminal, Brisbane City	8:42 am	9:12 am	9:42 am	10:12 am	10:42 am

8. a. If you left Sydney Street at 8:30 am, what time would you arrive at South Bank 3?
   b. If you wanted to arrive at the Maritime Museum at 10:15 am, what time would the ferry depart Holman Street?

9. How long does the ferry take to travel from Holman Street to North Quay 2?

Questions 10 and 11 refer to the City to Doomben train timetable.

☞ Queensland Rail

City to Doomben outbound

Monday to Thursday

Comes from / Station	am	am	am	am	VYS am	VYS am	am	MNY am	am	am	am	am	am	pm	pm	pm	pm	pm
Park Road					7:09	7:39		8:37				10:39						
South Bank					7:12	7:42		8:40				10:42						
South Brisbane					7:14	7:44	8:12	8:42	9:14	9:44	10:16	10:44	11:16	11:46	12:16	12:46	1:16	1:46
Roma Street	5:19	5:49	6:19	6:49	7:19	7:49	8:17	8:47	9:19	9:49	10:19	10:49	11:19	11:49	12:19	12:49	1:19	1:49
Central arrive	5:21	5:51	6:21	6:51	7:21	7:51	8:19	8:49	9:21	9:51	10:21	10:51	11:21	11:51	12:21	12:51	1:21	1:51
Central depart	5:23	5:53	6:23	6:53	7:23	7:53	8:23	8:53	9:23	9:53	10:23	10:53	11:23	11:53	12:23	12:53	1:23	1:53
Fortitude Valley	5:25	5:55	6:25	6:55	7:25	7:55	8:25	8:55	9:25	9:55	10:25	10:55	11:25	11:55	12:25	12:55	1:25	1:55
Bowen Hills	5:28	5:58	6:28	6:58	7:28	7:58	8:28	8:58	9:28	9:58	10:28	10:58	11:28	11:58	12:28	12:58	1:28	1:58
Albion	5:32	6:02	6:32	7:02	7:32	8:02	8:32	9:02	9:32	10:02	10:32	11:02	11:32	12:02	12:32	1:02	1:32	2:02
Wooloowin	5:34	6:04	6:34	7:04	7:34	8:04	8:34	9:04	9:34	10:04	10:34	11:04	11:34	12:04	12:34	1:04	1:34	2:04
Eagle Junction	5:37	6:07	6:37	7:07	7:37	8:07	8:37	9:07	9:37	10:07	10:37	11:07	11:37	12:07	12:37	1:07	1:37	2:07
Clayfield	5:40	6:10	6:40	7:10	7:40	8:10	8:40	9:10	9:40	10:10	10:40	11:10	11:40	12:10	12:40	1:10	1:40	2:10
Hendra	5:42	6:12	6:42	7:12	7:42	8:12	8:42	9:12	9:42	10:12	10:42	11:12	11:42	12:12	12:42	1:12	1:42	2:12
Ascot	5:44	6:14	6:44	7:14	7:44	8:14	8:44	9:14	9:44	10:14	10:44	11:14	11:44	12:14	12:44	1:14	1:44	2:14
Doomben	5:46	6:16	6:46	7:16	7:46	8:16	8:46	9:16	9:46	10:16	10:46	11:16	11:46	12:16	12:46	1:16	1:46	2:16

Monday to Thursday (continued)

Comes from / Station	pm	pm	SFC pm	IPS pm	pm	pm	pm	pm	pm	pm
Park Road			3:39					6:09		7:09
South Bank			3:42					6:12		7:12
South Brisbane			3:44					6:14		7:14
Roma Street	2:49	3:19	3:49	4:19	4:49	5:19	5:49	6:19	6:49	7:19
Central arrive	2:51	3:21	3:51	4:21	4:51	5:21	5:51	6:21	6:51	7:21
Central depart	2:53	3:23	3:53	4:23	4:53	5:23	5:53	6:23	6:53	7:23
Fortitude Valley	2:55	3:25	3:55	4:25	4:55	5:25	5:55	6:25	6:55	7:25
Bowen Hills	2:58	3:28	3:58	4:28	4:58	5:28	5:58	6:28	6:58	7:28
Albion	3:02	3:32	4:02	4:32	5:02	5:32	6:02	6:32	7:02	7:32
Wooloowin	3:04	3:34	4:04	4:34	5:04	5:34	6:04	6:34	7:04	7:34
Eagle Junction	3:07	3:37	4:07	4:37	5:07	5:37	6:07	6:37	7:07	7:37
Clayfield	3:10	3:40	4:10	4:40	5:10	5:40	6:10	6:40	7:10	7:40
Hendra	3:12	3:42	4:12	4:42	5:12	5:42	6:12	6:42	7:12	7:42
Ascot	3:14	3:44	4:14	4:44	5:14	5:44	6:14	6:44	7:14	7:44
Doomben	3:16	3:46	4:16	4:46	5:16	5:46	6:16	6:46	7:16	7:46

10. **a.** Is there a train leaving Roma Street station at 7:49 pm?

   b. If you needed to get to Doomben at 9:30 am, what time would you need to catch the train from Fortitude Valley?

11. **a.** How many trains leave from Park Road station between 9:00 am and 10:00 am?

   b. If I needed to be at a friend's home, which is a 15 minute walk from Eagle Junction station, for dinner at 6:00 pm, what is the latest train I could catch from South Brisbane?

   c. If you caught the 12:53 am train from Central, how long would it take to travel to Doomben station?

12. Fix the table shown by matching the historical events (in the right column) in Australia's history with the year it occurred (the left column) and link them to the timeline. You might need to research some of the answers.

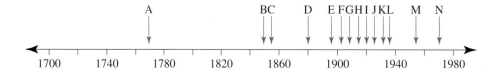

1902	Referendum to include Indigenous Australians in the census is successful
1770	First university in Australia (Sydney)
1880	Eureka Stockade
1915	Captain James Cook lands at Botany Bay
1967	Australian women gain the right to vote and stand for Parliament
1850	Ned Kelly hanged
1956	Banjo Patterson publishes *The man from Snowy River*
1930	Australians land at ANZAC Cove
1890	Australia becomes a federation
1932	Sydney Harbour Bridge opens
1923	Vegemite produced
1854	Phar Lap wins Melbourne Cup
1901	QANTAS is founded
1920	Melbourne Olympics/TV in Australia

Questions **13** and **14** refer to the timetable on moon phases in 2015.

Moon phases 2015

Please note: All times are Australian Eastern Standard Time. Add one hour for daylight savings (first Sunday in October to first Sunday in April).

New moon		First quarter		Full moon		Last quarter	
Date	**Time**	**Date**	**Time**	**Date**	**Time**	**Date**	**Time**
				Mon Jan 5	14:53	Tues Jan 13	19:47
Tues Jan 20	23:14	Tues Jan 27	14:48	Wed Feb 4	09:09	Thur Feb 12	13:50
Thur Feb 19	09:47	Thur Feb 26	03:14	Fri Mar 6	04:06	Sat Mar 14	03:48
Fri Mar 20	19:36	Fri Mar 27	17:43	Sat Apr 4	22:06	Sun Apr 12	13:44
Sun Apr 19	04:57	Sun Apr 26	09:55	Mon May 4	13:42	Mon May 11	20:36
Mon May 18	14:13	Tues May 26	03:19	Wed Jun 3	02:19	Wed Jun 10	01:42
Wed Jun 17	00:05	Wed Jun 24	21:03	Thur Jul 2	12:20	Thur Jul 9	06:24
Thur Jul 16	11:24	Fri Jul 24	14:04	Fri Jul 31	20:43	Fri Aug 7	12:03
Sat Aug 15	00:54	Sun Aug 23	05:31	Sun Aug 30	04:35	Sat Sep 5	19:54
Sun Sep 13	16:41	Mon Sep 21	18:59	Mon Sep 28	12:50	Mon Oct 5	07:06
Tues Oct 13	10:06	Wed Oct 21	08:31	Tues Oct 27	22:05	Tues Nov 3	22:24
Thur Nov 12	03:47	Thur Nov 19	18:27	Thur Nov 26	08:44	Thur Dec 3	17:40
Fri Dec 11	20:29	Sat Dec 19	01:14	Fri Dec 25	21:11		

13. a. When was the first new moon in 2015 (date and time)?
   b. How many full moons were there in 2015?
   c. Were there any months in 2015 that had more than one full moon? If so, which month?
14. a. How many days were there between the new moon and the full moon in September 2015?
   b. How many hours and minutes were there between the new moon and its first quarter in February?
   c. How many hours and minutes were there between the new moon in October and its last quarter?

Tide tables tell you when tides will be high or low at a particular location. An example of this type of table is shown below. Questions **15** and **16** refer to the tide predictions for Western Port (Stony Point) in September.

Mon. 1		Tues. 2		Wed. 3		Thurs. 4		Fri. 5		Sat. 6		Sun. 7	
Time	*Ht*	*Time*	*Ht*	*Time*	*Ht*	*Time*	*Ht*	*Time*	*Ht*	*Time*	*Ht*	*Time*	*Ht*
0110	2.49	0217	2.64	0314	2.75	0400	2.81	0443	2.82	0519	2.79	0553	2.74
0702	0.57	0801	0.56	0853	0.57	0937	0.61	1017	0.68	1054	0.75	1127	0.82
1359	2.71	1447	2.77	1529	2.77	1604	2.74	1634	2.70	1702	2.65	1730	2.60
1944	0.71	2032	0.52	2115	0.39	2154	0.32	2230	0.29	2302	0.30	2335	0.34

The tide heights (Ht) are measured in metres — **red** indicates low tide and **blue** indicates high tide. Times stated are Australian Eastern Standard Time (24-hour clock). During daylight-saving time (when in force) 1 hour needs to be added to the times stated.

15. **a.** What are the differences between the blue and the red numbers in the table?
    b. How many low tides are there each day?
    c. What is the lowest low tide in the table?
    d. When does the lowest low tide occur?

16. **a.** How much time is there between high tides on Saturday 6 September?
    b. Is the time difference between high tides the same each day?
    c. List three examples of how a tide table could be useful.

10.4 Time travel

10.4.1 Time zones

The world is divided into 24 different **time zones.**

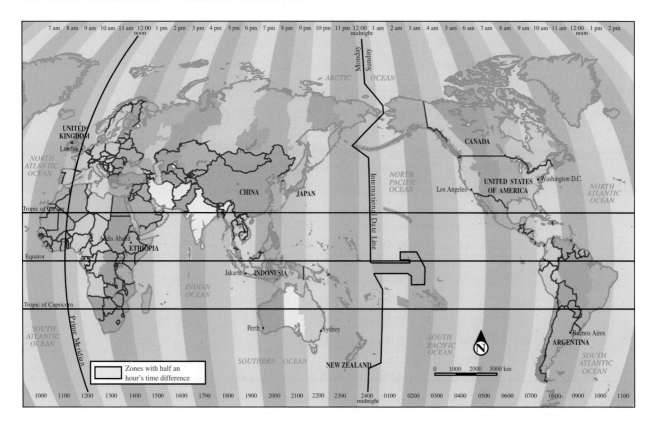

If you move from one time zone to another in an easterly direction, you need to move your clock forward. If you move from one time zone to another in a westerly direction, you need to move your clock back.

The International Date Line is the line on the map of Earth where a calendar day officially starts and ends. If you cross the International Date Line in an easterly direction, your calendar goes back a day. If you cross the International Date Line in a westerly direction, your calendar goes forward one day.

The time in England that used to be called Greenwich Mean Time (GMT) is now known as Coordinated Universal Time (UTC). It is referred to as Zulu time in the military.

UTC is the reference point for every time zone in the world. Time zones are either ahead of or behind UTC. For example, the time difference between Melbourne and Los Angeles is 18 hours, not taking into account Pacific Daylight Time (PDT). This is because Melbourne is 10 hours ahead of UTC (UTC + 10) and Los Angeles is 8 hours behind UTC (UTC − 8).

What time is it in Melbourne if the time in London is 9:00 am UTC?

THINK	WRITE
1. Melbourne is 10 hours ahead of London. Add 10 hours to the London time.	The time in London is 9:00 am. The time in Melbourne is 9:00 am + 10 hours.
2. Calculate the time in Melbourne.	The time in Melbourne is 7:00 pm.

on Resources

Interactivity: Time zones (int-6911)

Interactivity: Time zones (int-0008)

Interactivity: World time zones (int-3798)

10.4.2 Australian time zones

Australia is divided into three different standard time zones: Western Standard Time (WST), Central Standard Time (CST) and Eastern Standard Time (EST).

Western Australia is 2 hours behind Eastern Standard Time.

The Northern Territory and South Australia (CST) are half an hour behind Eastern Standard Time.

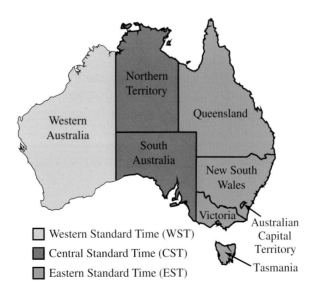

☐ Western Standard Time (WST)

☐ Central Standard Time (CST)

☐ Eastern Standard Time (EST)

If the time in Sydney is 9:30 pm EST, what time is it in Perth?

THINK	WRITE
1. Perth is 2 hours behind Sydney. Subtract 2 hours from Sydney time.	The time in Sydney is 9:30 pm. Therefore the time in Perth is 9:30 pm – 2 hours.
2. Calculate the time in Perth.	It is 7:30 pm in Perth.

10.4.3 Daylight-saving time

Many countries around the world have daylight-saving time during summer to make the most of the warmer weather.

Daylight-saving time starts on the first Sunday in October, when clocks go forward 1 hour at 2:00 am. Daylight-saving time finishes on the first Sunday in April, when clocks are turned back 1 hour at 3:00 am. Queensland, Western Australia and the Northern Territory do not observe daylight-saving time.

Daylight-saving time was first used in Australia in 1917 during World War I to cut down on fuel required for artificial lighting at night. It was re-introduced during World War II and by 1971, daylight saving was adopted by most eastern states and has been in place ever since.

When daylight-saving time is in operation:
- Tasmania, Victoria and New South Wales have the same time. The time in these states is called Eastern Daylight-Saving Time (EDT).
- Queensland is 1 hour behind Melbourne time (EDT).
- the Northern Territory is one-and-a-half hours behind Melbourne time (EDT).
- South Australia is half an hour behind Melbourne time (EDT).
- Western Australia is 2 hours behind Melbourne time (EDT).

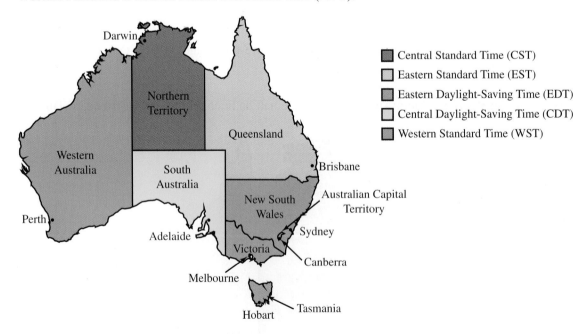

WORKED EXAMPLE 9

You are visiting Melbourne and want to call your friend in Darwin to wish her a happy New Year. If it is 8:00 pm in Melbourne, what time is it in Darwin?

THINK	WRITE
1. There is normally half an hour between the time in Darwin and the time in Melbourne.	8:00 pm – 30 minutes = 7:30 pm in Darwin.
2. On New Year's Day, Melbourne observes daylight-saving time because it is summer. This means Melbourne is another hour ahead.	7:30 pm – 1 hour = 6:30 pm
3. Write the answer.	It would be 6:30 pm in Darwin.

 Resources

Interactivity: Daylight saving (int-3799)

10.4.4 Comparing travel time

Australians are known for their long holiday drives, whether it is to go camping, take the caravan or just explore the magnificent country we live in. However, many means of transport are available to travel from one destination to another, depending on the distance of the journey. You can travel by bike, car, train, bus and plane, and each of these modes of transport will vary the time it will take to reach your destination.

> **To compare travel times use:**
>
> $$\text{Speed} = \frac{\text{Distance}}{\text{Time}}$$
>
> $$\therefore \text{Time} = \frac{\text{Distance}}{\text{Speed}}$$

WORKED EXAMPLE 10

If a plane trip from Brisbane to Sydney takes 1 hour, how much longer would it take to drive the 921 km at an average speed of 100 km/h?

THINK	WRITE
1. To calculate the time it takes to travel by car you need the distance, which is 921 km and the average speed, which is 100 km/h. Make sure the units of distance match.	Distance = 921 km Average speed = 100 km/h
2. Substitute the values into the formula for time: $\text{Time} = \dfrac{\text{Distance}}{\text{Speed}}$	$\begin{aligned} \text{Time} &= \dfrac{\text{Distance}}{\text{Speed}} \\ &= \dfrac{921}{100} \\ &= 9.21 \text{ hours} \\ &= 9 \text{ hours and } 13 \text{ minutes} \end{aligned}$

3. Calculate the difference in time by subtracting the time taken to fly.

4. Write the answer.

Time difference = 9 hr 13 min − 1 hr
= 8 hr 13 min

The time difference for driving compared to flying to Sydney from Brisbane is 8 hours and 13 minutes

Exercise 10.4 Time travel

1. **WE7** What time is it in Melbourne if the time in London is 11:45 am UTC?
2. What time is it in Los Angeles if the time in Sydney is 3:35 pm UTC?
3. a. How does the local time change as you move around the world in a westerly direction?
   b. How does the local time change as you move around the world in an easterly direction?
4. Complete the following table to show the times in different cities in June [Note: Auckland (New Zealand) is 2 hours ahead of EST].

Time in Perth	Time in Adelaide	Time in Melbourne	Time in Auckland
4:00 pm		6:00 pm	
			9:30 pm
	9:30 am		
10:45 pm			

5. What is the time in Brisbane (EST) if the time in London (UTC) is:
   a. 11:33 am?   b. noon?   c. 2:55 pm?   d. 7:14 am?
6. What is the time in Los Angeles (ignoring PDT) if the time Brisbane (UTC) is:
   a. 2:30 pm?   b. noon?   c. 6:05 am?   d. midnight?
7. **WE8** If the time in Sydney is 11:15 pm EST, what time is it in Perth?
8. If the time is 12:15 pm EST in Melbourne, what time is it in Adelaide?
9. Change 9:00 am Eastern Standard Time (EST) to Eastern Daylight-Saving Time (EDT) in Sydney.
10. **WE9** If you want to Skype with a friend at 7:30 pm on Christmas Day in Perth, what time should you tell them to be on their computer in Melbourne?
11. During summer, what time is it in Brisbane if it is:
    a. noon in the Australian Capital Territory?   b. 10:00 am in Sydney?
    c. 7:45 am in Hobart?   d. 5:15 pm in Darwin?
    e. 8:30 pm in Perth?   f. 3:25 pm in Adelaide?
12. **WE10** If a plane trip from Melbourne to Sydney takes 1 hour, how much longer would it take to drive the 877 km at an average speed of 85 km/h?
13. Robert and Anthony drive from Parkhurst to Brisbane which is 644 km. Anthony averages 92 km/h, while Robert has to stop to meet some friends on the way so he averages a speed of 84 km/h for the trip.
    a. How long did Anthony take to get to Brisbane?
    b. How long did Robert take to get to Brisbane?
    c. If they both left at 6:00 am, what time did each of them arrive at Brisbane?
14. Consider workers who live in northern New South Wales and drive for 1 hour to the Gold Coast in Queensland each day for work during daylight-saving time.
    a. What time (EST) do they need to leave home to be at work by 8:30 am?
    b. If they leave work at 5:00 pm, what time will it be when they arrive home?

15. In an Australian reality television show, a contestant from Perth was voted off the show as a result of a 20-minute voting period during the 1-hour show. The show was telecast live in the eastern states but delayed to the corresponding time in all other states. Use your knowledge of time zones to explain why her family may have been upset.

16. Your friend in New York wishes to call you in Toowoomba. If you wish to receive the telephone call at 6:00 pm EST, at what time in New York (ignoring PDT) should your friend make the call?

17. It takes around 10 times longer to drive from Brisbane to Proserpine compared to flying. Lucas took the plane and arrived at 2 pm after a 1 hour flight. What time would Zoe arrive in Proserpine if she left at the same time as Lucas but she travelled by car?

18. To have the time zones work in Santa's favour:
   a. what country should he deliver his presents to first?
   b. what direction should he travel from his first country?

19. A plane flew from Brisbane to London. Some of the details of the trip are listed.
   • Departure date: Friday 5 August
   • Departure time: 8:00 am
   • Travel time: 25 hours

   a. What was the time and date in Brisbane when the plane arrived in London?
   b. What was the local time (UTC) when the plane arrived?

20. If a plane trip from Perth to Darwin takes 4 hours:
   a. how long would it take to drive the 4000 km at an average speed of 100 km/h?
   b. how much longer would it take to drive the 4000 km at an average speed of 100 km/h compared to flying?
   c. It is too far to drive without a break so the travellers decided to have a 30 minute break after every 4 hours of driving. If they left at 0600 on 26 December, when did they arrive in Darwin?

10.5 Distance: scales and maps

10.5.1 Map distances

Maps are similar to the land they represent; they have exactly the same shape but very different sizes.

Maps are drawn using a scale, which is displayed on the map. The **scale** describes the ratio between the distance on the map and the actual distance. Maps also have an icon depicting north.

A scale may be written as:

• a ratio of distance on a map to actual distance, such as 1:100 000.
• a statement, such as '1 cm represents 1 km'.
• a scale bar.

The scale on the map can be used to calculate the size of objects and the distance between objects on the map.

The scale on the map of King Island shown is 1:1 000 000.

This means that 1 unit on the map represents 1 000 000 units in real life, so 1 mm on the map represents 1 000 000 mm in real life and 1 m on the map represents 1 000 000 m (or 1000 km) in real life.

Walking from Currie to Grassy would be walking in a south easterly direction.

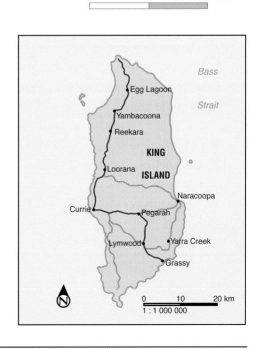

Use the numerical scale to determine the distance between Melbourne and Adelaide.
What direction would you be travelling when going from Melbourne to Adelaide?

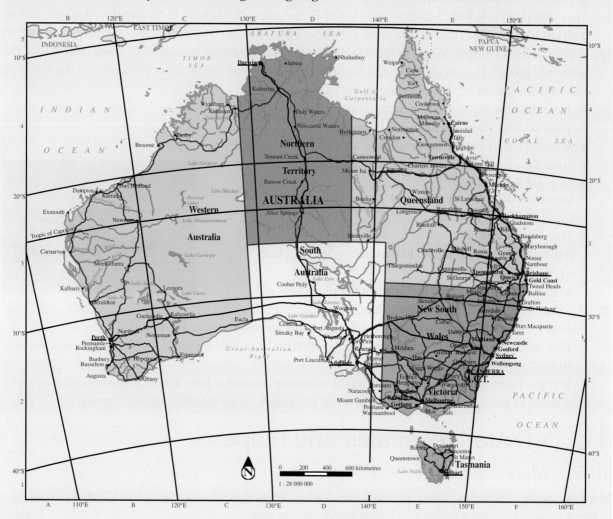

THINK

1. Use a ruler to measure the linear distance in millimetres between Melbourne and Adelaide.

2. Rewrite the numerical scale as a ratio using appropriate units.

3. Each millimetre represents 28 kilometres, so 23 millimetres represents 23 lots of 28 kilometres.

4. Write the answer.

WRITE

23 mm

1 mm : 28 000 000 mm
1 mm : 28 000 m
1 mm : 28 km

$23 \times 28 = 644$

The distance between Melbourne and Adelaide is 644 km.

Travelling from Melbourne to Adelaide would be in a north westerly direction.

Exercise 10.5 Distance: scales and maps

1. **WE11** Using the previous map of Australia with its numerical scale, determine the distance between Adelaide and Brisbane. What direction would you be travelling in from Adelaide to Brisbane?

2. Using the previous map of Australia with its numerical scale, determine the distance between Perth and Alice Springs. What direction would you be travelling in from Perth to Alice Springs?

3. What is the actual distance represented by 1 cm on the map for each of the following scale ratios? Write your answer in the unit specified in the brackets.
   a. 1:1000 (m)
   b. 1:50 000 (m)
   c. 1:250 000 (km)
   d. 1:7 000 000 (km)

4. Write a scale ratio for a map in which 1 cm on the map represents an actual distance of:
   a. 10 cm.
   b. 800 m.
   c. 6.5 km.
   d. 4000 km.

5. A scale map of Wilsons Promontory National Park is shown.

   a. What actual distance, in kilometres, is represented by 1 cm on the map?
   b. Find the actual distance in a straight line from:
      i. Tidal River to Sealers Cove.
      ii. Millers Landing to Tin Mine Cove.
      iii. Mount Oberon to the lighthouse.
   c. Estimate the length of the peninsula from its northernmost point near Tin Mine Cove to its southernmost point near the lighthouse.
   d. Mt Vereker is located in the middle of the park, 8.5 km from Tidal River. What distance should Mt Vereker appear from Tidal River on the map?

6. The map of Australia shown is drawn to a scale of 1 cm to 500 km.

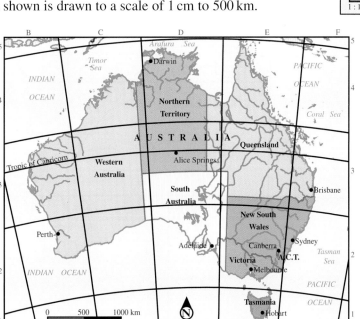

   a. How many kilometres are represented by 5 cm on this map?
   b. Use the map to estimate the straight line distance between:
      i. Melbourne and Perth.
      ii. Sydney and Canberra.
      iii. Darwin and Alice Springs.
      iv. Brisbane and Sydney.
      v. Melbourne and Adelaide.
      vi. Adelaide and Hobart.

7. This map of Tasmania is scaled at 1 : 3 000 000.

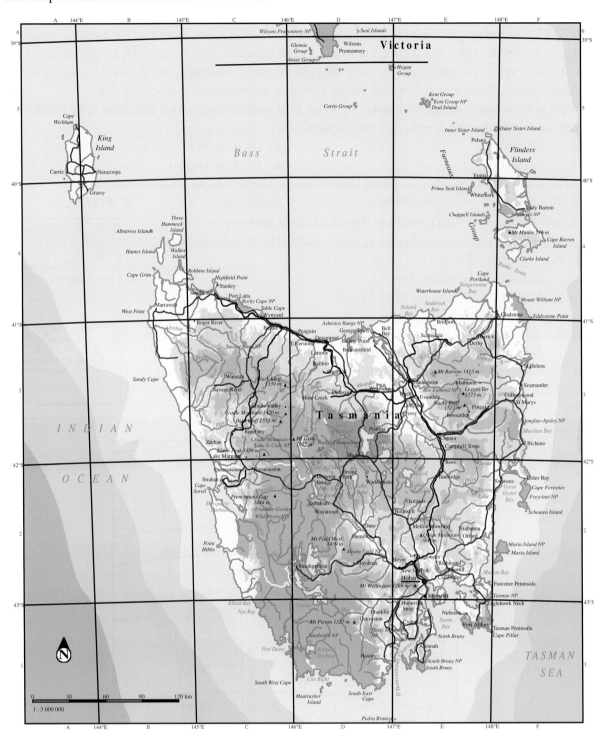

a. Write the scale as a ratio in the most appropriate units.
b. How many kilometres long and wide is King Island?
c. How many kilometres of ocean lie between Wilsons Promontory in Victoria and Devonport on Tasmania's central north coast?
d. How far is Hobart from Launceston?

8. Consider the map of north-eastern Victoria.

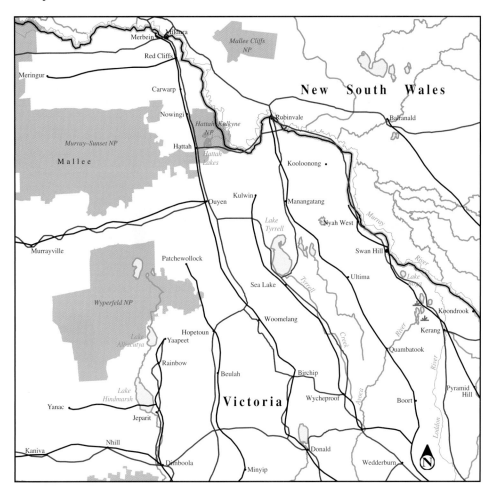

a. Measure the straight-line map distance between:
   i. Kaniva and Swan Hill.
   ii. Wedderburn and Meringur.
   iii. Mildura and Donald.
b. Given that the actual distances between the locations in part **a** are 236 km, 310 km and 255 km respectively, calculate the scale ratio for the map.
c. Do you get the same ratio for each of your calculations in part **b**? Discuss any variation.

9. Orienteering is a physically active and enjoyable pastime that requires map and compass reading skills and an ability to estimate distances. The aim of orienteering is to complete a given course that has been planned and mapped. There are many checkpoints in an orienteering course with competitors visiting each.

The following exercise is an example of a simple orienteering course conducted in a park. The directions that are listed below could be either:

• given to each participant prior to commencing the course, or
• left at each checkpoint.

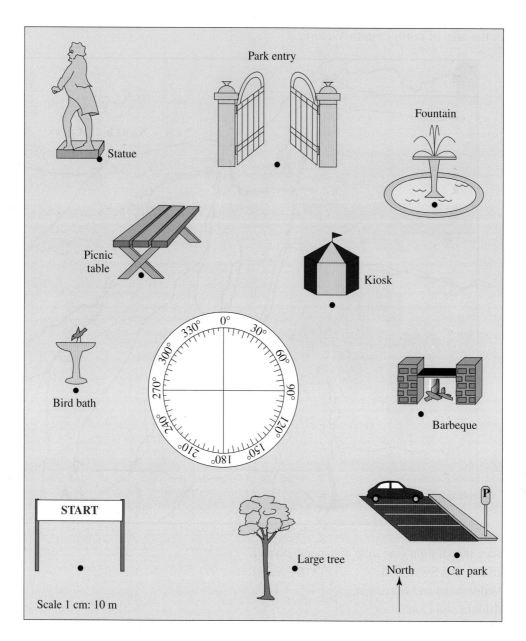

Use the compass rose or a protractor and your ruler to follow this course in the figure shown above.

- Find the point in the park labelled 'Start'.
- Proceed in an easterly direction for 60 m.
- Follow N20°E for 103 m.
- Now move 94 m on N83°W.
- Proceed along S6°W for 62 m.
- Now follow S86°E for 95 m.
  Where are you now?

10.6 Speed

10.6.1 Calculating speed

Average speed tells us how fast we are going by dividing the distance we travel by the time it takes to travel that distance.

$$\text{Speed} = \frac{\text{Distance}}{\text{Time}}$$

The most common unit for speed is metres per second (m/s). However speeds of objects will vary, for example the speed of a snail compared to a car will be very different and therefore different units may be more appropriate. The speed of a snail could be measured in millimetres per second (mm/s), whereas the speed of a car could be measured in kilometres per hour (km/h).

WORKED EXAMPLE 12

Calculate the average speed of a dragster that covers 400 m in 4.25 seconds, in:

a. m/s

b. km/h

THINK	WRITE
a. 1. Determine the distance and time in the correct units.	Distance $= 400\,\text{m}$ Time $= 4.25\,\text{s}$
2. Substitute values into the formula $\text{Speed} = \dfrac{\text{Distance}}{\text{Time}}$	$\text{Speed} = \dfrac{\text{Distance}}{\text{Time}}$ $\text{Speed} = \dfrac{400}{4.25}$ $\text{Speed} = 94.12\,\text{m/s}$
b. 1. Determine the distance and time in the correct units.	Distance $= 400\,\text{m} = 0.4\,\text{km}$ Time $= 4.25\,\text{s} = \dfrac{4.25}{60 \times 60}$ $= 1.18 \times 10^{-3}\,\text{hours}$
2. Substitute values into the formula $\text{Speed} = \dfrac{\text{Distance}}{\text{Time}}$	$\text{Speed} = \dfrac{\text{Distance}}{\text{Time}}$ $\text{Speed} = \dfrac{0.4}{1.18 \times 10^{-3}}$ $\text{Speed} = 338.82\,\text{km/h}$

10.6.2 Calculating distance

The speed equation can be rearranged to make distance the subject, allowing you to calculate the distance covered given the speed and time of a journey.

$$\textbf{Distance} = \textbf{Speed} \times \textbf{Time}$$

WORKED EXAMPLE 13

A car travels at a constant speed of 60 km/h for 1.5 hours. How far did the car travel in kilometres?

THINK	WRITE
1. To calculate the distance we need to know the speed and time in the correct units.	Speed = 60 km/h Time = 1.5 hours

2. Substitute into the distance equation.	Distance = Speed × Time
	   = 60 × 1.5
	   = 90 km

3. Understand the question.	Since the car travels at 60 km/h it covers 60 km in 1 hour, and therefore 30 km in half an hour, thus 90 km in 1.5 hours.
4. Write the answer.	The distance covered is 90 km.

10.6.3 Calculating time

The speed equation can also be rearranged to make time the subject, allowing you to calculate the time taken to cover a given distance at a certain speed.

$$\text{Time} = \frac{\text{Distance}}{\text{Speed}}$$

WORKED EXAMPLE 14

How long in hours would it take a person to walk 3 km at an average speed of 1.8 m/s?

THINK	WRITE
1. To calculate the time we need to know the distance and speed in appropriate units. We will use m/s and m and convert to hours at the end.	Distance = 3 km = 3000 m Speed = 1.8 m/s
2. Substitute into the equation $\text{Time} = \dfrac{\text{Distance}}{\text{Speed}}$.	$\text{Time} = \dfrac{\text{Distance}}{\text{Speed}}$ $= \dfrac{3000}{1.8}$ $= 1666.67 \text{ seconds}$
3. Convert seconds to hours by dividing by 60 (to convert to minutes) and then divided by 60 again (to convert to hours).	$\dfrac{1666.67}{60 \times 60} = 0.46 \text{ hour} \approx 28 \text{ minutes}$
4. Write the answer.	The time taken would be 0.46 hours.

Exercise 10.6 Speed

1. **WE12** Calculate the average speed of Sally who walks around her 2.5 km block in 1050 seconds in:
   a. m/s.
   b. km/h.
2. Calculate the average speed of a track cyclist who covered 1000 m in 1 minute and 25 seconds in km/h.

3. Select the appropriate unit for speed (mm/s, m/s or km/h) to measure the following:
   a. a sprinter running a 100 m race.
   b. a plane flying.
   c. the increasing snow depth.
   d. a surfer riding a wave.
4. Calculate the average speed (in m/s) of a skateboarder who travelled 50 m in 7 seconds.

5. Calculate the average speed (in km/h) of a boat that covered 2.6 km in 15 minutes.

6. **WE13** A car travels at a constant speed of 80 km/h for 3.25 hours. How far did the car travel in kilometres?
7. A truck travels at an average speed of 65 km/h for 335 minutes. How far did the truck travel in:

   a. kilometres?
   b. metres?

8. A kite surfer was moving at a constant speed of 11 m/s for 24 seconds. What distance (in metres) did she cover over this time?

9. Callum played a round of golf, averaging 5 shots per hole and a speed of 0.3 m/s. If it took him 4 hours to play his round of golf, how far did he walk during his round?

10. **WE14** How long in hours would it take a person to walk 10 km at an average speed of 1.6 m/s?

11. If Ahmed ran at an average speed of 5 m/s for a half marathon (21.1 km), how long would it take him to complete the half marathon in hours and to the nearest minute?

12. Haile Gebrselassie ran the Berlin marathon with an average speed of 5.67 m/s. How long did it take him to finish the marathon, given a marathon is 42.2 km, in:

   a. seconds?
   b. hours, minutes and seconds (to the nearest second)?

13. A cheetah can reach a maximum speed of 112 km/h. However they are only able to maintain this speed for a short period of time. If a cheetah ran at its maximum speed and covered a distance of 400 m, how long did it maintain its maximum speed for, in seconds?

14. Pigeons can cover vast distances to find their way home. If a pigeon covered 56.8 km in 1 hour and 45 minutes, find:

   a. its average speed in km/h.
   b. its average speed in m/s.

15. If a horse averages a speed of 25.6 km/h for 17 minutes and 36 seconds, find how far the horse travelled to two decimal places, in:
    a. kilometres.
    b. metres.

16. For Makybe Diva to win her third Melbourne Cup she averaged a speed of 57.9 km/h for the 3200 m race. What was her winning time in:

    a. minutes to two decimal places?
    b. minutes and seconds (to the nearest second)?

17. The speed of light is 3×10^8 m/s. Research the average distance from Earth to the following planets to calculate the time it would take light to get there from Earth. Note: use the average distance since it does vary.
    a. Saturn   b. Mercury   c. Jupiter   d. Mars

18. The following questions relate to reaction time of a driver before breaking.
    a. If a driver was travelling at 60 km/h and had a reaction time of 1.3 seconds before he applied the brakes, how far did he travel before he applied the brakes?
    b. If a driver had a reaction time of 1.5 seconds and travelled 30 m before applying the brakes, what speed was he travelling at?
    c. If the car travelled 42 m before the brakes were applied, what was the reaction time of the driver if they were travelling at 95 km/h?

10.7 Distance–time graphs

10.7.1 Time of journeys

We have learned how to estimate distances travelled via a map, however, if you are travelling somewhere you are often more interested in how long it will take to complete your journey. This is what a global positioning system (GPS) can do by estimating the distance of the trip, calculating the average speed of the trip depending on the roads travelled, and using this information to estimate the time the trip will take. Remember:

$$\text{Time} = \frac{\text{Distance}}{\text{Speed}}$$

To determine the time a journey takes:
- estimate the distance of the journey from the map
- estimate the average speed of the journey
- use these estimates to calculate time of the journey, using the formula $\text{Time} = \dfrac{\text{Distance}}{\text{Speed}}$.

Shane and Ravi went for a swim at Bondi Beach before planning to go to the cricket at the SCG. How long would they need to allow to get to the SCG, given the map of their journey shown and their estimate that they can average 50 km/h in the traffic?

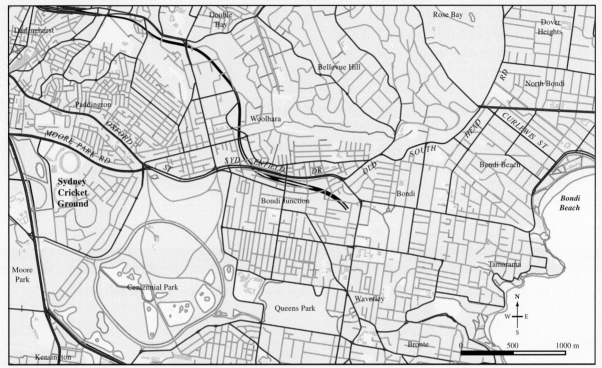

Source: N.S.W. Department of Finance

THINK	WRITE
1. Using the scale on the map estimate the distance. Measure the distance that represents 500 m and count how many 500 m are in the journey.	There appears to be 12 lots of 500 m. Distance $\quad$ 12 × 500 m $\qquad$ 6 km or 6000 m
2. The question tells you they estimate their speed as 50 km/h.	Speed = 50 km/h
3. Use the formula to calculate the time.	Time $= \dfrac{\text{Distance}}{\text{Speed}}$ $= \dfrac{6}{50}$ $= 0.12$ hours
4. Convert to minutes and seconds by multiplying the decimal by 60.	Time $= 0.12 \times 60$ $= 7.2$ minutes $= 7$ minutes 0.2×60 seconds $= 7$ minutes 12 seconds
5. Write the answer.	The estimated time for the trip from Bondi Beach to the SCG is 7 minutes and 12 seconds.

10.7.2 Distance versus time graphs

A journey doesn't always have a constant speed; it may vary at different periods of time. One way of displaying this is by a distance versus time graph. The distance is the *y*-axis (vertical) and the time is the *x*-axis (horizontal). The most common units are metres for distance and seconds for time.

You can get information about a journey that is represented in a distance versus time graph like the one shown. This distance versus time graph tells you the following.

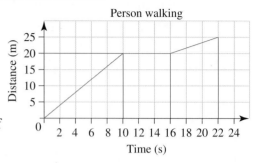

Person walking

- For the first 10 seconds the person travelled at a constant speed, since the graph has a straight line that goes 'uphill' from left to right (positive slope or gradient).
- Between 10 and 16 seconds the person stays at a distance of 20 m, therefore he did not move over this time, i.e. he is stationary since the graph is 'flat', (horizontal).
- Between 16 and 22 seconds the person travels at a constant speed, but his speed is less than his speed for the first 10 seconds since the slope is not as steep.

WORKED EXAMPLE 16

Use the distance versus time graph shown of a person on a walk to answer the following.

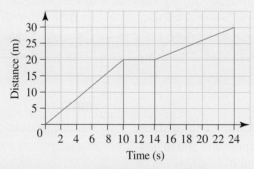

a. **State the time interval when the person is stationary.**

b. **State the times when the person is walking their fastest.**

THINK	WRITE
a. The person is stationary when the graph is 'flat' (horizontal)	The graph is flat so the person is stationary between 10 and 14 seconds.
b. The person walks the fastest when the slope of the graph is the steepest.	The person walks the fastest between 0 and 10 seconds.

Exercise 10.7 Distance–time graphs

1. **WE15** If the average speed travelling from Darwin to Uluru is 85 km/h, from the map shown how long would you expect it take to complete the journey?

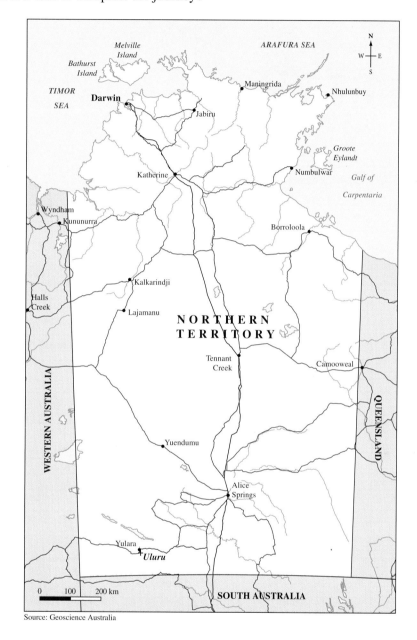

Source: Geoscience Australia

2. A family flew into Launceston and as a part of their holiday they wanted to visit Port Arthur by driving the hire car. They estimated they could average a speed of 70 km/h on the trip down. Using the map shown, how long to the nearest minute would it take them to travel to Port Arthur?

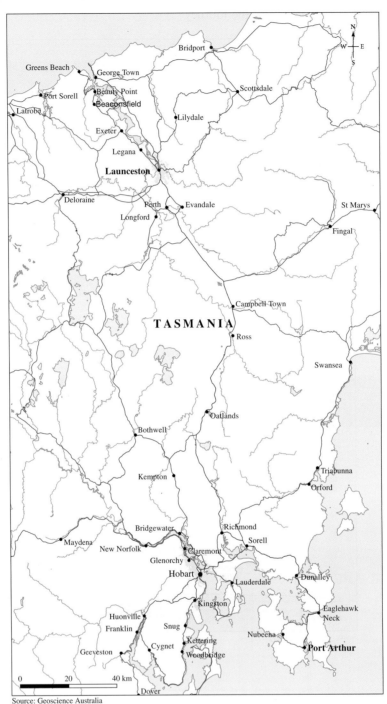

Source: Geoscience Australia

3. Janet walked 4000 m at an average speed of 2 m/s. How long did it take her to complete her walk in seconds?

4. Pedro went for a 7 km run at an average speed of 3.5 m/s. How long did it take him to complete his run in seconds?

5. Clancy rode her horse along the beach for 2.5 km at an average speed of 10 m/s. How long did it take her to cover the 2.5 km in minutes and seconds?

6. Cyril needs to drive from Perth to Geraldton, which is 434 km away by road. Taking traffic into account, Cyril estimates that he can average 82 km/h. How long would he expect it to take in hours, minutes and seconds?

7. Dermott decides to get up early from his Melbourne home and drive to Bell's Beach to go surfing. Estimate how long it will take him if he averages 75 km/h to get there, referring to the map shown.

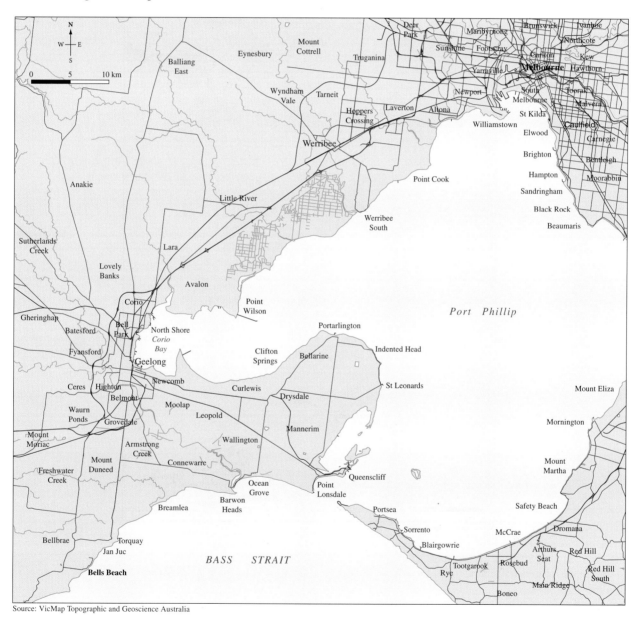

Source: VicMap Topographic and Geoscience Australia

8. A family is going for a holiday in Tasmania and decides to travel on the Spirit of Tasmania. The map shows the journey it takes. It travels at an average speed of 25 knots which is equivalent to 50 km/h. How long is the trip from Station Pier in Port Melbourne to Devonport in Tasmania?

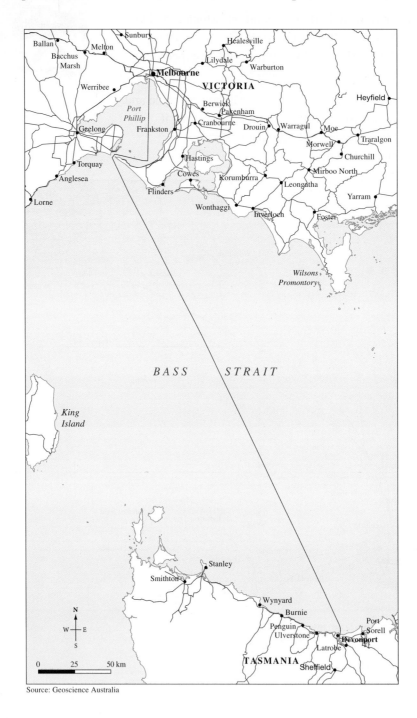

Source: Geoscience Australia

9. Johnny is looking forward to going to the AFL grand final at the MCG. He decides to drive to the grand final from his house in Frankston.

a. Taking traffic into consideration, Johnny estimates he will average 55 km/h for the journey. How long will it take him to get to the MCG given the map shown?

b. Johnny knows is costs him $1.55 per 10 km to drive his car. How much will the trip to the MCG cost him?

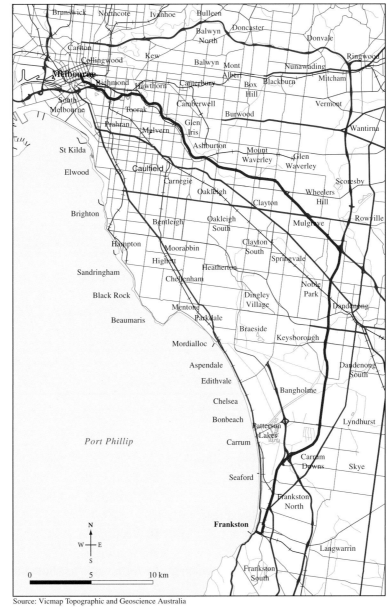

Source: Vicmap Topographic and Geoscience Australia

10. **WE16** Using the distance versus time graph shown of a person on a walk, answer the following.

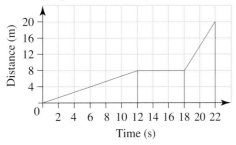

a. State the times when the person is stationary.

b. State the times when the person is walking their fastest.

11. Use the distance versus time graph shown of a train travelling from one station to another to answer the following questions.

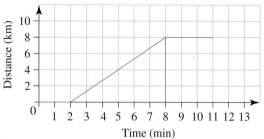

  a. What times was the train at the first station?
  b. How long after leaving the first station did it take to arrive at the next station?
  c. How long did the train wait for passengers at the first station?

12. Answer the following questions from the distance versus time graph shown.
  a. During what time interval is the object stationary?
  b. Between what times is the object moving fastest?

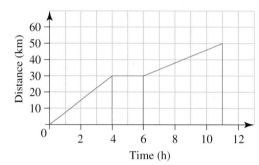

13. **MC** The distance versus time graph is of a person going for a walk.

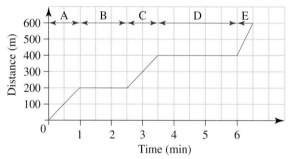

  a. In which sections has the person stopped walking?
     A. Section A
     C. Section D
     B. Section B
     D. Section B and D
  b. In which section is the person walking their fastest?
     A. Section B
     C. Section C
     B. Section A
     D. Section E

14. Briefly describe each section (A to D) of a person going for a walk, as shown in the distance versus time graph.

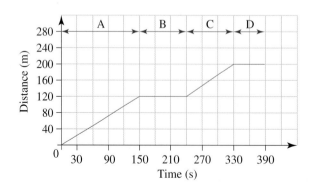

15. The map below shows the Phillip Island Grand Prix race track. Use this map with its scale to answer the following.

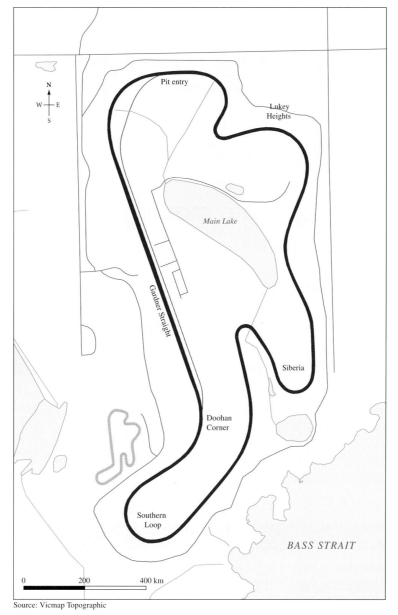

Source: Vicmap Topographic

a. Estimate the distance of one lap of the track.

b. If Casey Stoner could maintain his fastest lap time of 1 minute 30 seconds, how long would it take him to complete the 30 lap MotoGP race?

c. If each lap of the track is 4 km, what was Casey Stoner's average speed when he completed his fastest lap

  i. in m/s?

  ii. in km/h?

16. The distance time graph shown is that of a car driving through a section of Sydney.
    a. Explain what the car is doing in sections B and D and a possible reason for this.
    b. In which section is the car moving the fastest?
    c. Given that Speed $= \dfrac{\text{Distance}}{\text{Time}}$, calculate the maximum speed reached by the car on its journey
        i. in km/min.
        ii. in km/h.
    d. What is the car's average speed in km/h for section A?
    e. What is the car's average speed in m/s for section D?

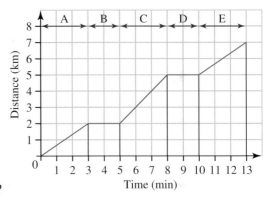

10.8 Calculations with motion

10.8.1 Motion in everyday life

Motion plays a part of our everyday living, whether it is how quickly we react on a sporting field, estimating the time required to travel to school, or knowing who the fastest runner in the school is.

To calculate speed, distance and time recall the equations shown:

$$\text{Speed} = \dfrac{\text{Distance}}{\text{Time}}$$

$$\text{Time} = \dfrac{\text{Distance}}{\text{Speed}}$$

$$\text{Distance} = \text{Speed} \times \text{Time}$$

Simple ways to convert speed

- To convert from km/h to m/s divide by 3.6.
- To convert from m/s to km/h multiply by 3.6.

WORKED EXAMPLE 17

The distance from home to school is 7.5 km and it takes twice as long to get home as it does to get to school due to traffic congestion. If it takes 10 minutes to get to school:
a. how long does it take to get home from school?
b. how long is the total trip to and from school?
c. what is the average speed of the total trip to and from school in:
    i. km/min   ii. km/h?

THINK	WRITE
a. It takes twice as long to get home from school compared to going in the morning.	Home from school $= 2 \times 10$ minutes $= 20$ minutes
b. Add the two times together.	To school $= 10$ minutes Total time $= 10 + 20$ $= 30$ minutes

---

c. i. To find the average speed of the total trip we need to use the total distance is 15 km and the total time is 30 minutes.

$$\text{Speed} = \frac{\text{Distance}}{\text{Time}}$$

$$= \frac{15}{30}$$

$$= 0.5 \text{ km/min}$$

d. ii. Use total distance is 15 km and time is 30 minutes is 0.5 hour.

$$\text{Speed} = \frac{\text{Distance}}{\text{Time}}$$

$$= \frac{15}{0.5}$$

$$= 30 \text{ km/h}$$

10.8.2 Energy in food and activity

Our bodies convert the food we eat into energy by using digestion and other chemical reactions that are part of human metabolism. Energy is required for basic functions such as breathing, keeping the heart beating and digesting food. It is also required for the performance of work and physical activity.

The kilojoule is a measure of energy. It can be used to tell us how much energy is contained in food and drink and can also measure the amount of energy used in physical activities and exercise. It is the standard metric measurement of energy. The calorie is another measure of energy and still used in the USA and the UK, which means many apps and discussions about energy related to food and activity use the calorie.

1 kilojoule = 0.24 calories
To convert calories to kilojoules ÷ 0.24
To convert kilojoules to calories × 0.24

The table below shows the number of minutes necessary to burn off particular food and highlights the need to combine exercise with a healthy diet.

Food	Number of minutes to burn kilojoules				
	Kilojoules	Walking	Riding bicycle	Swimming	Running
Apple, large	424	19	12	9	5
Bacon, 2 strips	403	18	12	9	5
Banana, small	397	17	11	8	4
Beans, green, 1serve	113	5	3	2	1
Beer, 1 glass	479	22	14	10	6
Bread and butter	319	15	10	7	4
Biscuit, plain	63	3	2	1	1
Biscuit, chocolate chip	214	10	6	5	3
Cake, 1 slice	1495	68	43	32	18

| Food | Number of minutes to burn kilojoules | | | | |
	Kilojoules	Walking	Riding bicycle	Swimming	Running
Carrot, raw	176	8	5	4	2
Cereal, $\frac{1}{2}$ cup with milk, sugar	840	38	24	18	10
Cheese, cottage, 1 tbsp.	113	5	3	2	1
Cheese, Cheddar 30 g	466	21	14	10	6
Chicken, fried, $\frac{1}{2}$ breast	974	45	28	21	12
Chicken dinner	2276	104	66	48	28
Doughnut	634	29	18	13	8
Drink, fizzy	445	20	13	9	5
Egg, fried	462	21	13	10	6
Egg, boiled	323	15	9	7	4
Fish steak, 110 g	861	39	25	18	11
French dressing, 1 tbsp.	248	11	7	5	3
Ham, 2 slices	701	32	20	15	9
Ice-cream, 1 serve	811	37	24	17	10
Malted milkshake	2108	97	61	45	26
Mayonnaise, 1 tbsp.	386	18	11	8	5
Milk, 1 glass	697	32	20	15	9
Milk, skim, 1 glass	340	16	10	7	4
Milkshake	1768	81	51	38	22
Orange, medium	286	13	8	6	4
Orange, juice, 1 glass	504	23	15	11	6
Pancake with syrup	521	24	15	11	6

WORKED EXAMPLE 18

Using the table above, answer the following questions.

a. On her way to swimming squad training after school, Benita bought a doughnut and a soft drink for afternoon tea. How many minutes of swimming would Benita need to do that afternoon to equal the amount of energy in kilojoules equivalent to her afternoon snack?

b. Kevin ate a boiled egg, a piece of toast with butter and an orange for breakfast, then cycled to school. How long would Kevin's trip to school need to be to use the same amount of energy as that contained in his breakfast?

THINK	WRITE
a. 1. Locate doughnuts and soft drinks in the table above and record their kilojoule content and equivalent swimming time	Doughnut contains 634 kj and takes 13 minutes to burn through swimming. Fizzy drink contains 445 kj and takes 9 minutes to burn through swimming.
2. Add the times together and write the answer.	$13 + 9 = 32$ minutes Benita would need to swim for 32 minutes to use the same amount of energy as that contained in her afternoon snack.
b. 1. Locate a boiled egg, bread and butter and an orange in the table on pages 419–20 and record their kilojoule content and equivalent cycling time.	Boiled egg contains 323 kj and takes 9 minutes to burn through cycling. Bread and butter contains 319 kj and takes 10 minutes to burn through cycling. Orange contains 286 kj and takes 8 minutes to burn through cycling.
2. Add the times together and write the answer.	$9 + 10 + 8 = 27$ minutes Kevin would need to cycle for 27 minutes to use the same amount of energy as he consumed for breakfast.

Exercise 10.8 Calculations with motion

1. **WE17** The distance from home to netball training is 6 km and it takes three quarters of the time to return to home as it does to get there. If it takes 8 minutes to get to netball training:
   a. how long does it take to get home from netball training?
   b. how long is the total trip to and from netball training?
   c. what is the average speed of the total trip to and from netball training in:
      i. km/min?
      ii. km/h?

2. Jane went for a 6.8 km hike which took her 1 hour and 3 minutes. What was Jane's average speed for her hike in m/s?

3. Calculate the average speed in the units shown in brackets, of the following:
   a. distance covered was 100 m in 25 seconds (m/s).
   b. distance covered was 4.8 km in 2 minutes (m/s).
   c. distance covered was 3800 m in 2 hours (km/h).
   d. distance covered was 14.6 km in 400 minutes (km/h).

4. Calculate the distance covered in the units shown in the brackets, of the following:
   a. travelled at 10 m/s for 250 seconds (m).
   b. travelled at 60 km/h for 3 hours (km).
   c. travelled at 25 m/s for 45 minutes (m).
   d. travelled at 100 km/h for 2 hours and 45 minutes (km).
5. Covert the following to m/s and to one decimal place.
   a. 50 km/h   b. 60 km/h   c. 80 km/h   d. 100 km/h
6. Convert the following to km/h and to one decimal place.
   a. 5 m/s   b. 15 m/s   c. 18 m/s   d. 30 m/s
7. **WE18** Using the table on pages 419–20, answer the following questions.
   a. Brigitte ordered a piece of cake and a milkshake for her morning tea. Her friend asked her to go for a walk along the beach after work that evening. How long would Brigitte need to walk to use the same number of kilojoules that she had eaten for morning tea? Give your answer in hours and minutes.
   b. Myvan ate cereal and a piece of toast with butter for breakfast, then walked to school. If it takes Myvan 20 minutes to walk to school, will she have used up all the energy contained in her breakfast?
8. **MC** 1200 calories is equivalent to
   A. 240 kj.   **B.** 5000 kj.   **C.** 288 kj.   **D.** 6000 kj.
9. Maurice flew from Melbourne to the Gold Coast, which took 2 hours and 6 minutes. On the flight back home, due to wind conditions it took 8 minutes longer. Calculate the average speed of the plane for the entire trip given the distance from Melbourne to the Gold Coast is 1345 km.

10. Kelly and her friend Kerry drove to an Ed Sheeran concert. It took them 50 minutes to get to the concert, but since it took them a lot of time to get out of the car park it took them 1:5 times longer to get home. The distance from their house to the concert is 50 km.
    a. How long did it take to get home from the concert?
    b. How long did the total trip take in hours and minutes?
    c. What was their average speed for the entire trip in km/h?

11. Brian intends to take the family on a holiday with the caravan to watch Bathurst (a car race). The petrol for his car costs $1.60 per 10 km when he is not towing, however, when he tows the cost is one and a half times the original cost. The distance from their house in Wagga Wagga to Bathurst is 315 km, therefore how much would it cost to tow the caravan to Bathurst and back home to Wagga Wagga?

12. A car travelling at 60 km/h in the dry has a braking distance of 20 m. However, if the road is wet its breaking distance is 1:4 times of that when dry. Given a car travels 25 m during the driver's reaction time, what is the total distance covered before the car comes to rest from when the driver initially sees the obstacle in front of them, given they are travelling at 60 km/h in the wet?

13. A batsman hits a ball at 35 m/s to a fieldsman, 7 m away. How long does the fieldsman have to react before he tries to catch it?

14. Robbie rides his bike to and from school each day for a week and times how long it takes. The distance to school from home is 6 km and his times are shown in the table.

Day	Time to school	Time from school
Monday	15 min 26 sec	14 min 24 sec
Tuesday	14 min 45 sec	15 min 09 sec
Wednesday	16 min 02 sec	15 min 03 sec
Thursday	15 min 13 sec	14 min 44 sec
Friday	14 min 38 sec	14 min 08 sec

   What was Robbie's average speed for the week?

15. A driver is travelling at 70 km/h when they see a kangaroo jump out in front of them. They take 1.5 seconds to react before braking and once the brakes are applied it takes 28 m to stop.
   a. How far does the car travel once the driver sees the kangaroo?
   b. If the kangaroo is 50 m in front of the car, what is the maximum speed, to the nearest km/h, they could be travelling at so they stop before the kangaroo?

16. On wet roads bald tyres are unsafe due to the increase in stopping distances. If a car is travelling at 80 km/h, on good tyres and dry roads, its braking distance is 35 m. Given the driver has a reaction time of 1.4 seconds before braking and it was a wet day with bald tyres, the braking distance increased by a further 70% compared to a dry day with good tyres.

   a. How far did the car travel before the brakes were applied, in metres to one decimal place?
   b. How far was the braking distance, in metres to one decimal place?
   c. What was the total distance travelled in the braking process to one decimal place?

17. Ruth liked to run each morning before breakfast for 30 minutes. She calculated that each kilometre she ran used 250 kj of energy. Her aim was to burn 1500 kj of energy each morning. At what speed does Ruth need to run to burn her 1500 kj?

10.9 Review: exam practice

10.9.1 Time, distance and speed: summary

Time

- Time can be expressed in 12-hour time or 24-hour time.
- 12-hour time is separated into two 12-hour sections.
- Midnight to midday is am (ante meridiem — before noon).
- Midday to midnight is pm (post meridiem — after noon).
- If the time is less than 1200, the time is am. If the time is greater than 1200, the time is pm.

TIME CONVERSIONS	
60 seconds	1 minute
60 minutes	1 hour
24 hours	1 day
7 days	1 week
14 days	1 fortnight
12 months	1 year
365 days	1 year
366 days	1 leap year
10 years	1 decade
100 years	1 century
1000 years	1 millennium

Interpret timetables

- A timeline can be used to record a series of events in the order that they occur.
- A scale must be used when drawing a timeline so that the time elapsed between events can be determined.
- A timetable is a list of events that are scheduled to happen and the timetable usually makes these events easier to read. This is because they allow you to quickly see a lot of information.
- When reading a timetable, be careful to follow it horizontally and vertically so that you stay on the same line.

Time travel

- If you move from one time zone to another in an easterly direction, you need to move your clock forward. If you move from one time zone to another in a westerly direction, you need to move your clock back.
- The International Date Line is the theoretical line on Earth where a calendar day officially starts and ends. If you cross the International Date Line in an easterly direction, your calendar goes back a day. If you cross the International Date Line in a westerly direction, your calendar goes forward one day.
- Australia is divided into three different standard time zones: Western Standard Time (WST), Central Standard Time (CST) and Eastern Standard Time (EST).
- Western Australia is 2 hours behind Eastern Standard Time.
- The Northern Territory and South Australia (CST) are half an hour behind Eastern Standard Time.
- Daylight-saving time starts on the first Sunday in October, when clocks go forward 1 hour at 2:00 am.
- Daylight-saving time finishes on the first Sunday in April, when clocks are turned back 1 hour at 3:00 am.
- Queensland, Western Australia and the Northern Territory do not observe daylight-saving time.
- Tasmania, Victoria and New South Wales always have the same time. The time in these states is called Eastern Daylight-Saving Time (EDT).
- Queensland is 1 hour behind Melbourne daylight-saving time (EDT).

- The Northern Territory is one-and-a-half hours behind Melbourne daylight saving time (EDT).
- South Australia is half an hour behind Melbourne time.
- Western Australia is 3 hours behind Melbourne daylight saving time (EDT).

To compare travel times use $\text{Speed} = \dfrac{\text{Distance}}{\text{Time}} \quad \therefore \text{Time} = \dfrac{\text{Distance}}{\text{Speed}}$

Distance: scales and maps

- A scale on a map describes the ratio between the distance on the map and the actual distance.
- The scale on the map can be used to calculate the size of objects and the distance between objects on the map.
- Maps include a north direction icon.

Speed

- Speed tells us how fast we are going by dividing the distance we travel by the time it takes to travel that distance.

$$\text{Speed} = \dfrac{\text{Distance}}{\text{Time}}$$

- The most common unit for speed is metres per second (m/s).

$$\text{Distance} = \text{Speed} \times \text{Time}$$

$$\text{Time} = \dfrac{\text{Distance}}{\text{Speed}}$$

- Kilojoules and calories are units of energy. 1 kilojoule = 0.24 calories.

Distance–time graphs

To find the time of a journey:
- estimate the distance of the journey from the map.
- estimate the average speed of the journey.
- use these estimates to calculate time of the journey, using $\text{Time} = \dfrac{\text{Distance}}{\text{Speed}}$.

distance versus time graphs tell you:
- if the graph has an increasing slope (gradient) then the object is moving with a constant speed.
- the steeper the graph (gradient), the faster the speed.
- if the graph is 'flat' (horizontal) the object is not moving (stationary).

Calculations with motion

To calculate speed, distance and time use the equations:

$$\text{Speed} = \dfrac{\text{Distance}}{\text{Time}}$$

$$\text{Time} = \dfrac{\text{Distance}}{\text{Speed}}$$

$$\text{Distance} = \text{Speed} \times \text{Time}$$

- To convert from km/h to m/s divide by 3.6.
- To convert from m/s to km/h multiply by 3.6.

Exercise 10.9 Review: exam practice

Simple familiar

1. **MC** The time 8:35 pm in 24 hour time is:

   A. 0835. **B.** 1835. **C.** 835. **D.** 2035.

2. **MC** Fredrick estimates that it will take him 1 hour and 25 minutes to walk to the lookout at the National Park. If he leaves at 10.38 am, he will get to the lookout at:

   A. 11:38 am. **B.** 12:13 pm. **C.** 12:03 pm. **D.** 11:03 am.

The shown train timetable relates to questions **3** and **4**.

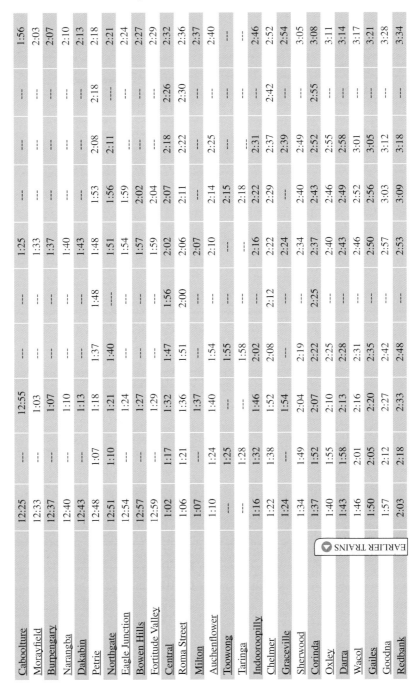

Station									
Caboolture	12:25	—	12:55	—	1:25	—	—	—	1:56
Morayfield	12:33	—	1:03	—	1:33	—	—	—	2:03
Burpengary	12:37	—	1:07	—	1:37	—	—	—	2:07
Narangba	12:40	—	1:10	—	1:40	—	—	—	2:10
Dakabin	12:43	—	1:13	—	1:43	—	—	—	2:13
Petrie	12:48	1:07	1:18	1:37	1:48	1:53	2:08	2:18	2:18
Northgate	12:51	1:10	1:21	1:40	1:51	1:56	2:11	—	2:21
Eagle Junction	12:54	—	1:24	—	1:54	1:59	—	—	2:24
Bowen Hills	12:57	—	1:27	—	1:57	2:02	—	—	2:27
Fortitude Valley	12:59	—	1:29	—	1:59	2:04	—	—	2:29
Central	1:02	1:17	1:32	1:47	2:02	2:07	2:18	2:26	2:32
Roma Street	1:06	1:21	1:36	1:51	2:06	2:11	2:22	2:30	2:36
Milton	1:07	—	1:37	—	2:07	—	—	—	2:37
Auchenflower	1:10	1:24	1:40	1:54	2:10	2:14	2:25	—	2:40
Toowong	—	1:25	—	1:55	—	2:15	—	—	—
Taringa	—	1:28	—	1:58	—	2:18	2:31	—	—
Indooroopilly	1:16	1:32	1:46	2:02	2:16	2:22	—	—	2:46
Chelmer	1:22	1:38	1:52	2:08	2:22	2:29	2:37	2:42	2:52
Graceville	1:24	—	1:54	—	2:24	—	2:39	—	2:54
Sherwood	1:34	1:49	2:04	2:19	2:34	2:40	2:49	—	3:05
Corinda	1:37	1:52	2:07	2:22	2:37	2:43	2:52	2:55	3:08
Oxley	1:40	1:55	2:10	2:25	2:40	2:46	2:55	—	3:11
Darra	1:43	1:58	2:13	2:28	2:43	2:49	2:58	—	3:14
Wacol	1:46	2:01	2:16	2:31	2:46	2:52	3:01	—	3:17
Gailes	1:50	2:05	2:20	2:35	2:50	2:56	3:05	—	3:21
Goodna	1:57	2:12	2:27	2:42	2:57	3:03	3:12	—	3:28
Redbank	2:03	2:18	2:33	2:48	2:53	3:09	3:18	—	3:34

EARLIER TRAINS ⌄

3. **MC** If you caught the 1:18 pm train from Petrie, you would arrive at Gailes at:

A. 1:50 pm. **B.** 2:05 pm.

C. 2:20 pm. **D.** 2:35 pm.

4. **MC** The time you need to catch the train from Roma Street if you want to arrive at Oxley at 2:40 pm is:

A. 2:06 am. **B.** 2:00 pm.

C. 2:11 pm. **D.** 2:06 pm.

5. **MC** If the time in Brisbane is 1:24 pm, the time in Perth in May is:

A. 3:34 pm. **B.** 11:24 am.

C. 2:24 pm. **D.** 12:24 pm.

6. **MC** The distance from Cairns to Port Douglas, as shown in the map and using the scale, is closest to:

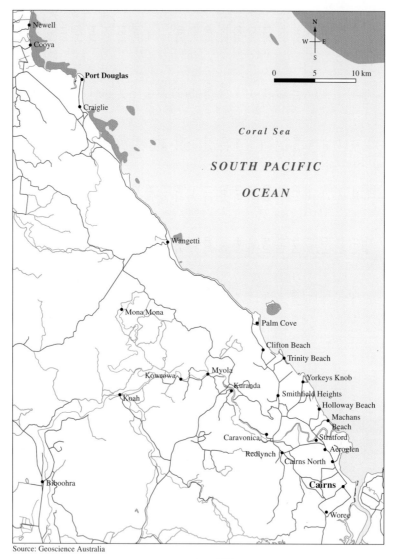

Source: Geoscience Australia

A. 7 km. **B.** 50 km.

C. 70 km. **D.** 100 km.

7. **MC** If Mark Webber completed one lap of the 5.303 km Australian Grand Prix circuit in 1 minute and 29.5 seconds, his average speed in m/s is closest to:

    A. 4.1 m/s.   **B.** 59.3 m/s.   **C.** 41.0 m/s.   **D.** 55.5 m/s.

8. **MC** From a distance versus time graph, the object is stationary when:
    A. the graph is sloping uphill from left to right.
    B. the graph is sloping downhill from left to right.
    C. the graph is horizontal.
    D. the graph is its steepest.

9. **MC** If petrol costs $1.49 per litre and your car uses 9.8 litres per 100 km, the cost of petrol to cover 400 km is closest to:
    A. $58.41.   **B.** $50.   **C.** $14.60.   **D.** $45.65.

10. **MC** If you drove for 2 hours and 30 minutes at an average speed of 70 km/h, the distance you would cover is closest to:
    A. 100 km.   **B.** 125 km.   **C.** 150 km.   **D.** 175 km.

11. **MC** 80 km/h converted to m/s is closest to:
    A. 52 m/s.   **B.** 12 m/s.   **C.** 80 m/s.   **D.** 22 m/s.

Complex familiar

12. Britney travels to school by various means. What time will she arrive at school if she leaves home at 7:48 am and travels by the following means?
    a. Walking which takes 1 hour and 5 minutes.
    b. Cycling which takes 37 minutes.
    c. Car which takes 17 minutes.
    d. Bus which takes 25 minutes.

13. Trams are used throughout Melbourne to assist commuters getting around the inner city suburbs. The following questions refer to the Melbourne tram timetable shown.
    a. How many trams leave the University of Melbourne between 8 am and 9 am?
    b. If you wanted to go shopping at Chapel Street and be there when the shops open at 9:00 am, what time tram would you need to catch from Bourke Street?
    c. How long does it take to get from Melbourne Central to the Alfred Hospital by tram, if you catch the 8:06 am from Melbourne Central?
    d. If you caught the tram from the University of Melbourne at 11:06 am, what time would you get to Toorak Road?

Monday to Thursday

Morning (am) / Afternoon (pm)	am	am	am	am	am	am	am	am	am	am	am	am	am	am	am	am	am
1–University of Melbourne/Swanston St (Carlton)	7:59	8:09	8:20	8:30	8:42	8:54	9:05	9:16	9:29	9:41	9:54	10:06	10:18	10:30	10:42	10:54	11:06
8–Melbourne Central Station/Swanston St (Melbourne City)	8:06	8:16	8:27	8:37	8:49	9:01	9:12	9:23	9:36	9:48	10:01	10:13	10:25	10:37	10:49	11:01	11:13
10–Bourke Street Mall/Swanston St (Melbourne City)	8:09	8:19	8:30	8:40	8:52	9:04	9:15	9:26	9:39	9:51	10:04	10:16	10:28	10:40	10:52	11:04	11:16
11–City Square/Swanston St (Melbourne City)	8:10	8:20	8:31	8:41	8:53	9:05	9:16	9:27	9:40	9:52	10:05	10:17	10:29	10:41	10:53	11:05	11:17
13–Flinders Street Railway Station/Swanston St (Melbourne City)	8:12	8:22	8:33	8:43	8:55	9:07	9:18	9:29	9:42	9:54	10:07	10:19	10:31	10:43	10:55	11:07	11:19
14–Arts Centre/St Kilda Rd (Southbank)	8:14	8:24	8:35	8:45	8:57	9:09	9:20	9:31	9:44	9:56	10:09	10:21	10:33	10:45	10:57	11:09	11:21
20–Domain Interchange/St Kilda Rd (Melbourne City)	8:21	8:31	8:42	8:52	9:03	9:15	9:26	9:37	9:50	10:02	10:15	10:27	10:39	10:51	11:03	11:15	11:27
25–Commercial Rd/St Kilda Rd (South Melbourne)	8:28	8:38	8:49	8:59	9:09	9:21	9:32	9:43	9:55	10:07	10:20	10:32	10:44	10:56	11:08	11:20	11:32
26–Alfred Hospital/Commercial Rd (South Yarra)	8:29	8:39	8:50	9:00	9:10	9:22	9:33	9:44	9:55	10:08	10:21	10:33	10:45	10:57	11:09	11:21	11:33
31–Chapel St/Commercial Rd (South Yarra)	8:34	8:44	8:55	9:05	9:15	9:27	9:38	9:49	10:01	10:13	10:26	10:38	10:50	11:02	11:14	11:26	11:38
37–Orrong Rd/Malvern Rd (Toorak)	8:40	8:50	9:01	9:10	9:20	9:32	9:43	9:54	10:06	10:18	10:31	10:43	10:55	11:07	11:19	11:31	11:43
43–Glenferrie Rd/Malvern Rd (Toorak)	8:46	8:56	9:05	9:14	9:24	9:36	9:48	9:59	10:11	10:23	10:36	10:48	11:00	11:12	11:24	11:36	11:48
44–Plant St/Malvern Rd (Malvern)	8:47	8:57	9:06	9:15	9:25	9:37	9:49	10:00	10:12	10:24	10:37	10:49	11:01	11:13	11:25	11:37	11:49
51–Gardiner Railway Station/Burke Rd (Hawthorn East)	8:52	9:02	9:10	9:19	9:29	9:41	9:53	10:04	10:16	10:28	10:41	10:53	11:05	11:17	11:29	11:41	11:53
54–Toorak Rd/Burke Rd (Hawthorn East)	8:56	9:06	9:14	9:23	9:33	9:45	9:57	10:08	10:20	10:32	10:45	10:57	11:09	11:21	11:33	11:45	11:57
61–Camberwell Junction/Burke Rd (Hawthorn East)	9:04	9:13	9:21	9:30	9:40	9:52	10:04	10:15	10:27	10:39	10:52	11:04	11:16	11:28	11:40	11:52	12:04
64–Camberwell Station/Burke Rd (Hawthorn East)	9:07	9:16	9:24	9:33	9:43	9:55	10:07	10:18	10:30	10:42	10:55	11:07	11:19	11:31	11:43	11:55	12:07
70–Cotham Rd/Burke Rd (Kew)	9:13	9:22	9:30	9:38	9:48	10:00	10:12	10:23	10:35	10:47	11:00	11:12	11:24	11:36	11:48	12:00	12:12

14. To travel from Coolaroo to Roma Street by train in the morning takes 1 hour and 20 minutes. Due to heavy traffic it takes 1:5 times this to drive by car compared to going by train.
   a. If you travelled by train and left Coolaroo at 7:30 am, what time would you arrive at Roma Street?
   b. If you travelled by car and left at 7:30 am, what time would you arrive at Roma Street?
   c. How much longer does it take to travel by car compared to going by train from Coolaroo to Roma Street?

15. Calculate the following.
   a. The average speed (in m/s) when covering 275 m in 32 seconds
   b. The distance covered when travelling at an average speed of 40 km/h for 35 seconds
   c. The time it takes to travel 40 km travelling at an average speed of 15 m/s
   d. The average speed (in km/h) when covering 562 m in 28 seconds

16. The Sydney to Hobart yacht race is a traditional race that is 1170 km (630 nautical miles) long that starts on Boxing Day. What is the average speed of a yacht that takes 1 day 19 hours and 30 minutes to complete the course?

Complex unfamiliar

17. The following table shows the schedule for Chung who left for work at 6:45 am from his home in Melbourne.

Activity	Time
Drive to the airport	48 minutes
Wait at airport	36 minutes
Fly to Adelaide	1 hour 4 minutes
Taxi to meeting	27 minutes
Meeting time	2 hours 38 minutes
Lunch	1 hour 22 minutes
Taxi to airport	23 minutes
Wait at airport	1 hour 31 minutes
Fly to Melbourne	57 minutes
Drive home	34 minutes

a. What time did Chung's meeting start?
b. What time did he start lunch?
c. What time was his flight back home?
d. What time (24 hour) did he arrive back in Melbourne?
e. What time did he arrive back home (12 hour)?
f. How long was Chung away from home?

18. The table shows the flight timetable of Qantas, flying from Melbourne. The following questions refer to the timetable.

QANTAS — Worldwide Timetable

< from Melbourne   **from Melbourne >**

Flight	Depart	Arrive	Freq.	Stops	Effective
"	1610	1720	1·2·3···7	-	from 07 Oct
QF0628	1720	1930	Daily	-	
"	1720	1830	1·2·3···7	-	from 07 Oct
QF0632	1815	2025	···4·5··	-	
"	1815	1925	1·2·3···7	-	from 07 Oct
"	1820	2030	1·2·3···	-	until 03 Oct
QF0634	1910	2120	1·2·3·4·5·7	-	
"	1910	2120	·····6·	-	15 Sep only
"	1910	2120	·····6·	-	from 06 Oct
"	1910	2020	1·2·3···7	-	from 07 Oct
QF1746	1910	2120	·····6·	-	22 Sep only
QF1748	2005	2215	·····6·	-	29 Sep only
QF0636	2010	2220	1·2·3·4·5·7	-	
"	2010	2120	1·2·3···7	-	from 07 Oct
Broome					
QF1050	0830	1105	··3···7	-	12 Sep - 03 Oct
"	0835	1010	······7	-	from 07 Oct
Cairns					
JQ0944 †	0600	0920	Daily	-	
"	0600	0820	1·2·3···7	-	from 07 Oct
JQ0249 †	0700	1020	····5··	-	from 21 Sep
QF0702	0910	1235	1·2·3·4·5··	-	
"	0910	1235	·····67	-	15 Sep - 16 Sep
"	0910	1135	1·2·3····	-	from 08 Oct
JQ0946 †	1020	1340	Daily	-	
"	1120	1340	1·2·3···7	-	from 07 Oct
QF0704	1210	1420	······7	-	from 07 Oct
"	1255	1620	1······	-	until 01 Oct
"	1255	1620	·····67	-	15 Sep - 16 Sep
"	1255	1620	··3····	-	19 Sep - 03 Oct
"	1255	1605	·····67	-	from 22 Sep
"	1255	1520	1·3····	-	from 08 Oct
JQ0948 †	1950	2310	····5··	-	
"	1950	2310	1··4··7	-	23 Sep - 04 Oct
"	2050	2310	1··45·7	-	from 07 Oct
Canberra					
QF0804	0700	0805	1·2·3·4·5··	-	
QF0814	0830	0935	·2·3·4·5·6·7	-	
"	0915	1020	1······	-	
QF0816	1040	1145	1·2·3·4·5··	-	
QF0848	1200	1305	1·2·4·5·6·	-	
"	1205	1310	·2·····	-	from 09 Oct
QF2130	1235	1345	Daily	-	
QF0812	1320	1425	Daily	-	
QF0820	1420	1525	1·3·4·5··	-	
QF0818	1530	1635	1·3·4·5·6·	-	
QF0826	1620	1725	1·2·3·4·5·7	-	
QF0834	1710	1815	1·2·3·4·5··	-	
"	1710	1815	·····6·	-	15 Sep - 22 Sep
"	1710	1815	·····6·	-	from 06 Oct
QF2132	1805	1915	··3·4·5··	-	
QF0822	1940	2045	1·2·3·4·5·7	-	
QF0834	2015	2120	·····6·	-	29 Sep only
Christchurch					
QF0133	1155	1720	·2··6·	-	11 Sep - 29 Sep
"	1155	1820	·2···6·	-	from 02 Oct
"	1155	1720	·2·····	-	from 09 Oct
"	1830	2355	1·3·4·5·7	-	until 28 Sep
"	1830	0055+1	1·3·4·5·7	-	from 30 Sep

Flight	Depart	Arrive	Freq.	Stops	Effective
"	1830	2355	1·3···7	-	from 07 Oct
JQ0203 †	2255	0515+1	1·3·5·7	-	from 30 Sep
"	2355	0515+1	····5·7	-	14 Sep - 28 Sep
"	2355	0515+1	1·3····	-	24 Sep - 26 Sep
"	2355	0515+1	1·3···7	-	from 07 Oct
Darwin					
QF0838	0845	1245	Daily	-	
"	0845	1145	1·2·3···7	-	from 07 Oct
JQ0678 †	2125	0110+1	1·2·4·5·6·7	-	
"	2125	0110+1	··3····	-	26 Sep - 03 Oct
"	2125	0010+1	1·2·3···7	-	from 07 Oct
Devonport					
QF2051	0820	0925	1·2·3·4·5··	-	
"	0950	1055	·····67	-	
QF2053	1225	1330	1·2·3·4·5··	-	
"	1315	1420	·····6·	-	
QF2057	1540	1645	1·2·3·4·5·7	-	
"	1640	1745	·····6·	-	
QF2061	1900	2005	1·2·3·4·5·7	-	
Dubai					
EK8409*	0500	1300	Daily	-	
"	0600	1300	1·2·3···7	-	from 07 Oct
EK8405*	1800	0450+1	Daily	1	
"	1900	0450+1	1·2·3···7	1	from 07 Oct
EK8407*	2115	0515+1	Daily	-	
"	2215	0515+1	1·2·3···7	-	from 07 Oct
Gold Coast					
JQ0430 †	0600	0800	Daily	-	
"	0600	0700	1·2·3···7	-	from 07 Oct
QF0880	0950	1155	··3····	-	26 Sep - 03 Oct
"	0955	1100	1·2·3···7	-	from 07 Oct
"	1000	1205	1·2·4·5·6·7	-	
"	1000	1205	··3····	-	12 Sep - 19 Sep
JQ0432 †	1005	1205	·2·3·4·5·6·7	-	
"	1005	1205	1······	-	24 Sep - 01 Oct
"	1005	1105	1·2·3···7	-	from 07 Oct
JQ0436 †	1230	1430	Daily	-	
"	1230	1330	1·2·3···7	-	from 07 Oct
JQ0438 †	1430	1630	Daily	-	
"	1525	1625	1·2·3···7	-	from 07 Oct
JQ0444 †	1600	1800	1·3·4·5·6·7	-	
"	1600	1800	·2·····	-	25 Sep - 02 Oct
"	1630	1730	1·2·3···7	-	from 07 Oct
JQ0442 †	1930	2130	Daily	-	
JQ0446 †	2010	2210	1···5·6·7	-	
"	2010	2210	·2·3·4···	-	25 Sep - 04 Oct
JQ0442 †	2030	2130	1·2·3···7	-	from 07 Oct
JQ0446 †	2110	2210	1·2·3···7	-	from 07 Oct
Guangzhou					
CZ0329*	1030	1800	Daily	-	
"	1130	1800	1·2·3···7	-	from 07 Oct
CZ0331*	2230	0600+1	1·2·3·5·6·7	-	
"	2230	0600+1	···4···	-	27 Sep - 04 Oct
"	2330	0600+1	1·2·3···7	-	from 07 Oct
Hamilton Island					
QF0870	0750	1050	····5·7	-	14 Sep - 16 Sep
"	0845	1145	··3··6·	-	
"	0845	1145	····5·7	-	from 21 Sep
"	0845	1045	··3···7	-	from 07 Oct

Flight	Depart	Arrive	Freq.	Stops	Effective
Hobart					
JQ0701 †	0605	0720	·2·3·4·5·6·7	-	
"	0605	0720	1······	-	from 24 Sep
QF1011	0815	0930	Daily	-	
JQ0705 †	0830	0945	Daily	-	
JQ0707 †	1040	1155	Daily	-	
QF1013	1210	1325	······7	-	
JQ0709 †	1245	1400	Daily	-	
QF1013	1255	1410	1·2·3·4·5·6·	-	
JQ0711 †	1445	1600	Daily	-	from 21 Sep
QF1015	1515	1630	1·2·4·5·7	-	
"	1540	1655	·····6·	-	15 Sep - 22 Sep
"	1540	1655	·····6·	-	from 06 Oct
QF1017	1920	2035	1·2·3·4·5·7	-	
QF1015	1950	2105	·····6·	-	29 Sep only
JQ0715 †	2045	2200	Daily	-	
Ho Chi Minh City					
JQ0279 †	1515	2105	···4···	-	13 Sep - 04 Oct
"	1840	0030+1	1····6·	-	
"	1940	0030+1	1······	-	from 08 Oct
Hong Kong					
QF0029	0935	1720	Daily	-	
"	1035	1720	1·2·3···7	-	from 07 Oct
Honolulu					
JQ0285 †	1545	0615	··3·5·7	-	
"	1545	0615	·2·····	-	25 Sep - 02 Oct
"	1645	0615	·2·3···7	-	from 07 Oct
Launceston					
JQ0731 †	0635	0740	1·3·5·6·7	-	
"	0635	0740	·2·4···	-	from 25 Sep
QF2281	0830	0945	1·2·3·4·5··	-	
"	0845	1000	·····67	-	
JQ0733 †	0910	1015	Daily	-	from 21 Sep
"	1010	1115	1·2·3···7	-	from 07 Oct
JQ0735 †	1025	1130	Daily	-	
"	1125	1230	1·2·3···7	-	from 07 Oct
QF2283	1145	1300	1·2·3···	-	
"	1220	1335	·····67	-	
QF2285	1535	1650	1·2·3·4·5·7	-	
"	1605	1720	·····6·	-	
JQ0739 †	1640	1745	Daily	-	
QF2289	1905	2020	1·2·3·4·5·7	-	
JQ0743 †	1905	2010	1··45·7	-	from 21 Sep
London–Heathrow					
QF0009	1515	0505+1	Daily	1	
"	1615	0505+1	1·2·3···7	1	from 07 Oct
Los Angeles					
QF0093	0900	0625	Daily	-	
"	1000	0625	1·2·3···7	-	from 07 Oct
QF0095	2140	1900	1···5··	-	
"	2240	1900	1······	-	from 08 Oct
Mildura					
QF2078	0825	0935	1·2·3·4·5··	-	
"	0850	1000	·····67	-	
QF2080	1200	1310	···4·5··	-	
QF2084	1545	1655	1·2·3·4·5·7	-	
"	1645	1755	·····6·	-	
QF2086	1910	2020	1·2·3·4·5·7	-	

*Indicates a codeshare flight operated by another carrier and QF flight number.

†Indicates a Jetstar operated flight and flight number.

1 = Monday 2 = Tuesday 3 = Wednesday 4 = Thursday 5 = Friday 6 = Saturday 7 = Sunday

a. What time does the flight leave for London in 12 hour time?

b. If you wanted to arrive at the Gold Coast at 2 pm what time flight would you catch from Melbourne?

c. How many times a week does the QF0850 travel to Canberra each week?

d. Use the timetable and your knowledge of time differences to calculate how long it takes to fly to Ho Chi Minh City.

19. Using the scaled map of Australia answer the following questions.

Source: Geoscience Australia

a. Approximately how far apart is Perth to Sydney?

b. Approximately how far is Adelaide from Canberra?

c. Approximately how far is Melbourne from Sydney?

d. Which of the cities Brisbane, Sydney or Hobart is closest to Darwin?

e. If you drove from Melbourne to Adelaide at an average speed of 85 km/h, approximately how long would it take?

f. If you drove from Melbourne to Canberra and it took you 7 hours and 20 minutes, approximately what was your average speed?

20. Amber leaves home at 7:15 am and drives from Mentone to Tullamarine Airport (Melbourne) at an average speed of 63 km/h. She then catches the first Qantas flight to Brisbane. She has lunch and meetings with work colleagues before catching the earliest Qantas flight back to Melbourne after 1600.
  a. Research to find the distance between Mentone and Tullamarine Airport.
  b. Knowing the distance, how long did it take Amber to get to the airport?
  c. What time did Amber get to the airport?
  d. Research what is the earliest flight she can catch assuming that there is no check-in time. What time does it leave?
  e. How long does the flight take to get to Brisbane?
  f. Research the flight she can catch home. What time does it leave?
  g. What time does she land at Tullamarine Airport?
  h. If it takes Amber 42 minutes to get her luggage and get back to her car and then another 40 minutes to drive home, what time does she arrive home at Mentone?

Answers

Chapter 10 Time, distance and speed

Exercise 10.2 Time

1. a. 5 hours 48 minutes
   b. 9 hours 52 minutes 30 seconds

2. a. 12 hours 9 minutes 21.6 seconds
   b. 3 hours 13 minutes 20 seconds

3. a. Five o'clock
   b. Twenty-three minutes past seven
   c. Thirteen minutes to 7 o'clock
   d. Quarter past two
   e. Five minutes to eight
   f. Ten minutes past midnight
   g. Twelve o'clock
   h. Twenty-five to six

4. a. 1025   b. 0733
   c. 1345   d. 2012
   e. 2206   f. 2345

5. 2 hours 36 minutes

6. 1 hour 25 minutes

7. a. 3:51 pm   b. 8:22 pm
   c. 3:15 am   d. 11:31 am
   e. 9:02 am   f. 10:15 am

8. B

9. a. 180 minutes   b. 315 minutes
   c. 1 440 minutes   d. 45 minutes
   e. 198 minutes   f. 394 minutes

10. They are not the same length of time, since 2.25 hours (= 2 hours 15 minutes) is 10 minutes less than 2 hours and 25 minutes.

11. a. 3 hours 30 minutes
    b. 5 hours 5 minutes
    c. 1 hour 17 minutes
    d. 2 hours 27 minutes
    e. 9 hours 33 minutes
    f. 140 hours 27 minutes
    g. 8 hours 48 minutes 54 seconds
    h. 153 hours 51 minutes 54 seconds

12. 4 hours 15 minutes

13. 4 hours 57 minutes

14. a. 1 hour 25 minutes   b. 8 hours 15 minutes
    c. 35 hours 10 minutes   d. 43 hours 49 minutes
    e. 22 hours 20 minutes   f. 8 hours 35 minutes

15. a. 1 pm   b. 4:15 am
    c. 9:50 am   d. 5:15 pm
    e. 5:22 pm   f. 11:33 am

16. 106 days

17. 115 days

18. 88 days

19. a. 80 minutes or 1 hour 20 minutes
    b. 8:45 am
    c. Yes, just on time.

20. a. 3600 seconds   b. 86 400 seconds
    c. 1440 minutes   d. 8760 hours

Exercise 10.3 Interpret timetables

1. a. 2006   b. 1998

2. a. 10 years apart   b. Separated by 13 years

3. a. 1969   b. 1963

4. a. 1958   b. 11 years

5. a. 20 years   b. 4 years

6. a. 14:19 or 2:19 pm   b. 8:05 am

7. a. 7:01 am the next day   b. 2 hours 40 minutes

8. a. 9:07 am   b. 9:45 am

9. a. 27 minutes
   b. 11:30 am

10. a. No
    b. 8:55 am

11. a. 2 trains
    b. 5:14 pm from South Brisbane Station
    c. 23 minutes

12. A : (1770) Captain James Cook lands at Botany Bay
    B : (1850) First University in Australia (Sydney)
    C : (1854) Eureka Stockade
    D : (1880) Ned Kelly hanged
    E : (1890) Banjo Patterson publishes *The man from Snowy River*
    F : (1901) Australia becomes a federation
    G : (1902) Australian women gain the right to vote and stand for Parliament
    H : (1915) Australians land at ANZAC Cove
    I : (1920) QANTAS is founded
    J : (1923) Vegemite produced
    K : (1930) Phar Lap wins Melbourne Cup
    L : (1932) Sydney Harbour Bridge opens
    M : (1956) Melbourne Olympics / TV in Australia
    N : (1967) Referendum to include Indigenous Australians in the census is successful

13. a. 20th January 11:14
    b. 13
    c. July has 2 full moons.

14. a. 15 days
    b. 161 hours 27 minutes
    c. 516 hours 18 minutes

15. a. Red indicates low tide and blue indicates high tide
    b. 2 per day.
    c. 0.29 metres
    d. Friday 10:30 pm

16. a. 11 hours 43 minutes
    b. No
    c. 1. Getting a ship or boat through a channel
       2. Knowing when rocks are exposed to go looking for crabs
       3. Knowing when the tide is out and good for skim boarding
       There could be many other reasons.

Exercise 10.4 Time travel

1. 9:45 pm

2. 9:35 pm the day before

3. a. You need to move your clock backwards.
   b. You need to move your clock forwards.

4.

Time in Perth	Time in Adelaide	Time in Melbourne	Time in Auckland
4:00 pm	5:30 pm	6:00 pm	8:00 pm
5:30 pm	7:00 pm	7:30 pm	9:30 pm
8:00 am	9:30 am	10:00 am	12 midday
10:45 pm	12:15 pm	12:45 am	2:45 am

5. a. 9:33 pm   b. 10:00 pm
    c. 12:55 am   d. 5:14 pm

6. a. 8:30 pm the day before
    b. 6:00 pm the day before
    c. 12:05 pm the day before
    d. 6:00 am

7. 9:15 pm

8. 11:45 am

9. 10:00 am

10. 10:30 pm

11. a. 11 am   b. 9:00 am
    c. 6:45 am   d. 5:45 pm
    e. 10:30 pm   f. 2:55 pm

12. 9 hours 19 minutes

13. a. 7 hours   b. 7 hours 40 minutes
    c. Anthony: 1:00 pm
       Robert: 1:40 pm

14. a. 8:30 am   b. 7:00 pm

15. Perth is 2 hours behind and the show was delayed, so the voting could have been closed before the show was even screened in Perth. Thus it would have been too late to vote if they wanted to support the Perth contestant.

16. 3:00 am

17. 11:00 pm

18. a. New Zealand.
    b. Santa should travel in a westerly direction.

19. a. 9:00 am Saturday 6 August
    b. 11:00 pm Friday 5 August

20. a. 40 hours
    b. 36 hours longer to drive compared to fly.
    c. 2:30 am on 28 December

Exercise 10.5 Distance: scales and maps

1. 1428 km, north easterly (NE)

2. 1760 km, north easterly (NE)

3. a. 10 m   b. 500 m
    c. 2.5 km   d. 70 km

4. a. 1:10   b. 1:80 000
    c. 1:650 000   d. 1:400 000 000

5. a. 1 cm on the map represents 10 km.
    b. i. 11 km   ii. 16 km   iii. 12 km
    c. 40 km
    d. 8.5 mm

6. a. 2500 km
    b. i. 2650 km   ii. 250 km   iii. 1250 km
       iv. 750 km   v. 650 km   vi. 1125 km

7. a. 1 cm: 30 km
    b. Wide: 24 km, long: 63 km
    c. 225 km
    d. 159 km

8. a. i. 107 mm   ii. 141 mm   iii. 116 mm
    b. i. 107 : 236 000 000
       ii. 141 : 310 000 000
       iii. 29 : 63 750 000
    c. There is slight variation in the calculations in part b, due to error involved in measuring, but these ratios are quite similar when checked by converting to decimals.

9. At the barbeque

Exercise 10.6 Speed

1. a. 2.38 m/s   b. 8.57 km/h

2. 42.35 km/h

3. a. m/s   b. km/h
    c. mm/s   d. m/s

4. 7.14 m/s

5. 10.4 km/h

6. 260 km

7. a. 362.917 km   b. 362 917 m

8. 264 m

9. 4320 m

10. 1 hour 44 minutes 10 seconds (1.736 hours)

11. 1 hour 10 minutes

12. a. 7442.6808 seconds
    b. 2 hours 4 minutes 3 seconds

13. 12.86 seconds

14. a. 32.46 km/h   b. 9.02 m/s

15. a. 7.509 km   b. 7509 m

16. a. 3.32 minutes
    b. 3 minutes 19 seconds

17. a. 1 hour 19 minutes
    b. 4.28 minutes
    c. 52.67 minutes
    d. 3.03 minutes

18. a. 21.67 m   b. 20 m/s or 72 km/h
    c. 1.59 seconds

Exercise 10.7 Distance–time graphs

1. Approximately 23 hours

2. Approximately 3 hours

3. 2000 seconds

4. 2000 seconds

5. 4 minutes 10 seconds

6. 5 hours 17 minutes 34 seconds

7. Approximately 1 hour 20 minutes

8. Approximately 9 hours

9. a. Approximately 53 minutes
    b. Approximately $7.50

10. a. 12 to 18 seconds   b. 18 to 22 seconds

11. a. 0 to 2 minutes   b. 6 minutes
   c. 2 minutes

12. a. 4 to 6 hours   b. 0 to 4 hours

13. a. D   b. D

14. Section A: Walking from 0 m to 120 m at a constant rate of
$$\frac{120}{150} = \frac{4}{5} \text{m/s} = 0.80 \text{ m/s}$$
Section B: Stationary, not moving.
Section C: Walking from 120 m to 200 m at a constant rate
of $\frac{80}{90} = \frac{8}{9}$ m/s = 0.89 m/s
Section D: Stationary, not moving.

15. a. Approximately 4.3 km
   b. 45 minutes
   c.   i. 44.44 m/s
       ii. 160 km/h

16. a. Since section B and D is flat therefore it is not moving, possible due to stopping at traffic lights (i.e. stationary).
   b. The car is travelling the fastest when it has the greatest gradient, therefore in section C.
   c.   i. 1 km/min   ii. 60 km/h
   d. 40 km/h
   e. 0 m/s

Exercise 10.8 Calculations with motion

1. a. 6 minutes
   b. 14 minutes
   c.   i. 0.86 km/min   ii. 51.43 km/h

2. 1.80 m/sec

3. a. 4 m/s   b. 40 m/s
   c. 1.9 km/h   d. 2.19 km/h

4. a. 2500 m   b. 180 km
   c. 67 500 m   d. 275 km

5. a. 13.9 m/s   b. 16.7 m/s
   c. 22.2 m/s   d. 27.8 m/s

6. a. 18.0 km/h   b. 54.0 km/h
   c. 64.8 km/h   d. 108.0 km/h

7. a. 2 hours 29 minutes
   b. No

8. B

9. 620.77 km/h

10. a. 75 minutes
   b. 2 hours 5 minutes
   c. 48 km/h

11. $151.20

12. 53 m

13. 0.2 seconds

14. 6.69 m/s

15. a. 29.17 m   b. 52.8 km/h

16. a. 31.1 m   b. 59.5 m   c. 90.6 m

17. 12 km/h

Exercise 10.9 Review: exam practice

1. D   2. C   3. C
4. D   5. B   6. C
7. B   8. C   9. A
10. D   11. D

12. a. 8:53 am   b. 8:25 am
   c. 8:05 am   d. 8:13 am

13. a. 5 trams   b. 8:30 am
   c. 23 minutes   d. 11:57

14. a. 8:50 am   b. 9:30 am
   c. 40 minutes

15. a. 8.59 m/s
   b. 388.89 m
   c. 44 minutes 27 seconds
   d. 72.26 km/h

16. 26.90 km/h

17. a. 9:40 am
   b. 12:18 pm
   c. 3:34 pm
   d. 1631
   e. 5:05 pm
   f. 10 hours 20 minutes

18. a. 4:40 pm
   b. 12:55 pm
   c. Five days a week (Monday to Friday).
   d. 7 hours 40 minutes

19. a. Approximately 3300 km
   b. Approximately 1000 km
   c. Approximately 750 km
   d. Brisbane
   e. Approximately 9 hours
   f. Approximately 68 km/h

20. a. Approximately 46.3 km
   b. 44 minutes
   c. 7:59 am
   d. QF0608 at 8:05 am. Answers may change over time.
   e. 2 hours 10 minutes
   f. Since it is after 4:00 pm catch QF 0631 at 5:00 pm.
   g. 7:25 pm
   h. 8:47 pm

PRACTICE ASSESSMENT 3

Essential Mathematics: Problem solving and modelling task

Unit
Unit 1: Money, travel and data

Topic
Topic 1: Time and motion

Conditions

Duration	Mode	Individual/group
5 weeks	Written report, up to 8 pages (maximum 1000 words) excluding appendix	Individual
Resources permitted		
The use of technology is required, for example: • calculator • spreadsheets • Internet • other mathematical software		

Criterion*	Grade
Formulate *Assessment objectives 1, 2, 5	
Solve *Assessment objectives 1, 6	
Evaluate and verify *Assessment objectives 4, 5	
Communicate *Assessment objective 3	
Milestones	
Week 1	
Week 2	
Week 3	
Week 4	
Week 5 (assessment submission)	

*Queensland Curriculum & Assessment Authority, *Essential Mathematics Applied Senior Syllabus 2019 v1.1*, Brisbane, 2018.

   For the most up to date assessment information, please see www.qcaa.qld.edu.au/senior.

Context

Australia has many beautiful places to visit and one of those is Perth in Western Australia. Unfortunately, we only get limited holiday time so it is important to plan ahead so we can see what we want to see in this limited time frame. To do this we need to consider how far we travel, how we want to travel, how fast we travel and how long this takes.

Task

This task requires you to plan a trip to Perth using different modes of transport. The trip is to start from your home and arrive in Perth city. When you arrive at Perth you must travel out to see at least one attraction. Your holiday will last one week.

Investigate the time requirements for your travel plan. You will need to investigate a combination of travel methods including plane, bus, car and train. Develop a logbook for each of your possible trips based on the results of your research, including the time of each trip, the average speed of each trip and the distance of each trip.

To complete this task you must:

- Present your findings as an investigative report based on the approach to problem-solving and mathematical modelling outlined in the Essential Mathematics syllabus, and on the flow chart on the following page of this instrument.
- Provide a response that highlights the real-life application of mathematics and demonstrates your knowledge of Unit 2, Topic 2 subject matter.
- Develop a unique response that is appropriately structured, and uses different data and assumptions to other students in your class and school.
- Show all working to support your response.

Approach to problem-solving and modelling

In this task you will investigate the time taken, average speed and distance in travelling to Perth for a trip. Your report should consider:
- different methods of transport
- the total distance of each section of the trip
- the time for each section of the trip
- the average speed for each section of the trip
- possible time differences.

Design a logbook that tracks each section of your trip and the times that you will be leaving and arriving for each part of the trip. Remember to state the necessary assumptions, variables and observations. You must also explain how you will make use of technology.

Develop a logbook to explain where you will be and at what time in your trip to Perth. Consider any necessary assumptions, variables and observations in your calculations. You will make further refinements in Stage 3, as necessary.

The logbook should include:
- times
- distances travelled
- average speeds of each section.

How long will it take you to take the trip, and how long will you have in Perth? What attraction will you visit when in Perth and how will you get there? You will need to show the time taken to travel, investigating different modes of transport such as plane, car, bus and train.

You must use technology efficiently and show detailed calculations demonstrating the procedures used to plan and budget for the car purchase.

Is it solved?

Evaluate the reasonableness of your original solution.

Based on your logbook and the assumptions made about your trip, consider what is the best way to travel and what times you need to depart and arrive each journey. Look at the strengths and limitations of your plan and make any necessary changes. Justify and explain all procedures you have used and decisions you have made. Considering the original task, how valid is your solution?

Is the solution verified?

Once you have completed all necessary calculations, you should consider how you have communicated all aspects of your report. Communicate using appropriate language that refers to the calculations and tables included in previous sections. Your response should be coherently and concisely organised.

Ensure you have:
- used mathematical, statistical and everyday language
- considered the strengths and limitations of your solution
- drawn conclusions by discussing your results
- included recommendations.

CHAPTER 11
Data collection

11.1 Overview

CONTENT

In this chapter, students will learn to:
- investigate the procedure for conducting a census
- investigate the advantages and disadvantages of conducting a census [complex]
- understand the purpose of sampling to provide an estimate of population values when a census is not used
- investigate the different kinds of samples [complex]
- investigate the advantages and disadvantages of these kinds of samples [complex]
- identify the target population to be surveyed
- investigate questionnaire design principles, including simple language, unambiguous questions, consideration of number of choices, issues of privacy and ethics, and freedom from bias [complex]
- describe the faults in the process of collecting data
- describe sources of error in surveys, including sampling error and measurement error
- investigate the possible misrepresentation of the results of a survey due to misunderstanding the procedure or the reliability of generalising the survey findings to the entire population [complex]
- investigate errors and misrepresentation in surveys, including examples of media misrepresentations of surveys [complex].

Fully worked solutions for this chapter are available in the Resources section of your eBookPLUS at www.jacplus.com.au.

11.2 Census and surveys

11.2.1 Data collection

Data collection is a process in which data is collected to obtain information and draw conclusions about issues of concern regarding a given population. A **population** consists of a complete group of people, objects, events, etc. with at least one common characteristic. Any subset of a population is called a **sample**. Data is collected using either a **census** of the entire population or a **survey** of a sample of the population.

11.2.2 Census

A census is conducted by official bodies at regular time intervals on a set date. In Australia, a census of the entire population and housing is conducted by the **Australian Bureau of Statistics (ABS)** every five years on **census night**. The next census in Australia is set to occur in August 2021.

Procedure for conducting a census

A census consists of three main stages:
- planning, when strategies and methods of conducting the census are established
- collecting data, when the required information is collected from each member of the population
- producing and releasing results, when the data collected is analysed and the results are released to the public.

Advantages and disadvantages of conducting a census

The information collected through a census has its advantages and disadvantages.

Advantages

The advantages of the information collected through a census is that it:
- represents a true measure of the entire population
- can be used for further studies
- provides detailed information about minority groups within the entire population.

Disadvantages

The disadvantages of the information collected through a census include the:
- high costs involved
- long period of time required to collect, analyse, produce and release the results
- data can become out of date quite quickly.

For example, a company manufacturing school shoes wanted to collect data on the shoe size of every student in the country. This process would be very costly and it would take a long time to collect, analyse and produce the results. The company should investigate a sample of the student population for a fast and cost-effective estimate.

WORKED EXAMPLE 1

For each of the following situations state whether the data should be collected using a census or a survey. Give reasons for your choice.

a. All students in a school are asked to state their method of travel to and from school.

▶

---

b. Data is collected from every third person leaving a shopping centre on the time spent shopping.

c. Data was collected from 500 households regarding their average monthly water usage.

THINK	WRITE/DRAW
a. 1. Determine whether the group considered represents an entire population or a sample.	All students in a school represent the whole student population of the school.
2. State whether a census or a survey is required.	As this represents the entire student population for the given school, a census is required.
b. 1. Determine whether the group considered represents an entire population or a survey.	As only every third person is involved in the data collection. This represents a sample of the whole population of people leaving the shopping centre.
2. State whether a census or a sample is required.	As this represents a sample, a survey is required.
c. 1. Determine whether the group considered represents an entire population or a survey.	Only 500 households are involved in the data collection, so this represents a sample of all the households.
2. State whether a census or a survey is required.	As this represents a sample, a survey is required.

 Resources

 Interactivity: Collecting data (int-3807)

11.2.3 Surveys and sampling

As already stated in the introduction of this chapter, a **survey** of a sample of a given population is considered when a census cannot be used.

Sampling is the process of selecting a sample of a population to provide an estimate of the entire population. A sample has to maintain all the characteristics of the population it represents. Four methods of sampling will be discussed in the following section. Regardless of the method of sampling used, keep in mind that the sample has to be **representative** of the population.

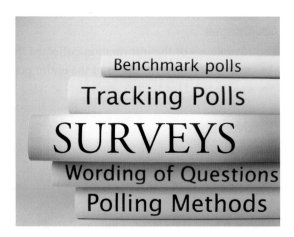

Simple random sampling

Simple random sampling is the basic method of selecting a sample. This type of sampling ensures that each individual of a population has an equal chance of being selected for the sample.

There are some complex formulae to calculate the **sample size**. For the purpose of this chapter, the sample size will be calculated using the formula

$$S = \sqrt{N}, \text{ where } N \text{ is the size of the population}$$

Consider a population of 25 Year 12 Essential Mathematics students and a sample of 5 students. A basic method of choosing the sample is by assigning each Year 12 student a unique number m 1 to 25, writing each

---

number on a piece of paper, placing all the papers in a box or a bowl, shaking them well and then choosing 5 pieces of paper. The students who correspond to the numbers drawn will be the sample of the Year 12 student population.

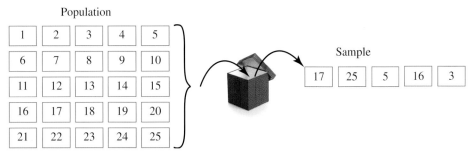

Calculators or computer software make the process of selecting a sample a lot easier by using a **random number generator**. An Excel worksheet generates random numbers using the **RAND( )** or **RANDBE-TWEEN (a,b)** command.

The **RAND( )** command generates a random number between 0 and 1. Depending on the size of the sample these generated numbers have to be multiplied by n, where n is the size of the population.

	A	B	C	D	E
1	0.433821				
2	0.78391				
3	0.547901				
4	0.892612				
5	0.390466				
6	0.264742				
7	0.690003				
8	0.070899				
9	0.876409				
10	0.976012				

RANDBETWEEN(a,b) generates a random number from **a** to **b**. For the population of Year 12 Essential Mathematics students, the sample of five students can be generated by using RANDBETWEEN(1, 25).

	A	B	C	D	E	F
1	9					
2	3					
3	25					
4	3					
5	17					

If the same number appears twice, a new number will need to be generated in its place to achieve five different numbers.

WORKED EXAMPLE 2

Select a sample of 12 days between 1 December and 31 January
a. **'by hand'.**
b. **using a random number generator.**

THINK	WRITE/DRAW
a. 1. Write every member of the population on a piece of paper.	There are 31 days in December and 31 days in January. This will require 62 pieces of paper to write down each day. 1/12, 2/12,30/01, ... ,31/01 *Note:* Alternatively, each day could be assigned a number, in order, from 1 to 62. ▸

2. Fold all papers and put them in a box.

Ensure that the papers are folded properly so no number can be seen.

3. Select the required sample. A sample of 12 days is required so randomly choose 12 papers from the box.

Sample: 15, 9, 41, 12, 1, 7, 36, 13, 26, 5, 50, 48

4. Convert the numbers into the data represented.

15 represents 15/12, 9 represents 9/12, 41 represents 10/01, 12 represents 12/12, 1 represents 1/12, 7 represents 7/12, 36 represents 5/01, 13 represents 13/12, 26 represents 26/12, 5 represents 5/12, 50 represents 19/01, 48 represents 17/01.

5. State the sample selected.

15/12, 9/12, 10/01, 12/12, 1/12, 7/12, 5/01, 13/12, 26/12, 5/12, 19/01, 17/01.

b. 1. Assign each member of the population a unique number.

Assign each day a number, in order, from 1 to 62: 1/12 = 1, 2/12 = 2, ... , 31/01 = 62.

2. Open a new Excel worksheet and use the RANDOMBETWEEN(1, 62) command to generate the random numbers required.

	A	B	C	D	E	F
1	29					
2	27					
3	17					
4	26					
5	4					
6	45					
7	41					
8	25					
9	10					
10	28					
11	16					
12	8					

Note: Some of the numbers may be repeated. For this reason, more than 12 numbers should be generated. Select the first 12 unique numbers.

3. Convert the numbers into the data represented.

29 represents 29/12, 27 represents 27/12, 17 represents 17/12, 26 represents 26/12, 4 represents 4/12, 45 represents 14/01, 41 represents 10/01, 25 represents 25/12, 10 represents 10/12, 28 represents 28/12, 16 represents16/12, 8 represents 8/12.

4. State the sample selected.

29/12, 27/12, 17/12, 26/12, 4/12, 14/01, 10/01, 25/12, 10/12, 28/12, 16/12, 8/12.

on Resources

Interactivity: Random numbers (int-0089)

Systematic sampling

Systematic sampling or **systematic random sampling** requires a starting point chosen at random with members of the population chosen at regular intervals.

For example, if a sample of 5 students out of 30 is needed for a survey, to ensure that each member of the population has an equal chance of being chosen, divide the whole population by the size of the sample. In this case, $30 \div 5 = 6$; so, the starting point will be a random number from 1 to 6 and then, every 6th student will be chosen.

Consider the case of choosing a sample of 5 students out of a population of 31 students. In this case, $31 \div 5 = 6.2$; so, should the sample be chosen using every 6th student or every 7th student? If every 6th student is chosen, the last student will never be chosen. For this reason, the starting point should be between 1 and 7. If the 7th student is chosen, the last chosen student might not exist. For this reason, the starting point should be between 1 and 3.

The starting point interval is chosen using the formula:

> **Maximum starting point** $= N - I(s - 1)$,
> where N is the size of the population
> s is the size of the sample
> I is the whole number of $\dfrac{N}{s}$.

Advantages

Advantages of using systematic sampling include
- it is easier to form than a randomly generated sample
- the sample is more evenly spread over the population.

Disadvantage

A disadvantage of using systematic sampling is that the population could have some hidden pattern; for example, every 6th student in the example discussed is a girl.

WORKED EXAMPLE 3

Using a systematic random sampling method, a researcher wants to choose 8 people from a population of 59 people. State one such sample.

THINK	WRITE/DRAW
1. Write a list of all the people in the population.	For the purpose of this example we are going to use the initials of the people. GF, AS, TG, YH, ID, BK, CD, YT, UE, OM, LP, HI, BT, SJ, FR, IV, BX, MN, IT, UM, WK, FA, FP, ST, VP, AA, AR, BT, OD, PM, LC, RP, PO, GV, BF, TP, AD, IP, CR, AN, OO, FC, LM, LC, AF, AK, PY, SF, GH, VS, MR, CB, FJ, RM, CS, KF, GS, WM, FX
2. Choose an interval.	$\dfrac{59}{8} = 7.37$. The researcher could choose either every 7th person or every 8th person.
3. Choose the starting point.	Let $I = 7$ Maximum starting point $= N - I(s - 1)$ $\qquad\qquad\qquad\qquad = 59 - 7 \times (8 - 1)$ $\qquad\qquad\qquad\qquad = 59 - 49$ $\qquad\qquad\qquad\qquad = 10$ The starting point will be a number from 1 to 10. Let this number be 4.
4. Form the sample.	The sample will be formed by the following people: 4th, 11th, 18th, 25th, 32nd, 39th, 46th, 53rd GF, AS, TG, **YH**, ID, BK, CD, YT, UE, OM, **LP**, HI, BT, SJ, FR, IV, BX, **MN**, IT, UM, WK, FA, FP, ST, **VP**, AA, AR, BT, OD, PM, LC, **MR**, PO, GV, BF, TP, AD, IP, **CR**, AN, OO, FC, LM, LC, AF, **AK**, PY, SF, GH, VS, RP, CB, **FJ**, RM, CS, KF, GS, WM, FX
5. State the sample selected.	YH, LP, MN, VP, MR, CR, AK, FJ

Self-selected sampling

Self-selected sampling is used when the members of a population are given the choice to participate in a research or not. In this type of sampling the researcher chooses the **sampling strategy**. Such strategies may include:

- asking for volunteers to participate in a trial of a drug
- a website asking customers to answer a short questionnaire.

Self-selected sampling requires two steps:

- publicising the needs of the study to potential participants
- accepting or rejecting the applicants offering to participate in the study.

For example, if you were asked to form a sample of 20 students out of 200 Year 11 students, the two steps would be:
- advertising your study around the school
- accept the Year 11 students that showed an interest in participating in your study and rejecting any other students who are not in Year 11.

Advantages

Advantages of using self-selected sampling include
- it is less time consuming to generate the sample
- participants would be more committed to the study.

Disadvantages

Disadvantages of using self-selected sampling are that
- it is likely to have some degree of bias
- it may not be representative of the population studied.

Although it does not benefit from the random selection provided by the simple random sampling, the self-selected sampling is very popular in studies that require comparisons between the mean performances of two or more groups.

WORKED EXAMPLE 4

A pharmaceutical company wants to trial a new flu vaccine. The study requires adults aged between 20 and 80 years. State the steps required to form a self-selected sample of 100 participants.

THINK	WRITE/DRAW
1. State the ways to publicise the study.	Possible ways to publicise the study are: • radio • TV • hospitals • general practitioners.
2. State the characteristics required for accepting or rejecting the applicants.	Accept 100 adults aged between 20 and 80 years old. Reject any others.

Stratified sampling

Stratified sampling is used when there are variations in the characteristics of a population. This method requires the population to be divided into sub-groups called **strata**. For example, if the Year 12 students of a secondary college were asked about their favourite movie, the preferences could be different for boys and girls. For this reason, the sample would have to reflect the same proportion of boys and girls as the actual population.

$$\text{Sample size for each subgroup} = \frac{\text{Sample size}}{\text{Population size}} \times \text{Subgroup size}$$

Once the sample size of each subgroup has been determined, random sampling or systematic sampling can be used to form the sample required.

Calculate the number of female students and male students required to be part of a sample of 25 students if the student population is 652 with 317 male students and 335 female students.

THINK	WRITE/DRAW
1. State the formula for determining the sample size.	$$\text{Sample size for each subgroup} = \frac{\text{Sample size}}{\text{Population size}} \times \text{Subgroup size}$$
2. Calculate the sample size for each subgroup.	$$\text{Sample size for females} = \frac{25}{652} \times 335$$ $$= 13$$ $$\text{Sample size for males} = \frac{25}{652} \times 317$$ $$= 12$$
3. State the answer.	The sample should contain 13 female students and 12 male students.

Exercise 11.2 Census and surveys

1. **WE1** For each of the following situations state whether the data should be collected using a census or a survey. Give reasons for your choice.
   a. An online business recorded the time spent by its customers searching for a product.
   b. 50% of the Year 12 students at a secondary college were asked to state their preference for either online tutorials or face-to-face tutorials.
   c. The heights of all the patients at a local hospital were recorded.

2. Which of the following represents a sample and which one represents a population? Give reasons for your choice.
   a. The number of absences of all year 12 students in a school and the number of absences of all students in the school.
   b. The number of passenger airplanes landing at an airport and the total number of airplanes landing at the same airport.

3. State whether a census or a survey is required to collect the data in the following statistical investigations. Give reasons for your answer.
   a. Water savings in 150 suburbs across the country.
   b. Highest educational level of all people in Australia.
   c. Roll marking at homeroom in a secondary college.
   d. Every fifth person leaving the theatre is asked whether they liked the play or not.
   e. Every customer in a car dealership is asked to fill in a questionnaire.

4. **WE2** Select a sample of 5 students to participate in a relay from a group of 26 students
   a. 'by hand'.
   b. using a random number generator.

5. A small business has 29 employees. The owner of the business has decided to survey 8 employees on their opinion about working conditions. Select the required sample
   a. 'by hand'.
   b. using a random number generator.

6. Calculate the approximate sample size of a population of 23 500 people.

7. Generate a sample of 10 members from a population of 100 people using a random number generator using an Excel worksheet and the command
   a. RAND ( )
   b. RANDBETWEEN(0, 100)

8. **WE3** A car manufacturer wants to test 5 cars from a lot of 32 cars. Form one possible sample by using a systematic random sampling method.

9. A quality control officer is conducting a survey of 46 products. She decides to sample 9 products using a systematic random sampling method. Form two possible samples by using systematic random sampling methods.

10. Stratified sampling is used in the following surveys. Calculate the number of members needed from each subgroup.
    a. A school has a population of 127 Year 12 students where 61 students are boys. A sample of 11 students is chosen to represent all the Year 12 students in an interstate competition.
    b. The local cinema is running a promotion movie screening. A sample of 10 people from those attending the screening will be chosen to receive a free pass to the next movie screening. If the total number of people to attend the promotion screening is 138 with 112 adults and 26 children, how many children and how many adults will receive a free movie pass in order to keep a proportional sampling?

11. **WE4** The local council is seeking volunteers to participate in a study on how to improve the local public library. State the steps required to form a self-selected sample of 50 participants.

12. A gym instructor is conducting a study on the effects of regular physical exercise on the wellbeing of teenagers. State the steps required to form a self-selected sample of 30 participants.

13. Census At School Australia is a project that collects information from Australian students using a questionnaire developed by The Australian Bureau of Statistics. It is not compulsory for students to participate in this census; however, students can volunteer to participate. What type of sampling is used in this project?

14. A farmer has 46 black sheep and 29 white sheep. He wants to try a new diet on a sample of 10 sheep.
    a. What type of sampling is the farmer using?
    b. Calculate the number of black sheep and white sheep required for this trial in order to give a proportional representation of both types of sheep.

15. A manufacturing company of mobile phones is testing two batches of 100 phones each for defects. In the first batch, every 5th phone is checked while in the second batch, every 7th phone is checked.

   a. What method of sampling is used for this testing?

   b. How many phones are being tested for defects in the two batches?

   c. What is the maximum starting point for each batch?

16. **WE5** Calculate the sample size of the subgroups in a sample of 33 people from a population of 568 males and 432 females.

17. Calculate the sample size of the subgroups in a sample of 15 cars from a population of 179 white cars and 215 black cars.

18. At Guntawang Secondary College, a survey is conducted to collect information from the 230 students enrolled.

   a. Calculate the appropriate sample size for this population.

   b. Explain how the sample could be chosen.

19. Generate a sample of 12 members from a population of 140 people using a random number generator using an Excel worksheet and the command

   a. RAND( )   b. RANDBETWEEN(0, 140)

20. For a population of 800 people,

   a. calculate the size of the sample required.

   b. use the command RAND(0, 800) to generate the sample.

11.3 Simple survey procedures

11.3.1 Surveys

Surveys, if not correctly done, might provide misleading results or results that are not representative of the whole population. Therefore the procedures and methods of conducting surveys are of great importance.

In a survey the data is collected from a sample of a given population. The characteristics of the population are estimated from the characteristics of the sample using statistical procedures.

Target population

The members of a population could be any entities such as people, animals, organisations and businesses. A survey is concerned with two populations: the **target population** and the **survey population**. The target population is the entire population while the survey population includes those members of the target population who are available to be selected in a sample.

For example, if you were interested in finding the students' opinion about a school policy, the target population would be all the students in the school. However, if you were interested in finding out the student's opinion about learning to drive, the target population would be all the students aged 16 or over.

A company is conducting three surveys in order to obtain information required to improve its employees' work conditions and life style. Define the target population in each of the surveys.
a. A survey about the time spent by its employees to travel to work.
b. A survey about the amount of money spent weekly on childcare.
c. A survey about the amount of time spent working on a computer.

THINK	WRITE/DRAW
a. Define the target population.	All the company's employees.
b. Define the target population.	All employees with children.
c. Define the target population.	All employees required to do work on a computer.

Questionnaires

Surveys can be administered using paper or electronic questionnaires, face-to-face interviews or by telephone. The most common type of survey is the traditional paper survey in the form of questionnaires. A **questionnaire** is a list of questions used to collect data from the survey's participants. Designing questionnaires is probably one of the most important tasks when administering a survey.

Some general guidelines for wording questions

- Use simple and easy to understand language.
- Avoid abbreviations and jargon. For example, VCE (Victorian Certificate of Education), ABS (Australian Bureau of Statistics) and HSC (Higher School Certificate) might be understood by some people but unknown to others.
- Use short sentences. For example, 'There are many types of movies available in cinemas that people watch. Which type do you think most represents your preference?' could be replaced by a simple, straightforward question such as 'What type of movie do you prefer?'
- Avoid vague or ambiguous questions by asking precise questions.
- Avoid double negatives. For example, 'Would you say that your mathematics teacher is not unqualified?' is an affirmation; however, some people might not be aware of this and the question might look confusing.
- Ask for one piece of information at a time and avoid double-barrel questions. 'How satisfied are you with your job and your boss?' is a double-barrel question putting two pieces of information together.
- Avoid leading questions. For example, 'Do you agree with the new school council structure? Yes or No.' could be replaced with a question such as 'Do you agree or disagree with the new school council structure? 1. Strongly agree, 2. Agree, 3. Disagree, 4. Strongly disagree.'
- Give consideration to sensitive issues like privacy and ethics. Questions about issues such as income, religious beliefs, political views and gender orientation are sensitive topics and they have to be stated in a sensitive manner.
- Avoid or eliminate bias. Consider the question, 'How would you rate your mathematics textbook: Excellent, Good or Average?' The bias is in the absence of a negative option. Questions with 'always' and 'never' can also bias participants.

Consider the following types of questions and explain why they are
not appropriate for a questionnaire. For every question state
a possible acceptable replacement.

a. How many TV sets do you own?

b. How many times did you purchase lunch from the school's
cafeteria last year?

c. What is your yearly income?

THINK	WRITE/DRAW
a. 1. State any concerns about the question.	The question assumes that all respondents own a TV set. Some respondents might have a TV set in their household but they don't own it.
2. State a possible replacement question or questions.	Do you own a TV set? If yes, how many TV sets do you own?
b. 1. State any concerns about the question.	It is hard to understand what period 'last year' covers. The question assumes that all students buy lunch from the school's cafeteria. It is hard to recall the exact number of times.
2. State a possible replacement question or questions.	Did you buy lunch at the school's cafeteria last school year (or over the last 12 months)? If yes, state the approximate number of times. Alternatively, provide categories like 0–5, 6–10, 10–15 etc.
c. 1. State any concerns about the question.	People usually don't like to state their exact income. Broad categories are more likely to be answered honestly.
2. State a possible replacement question or questions.	What category best describes your annual income? Less than $25 000, $25 000–$50 000, $50 000–$75 000, etc.

Types of questions

Questions have to be designed keeping in mind the relevance to the survey. The data collected has to be clear
and easy to analyse.

There are three main types of questions: **open-ended, closed** and **partially closed**.

In **open-ended** questions the participants have the opportunity to state their responses in their own words.
However, the participants have to spend more time writing the answer and the responses are harder to analyse.

Closed questions provide a list of answers from which the participants have to choose one or more options. The data collected through closed questions is easier to analyse and less time consuming. However, care has to be taken that all the possible options have been listed.

Partially closed questions provide a compromise between the open-ended and closed questions. Both answer choices and an option of creating own responses are provided. This type of question allows participants to state their own answer if their choice does not fit the options listed.

WORKED EXAMPLE 8

State whether the following questions are open-ended, partially closed or closed.

a. **What mode of transport do you use when travelling to school?**

b. **What mode of transport do you use to travel to school? Multiple options may be selected.**
   1. Walk   2. Bus   3. Tram   4. Bike   5. Car

c. **What mode of transport do you use to travel to school?**
   1. Walk   2. Bus   3. Tram   4. Bike   5. Car   6. Other ☐

THINK	WRITE/DRAW
a. 1. Consider if a written answer is required or possible options are available.	The question requires a written answer.
2. State the type of the question.	Open-ended
b. 1. Consider if a written answer is required or possible options are available.	The question has possible options available. However, it does not exhaust all possible options.
2. State the type of the question.	Closed
c. 1. Consider if a written answer is required or possible options are available.	The question has possible options available. However, it does allow the participant to write their own option.
2. State the type of the question.	Partially closed

Exercise 11.3 Simple survey procedures

1. **WE6** Define the target population in the following surveys:
   a. The time spent by the Year 12 students in your school completing homework.
   b. The time spent by the students in your school completing homework.
   c. The time spent by the students in Australia completing homework.

2. Define the target population of the following samples:
   a. A dairy farm has selected a sample of 20 bottles of milk to test for quality.
   b. An internet provider has selected a sample of 150 people to survey about the quality of their Internet service.

3. Maya is conducting a survey on the preferred way of doing research for statistical investigations. State the target population if she decides to survey
   a. a sample of 20 year 11 students.
   b. a sample of 30 students from years 11 and 12.
   c. a sample of 50 students from the whole school.

4. **WE7** Consider the following types of questions and explain why they are not appropriate for a questionnaire. For every question state a possible acceptable replacement.
   a. What brand of mobile phone do you have?
   b. Do you agree or disagree that drinking alcohol on the beach is not permitted but smoking is permitted?
   c. What church do you attend on Sundays?

5. Consider the following types of questions and explain why they are not appropriate for a questionnaire. For every question state a possible acceptable replacement.

 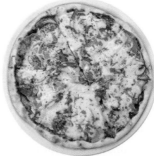

   a. Do you like pizza, pasta or both?
   b. Are your parents not unclear in their expectations of you?

6. **WE8** State whether the following questions are open-ended, partially closed or closed.

   a. How would you rate the quality of the meat bought from our supermarket?
      1. Excellent,   2. Good,   3. Average,   4. Poor
   b. What is your favourite brand of cereal?
   c. Can you improve your grades in mathematics? Yes or No.
   d. What steps have you taken to improve your Mathematics grades?
      1. Completing more homework.
      2. Asking the teacher for assistance more often.
      3. Spending more time understanding the concepts.
      4. Other.

7. Write one question of the following types.
   a. Open-ended   b. Closed   c. Partially closed
8. Design an open-ended question, a partially closed question and a closed question for a questionnaire surveying the Internet search engine used by the students in your year level.
9. The following questions are not properly worded for a questionnaire. State the reasoning behind this statement and rewrite them using the guidelines required.
   a. Do you catch a bus to work or a tram?
   b. Which of these is the most important issue facing teenagers today? Circle one answer.
      The environment, Binge drinking, Teenage pregnancy.
   c. Do you feel better now that you have stopped smoking?
   d. How much money do you earn per week?
10. State if the following questions are open-ended, partially closed or closed.
   a. What is your country of birth?
      ☐ Australia ☐ Other _____
   b. Do you have an account on a social networking site?
      ☐ Yes ☐ No
   c. What is your preferred way of spending weekends?
11. The ABS conducts a census every five years. To monitor changes that might occur between these times, surveys are conducted on samples of the population. The ABS selects a representative sample of the population and interviewers are allocated particular households. It is important that no substitutes occur in the sampling. The interviewer must persevere until the selected household supplies the information requested. It is a legal requirement that selected households cooperate.

    The following questionnaire is reproduced from the ABS website www.abs.gov.au. It illustrates the format and types of questions asked by an interviewer collecting data regarding employment from a sample.

Q.1.	**I WOULD LIKE TO ASK ABOUT LAST WEEK, THAT IS, THE WEEK STARTING MONDAY THE ... AND ENDING (LAST SUNDAY THE .../YESTERDAY).**
Q.2.	LAST WEEK DID ... DO ANY WORK AT ALL IN A JOB. BUSINESS OR FARM? Yes ☐ **Go to Q.5** No ☐ Permanently unable to work ☐ **No More Questions** Permanently not intending to work (if aged 65+ only) ☐ **No More Questions**
Q.3.	LAST WEEK DID ... DO ANY WORK WITHOUT PAY IN A FAMILY BUSINESS? Yes ☐ **Go to Q.5** No ☐ Permanently not intending to work (if aged 65+ only) ☐ **No More Questions**
Q.4.	DID ... HAVE A JOB, BUSINESS OR FARM THAT ... WAS AWAY FROM BECAUSE OF HOLIDAYS, SICKNESS OR ANY OTHER REASON? Yes ☐ No ☐ **Go to Q.13** Permanently not intending to work (if aged 65+ only) ☐ **No More Questions**
Q.5.	DID ... HAVE MORE THAN ONE JOB OR BUSINESS LAST WEEK? Yes ☐ No ☐ **Go to Q.7**

Q.6.	THE NEXT FEW **QUESTIONS** ARE ABOUT THE JOB OR BUSINESS IN WHICH ... USUALLY WORKS THE MOST HOURS.	
Q.7.	DOES ... WORK FOR AN EMPLOYER, OR IN ... OWN BUSINESS?	
	Employer	☐
	Own business	☐ **Go to Q.10**
	Other/Uncertain	☐ **Go to Q.9**
Q.8.	IS ... PAID A WAGE OR SALARY, <u>OR</u> SOME OTHER FORM OF PAYMENT?	
	Wage/Salary	☐ **Go to Q.12**
	Other/Uncertain	☐
Q.9.	WHAT ARE ... (WORKING/PAYMENT) ARRANGEMENTS?	
	Unpaid voluntary work	☐ **Go to Q.13**
	Contractor/Subcontractor	☐
	Own business/Partnership	☐
	Commission only	☐
	Commission with retainer	☐ **Go to Q.12**
	In a family business without pay	☐ **Go to Q.12**
	Payment in kind	☐ **Go to Q.12**
	Paid by the piece/item produced	☐ **Go to Q.12**
	Wage/salary earner	☐ **Go to Q.12**
	Other	☐ **Go to Q.12**
Q.10.	DOES ... HAVE EMPLOYEES (IN THAT BUSINESS)?	
	Yes	☐
	No	☐
Q.11.	IS THAT BUSINESS INCORPORATED?	
	Yes	☐
	No	☐
Q.12.	HOW MANY HOURS DOES ... USUALLY WORK EACH WEEK IN (THAT JOB/THAT BUSINESS/ALL ... JOBS)?	
	1 hours or more	☐ **No More Questions**
	Less than 1 hours/no hours	☐
	Insert occupation questions if required	
	Insert industry questions if required	
Q.13.	AT ANY TIME DURING THE LAST 4 WEEKS HAS ... BEEN LOOKING FOR FULL-TIME OR PART-TIME WORK?	
	Yes, full-time work	☐
	Yes, part-time work	☐
	No	☐ **No More Questions**

▶

Q.14.	AT ANY TIME IN THE LAST 4 WEEKS HAS …	
	Written, phoned or applied in person to an employer for work?	☐
	Answered an advertisement for a job?	☐
	Looked in newspapers?	☐
	Yes	☐
	No	☐
	Checked factory notice boards, or used the touchscreens at Centrelink offices?	
	AT ANY TIME IN THE LAST 4 WEEKS HAS …	
	Been registered with Centrelink as a jobseeker?	☐
	Checked or registered with an employment agency?	☐
	Done anything else to find a job?	☐
	Advertised or tendered for work	☐
	Contacted friends/relatives	☐
	Other	☐ **No More Questions**
	Only looked in newspapers	☐ **No More Questions**
	None of these	☐ **No More Questions**
Q.15.	IF … HAD FOUND A JOB COULD … HAVE STARTED WORK LAST WEEK?	
	Yes	☐
	No	☐
	Don't know	☐
	Remaining questions are only required if Duration of Unemployment is needed for output or to derive the long-term unemployed.	
Q.16.	WHEN DID … BEGIN LOOKING FOR WORK?	
	Enter Date	… ../ … ../ … …
	Less than 2 years ago	**DD MM YY**
	2 years or more ago	… ../ … ../ … …
		DD MM YY
	5 years or more ago	… ../ … ../ … …
		DD MM YY
	Did not look for work	☐
Q.17.	WHEN DID … LAST WORK FOR TWO WEEKS OR MORE?	
	Enter Date	… ../ … ../ … …
	Less than 2 years ago	**DD MM YY**
		… ../ … ../ … …
	2 years or more ago	**DD MM YY**
		… ../ … ../ … …
	5 years or more ago	**DD MM YY**
	Has never worked (for two weeks or more)	☐ **No More Questions**

Reading the questionnaire carefully you will note that, although the questions are labelled 1 to 17, there are only 15 questions requiring answers (two are introductory statements to be read by the interviewer). Because of directions to forward questions, no individual would be asked all 15 questions.

    a. How many questions would be asked of those who have a job?

    b. How many questions would unemployed individuals answer?

    c. How many questions apply to those not in the labour force?

Choose a topic of interest to you and conduct a survey

12. a. Design an interview questionnaire of a similar format to the ABS survey, using directions to forward questions.

    b. Decide on a technique to select a representative sample of the students in your class.

    c. Administer your questionnaire to this sample.

    d. Collate your results.

    e. Draw conclusions from your results.

    f. Prepare a report which details the:

        i. aim of your survey

        ii. design of the survey

        iii. sample selection technique

        iv. results of the survey collated in table format

        v. conclusions.

11.4 Sources of bias

11.4.1 Bias

Bias is a distortion that occurs during the data collection process which produces results that are not representative of the population. It can occur in different forms and at various stages during the process. Care has to be taken when planning for a survey to avoid and possibly eliminate bias.

Accurate data collection is essential in maintaining the integrity of any statistical study because its results are often used in decision making.

Sources of bias include:

- the sample is not representative of the population being studied
- questions that do not respect the guidelines required.

11.4.2 Sources of error

When forming a sample from the population, there are two main errors to look for: sampling error and measurement error.

Sampling error

This type of error is probably the most common type of error that happens in a survey. **Sampling errors** occurs due to the fact that surveys are conducted on one sample out of the large number of possible survey samples.

Some sampling errors occur when:

- the sample is too small
- subgroups of the population are unrepresented or overrepresented
- the sample was not selected at random.

The **'margin of error'** is the range of values that result when using a sample of a population. For example, if 60% of the people in a sample answered 'Yes' to a question then, it is assumed that, for a 5% margin of error, between 57.5% and 62.5% of the population answered 'Yes' to that question. The margins of error are calculated using the formula:

$$\text{Lower margin} = x\% - \frac{e\%}{2} \text{ and Upper margin} = x\% + \frac{e\%}{2}$$
**where $x\%$ is the percentage of the sample considered,
$e\%$ is the percentage error**

The **confidence level** is used to find the percentage of results that will not fall within the margin of error stated. If the confidence level for the situation described above is 95%, it means that there is a 5% risk that the results will not fall between 57.5% and 62.5%.

$$\text{Risk} = 100\% - \text{Confidence level}$$

This table provides recommended sample sizes for two different margins of error.

Population size	Sample size 5%	Sample size 10%	Population size	Sample size 5%	Sample size 10%
10	10		275	163	74
15	14		300	172	76
20	19		325	180	77
25	24		350	187	78
30	28		375	194	80
35	32		400	201	81
40	36		425	207	82
45	40		450	212	82
50	44		475	218	83
55	48		500	222	83
60	52		1000	286	91
65	56		2000	333	95
70	59		3000	353	97
75	63		4000	364	98
80	66		5000	370	98
85	70		6000	375	98
90	73		7000	378	99
95	76		8000	381	99
100	81	51	9000	383	99
125	96	56	10 000	385	99
150	110	61	15 000	390	99
175	122	64	20 000	392	100
200	134	67	25 000	394	100
225	144	70	50 000	397	100
250	154	72	100 000	398	100

Source: Isaac and Michael, 1981; Smith, M. F, 1983.

WORKED EXAMPLE 9

a. Calculate the lower and upper margins for a 10% margin of error if 24% of the members of a sample were found to be blue eyed.

b. If the confidence level is 85%, what is the risk of the results falling outside the margins calculated in part a?

THINK	WRITE/DRAW
a. 1. Write the formula for the lower margin.	Lower margin $= x\% - \dfrac{e\%}{2}$
2. Substitute the known values and simplify.	$= 24\% - \dfrac{10\%}{2}$ $= 24\% - 5\%$ $= 19\%$
3. Write the formula for the upper margin.	Upper margin $= x\% + \dfrac{e\%}{2}$
4. Substitute the known values and simplify.	$= 24\% + \dfrac{10\%}{2}$ $= 24\% + 5\%$ $= 29\%$
5. State the answer.	Between 19% and 29% of the population is blue eyed.
b. 1. Calculate the risk.	Risk $= 100\% - $ Confidence level $= 100\% - 85\%$ $= 15\%$
2. State the answer.	There is a 15% risk that the blue eyed members will fall outside the interval 19% – 29%.

Measurement error

This type of error is probably the second most common type of error that happens in a survey. A **measurement error** is, in general, the difference between the measured value and the real value. In statistics this error occurs when there are imperfections in the process of data collection. A poorly designed questionnaire is the most common type of measurement error.

11.4.3 Misrepresentation and misunderstanding

A **misrepresentation** of the results of a survey is an untrue statement that does not reflect the characteristics of the population. Sometimes the media misrepresents the results of a survey by stating conclusions that contradict a survey's report or by failing to mention the sponsors of the survey.

WORKED EXAMPLE 10

Consider the following excerpt from an article that reviews a survey conducted by the Environment Committee of the Association of Professional Engineers, Geologists and Geophysicists of Alberta, Canada, (APEGGA). The survey was conducted on the members of APEGGA 'to assess their beliefs and values about climate change'.

Peer-reviewed survey finds majority of scientists skeptical of global warming crisis

It is becoming clear that not only do many scientists dispute the asserted global warming crisis, but these skeptical scientists may indeed form a scientific consensus.

Don't look now, but maybe a scientific consensus exists concerning global warming after all. Only 36 per cent of geoscientists and engineers believe that humans are creating a global warming crisis, according to a survey reported in the peer-reviewed *Organization Studies*. By contrast, a strong majority of the 1077 respondents believe that nature is the primary cause of recent global warming and/or that future global warming will not be a very serious problem.[1]

The Actual Survey, Science or science fiction? professionals' discursive construction of climate chage'[2], states that 'Given the similarity of the survey respondents to APEGGA's general membership, this suggests [a 3.1% error] that responses may be generalizable to the membership as a whole' and '27.4% believe it is caused by primarily natural factors (natural variation, volcanoes, sunspots, lithosphere motions, etc.), 25.7% believe it is caused by primarily human factors (burning fossil fuels, changing land use, enhanced water evaporation due to irrigation), and 45.2% believe that climate change is caused by both human and natural factors'.[3]

1. *Source: Forbes*, http://www.forbes.com/sites/jamestaylor/2013/02/13/peer-reviewed-survey-finds-majority-of-scientists-skeptical-of-global-warming-crisis/
2. Lianne M. Lefsrud, University of Alberta, Canada; Renate E. Meyer, Vienna University of Economics and Business, Austria and Copenhagen Business School, Denmark.
3. *Source:* SAGE journals, *Organization Studies*, http://oss.sagepub.com/content/33/11/1477.full

State any misinterpretations or misunderstandings found in the article when discussing the survey.

THINK	WRITE/DRAW
1. Start with the title of the article and check the survey to check its reliability.	'Peer-reviewed survey finds majority of scientists skeptical of global warming crisis' The title generalises the findings of the survey to all scientists. The survey states ' ... generalizable to the membership as a whole.'
2. Comment on the misrepresentation of the results.	As only members of an organisation were surveyed the results can be generalised only to the organisation's scientific population and not all scientists.
3. Check each sentence and compare with the results stated in the survey.	Article: 'By contrast, a strong majority of the 1077 respondents believe that nature is the primary cause of recent global warming and/or that future global warming will not be a very serious problem.' Survey: '27.4% believe it is caused by primarily natural factors (natural variation, volcanoes, sunspots, lithosphere motions, etc.), 25.7% believe it is caused by primarily human factors (burning fossil fuels, changing land use, enhanced water evaporation due to irrigation), and 45.2% believe that climate change is caused by both human and natural factors.'

▶

4. Comment on the misrepresentation or misunderstanding of the results.	According to the survey only 27.4% of the respondents believe that global warming is caused primarily by natural factors, while the article states that ' ... a strong majority of the 1077 respondents believe that nature is the primary cause of recent global warming ... ' It is clear this article is presenting a biased view.

11.4.4 Statistical interpretation bias

Once the data have been collected, collated and subjected to statistical calculations, bias may still occur in the interpretation of the results.

Misleading graphs can be drawn, leading to a biased interpretation of the data. Graphical representations of a set of data can give a visual impression of 'little change' or 'major change' depending on the scales used on the axes (we learned about misleading graphs in Chapter 7).

The use of terms such as 'majority', 'almost all' and 'most' are open to interpretation. When we consider that 50.1% 'for' and 49.9% 'against' represents a 'majority for' an issue, the true figures have been hidden behind words with very broad meanings. Although we would probably not learn the real facts, we should be wary of statistical issues quoted in such terms.

WORKED EXAMPLE 11

Discuss why the following selected samples could provide bias in the statistics collected.

a. In order to determine the extent of unemployment in a community, a committee phoned two households (randomly selected) from each page of the local telephone directory during the day.

b. A newspaper ran a feature article on the use of animals to test cosmetics. A form beneath the article invited responses to the article.

THINK	WRITE
a. 1. Consider phone book selection.	a. Phoning two randomly selected households per page of the telephone directory is possibly a representative sample.
2. Consider those with no phone contact.	However, those without a home phone and those with unlisted numbers could not form part of the sample.
3. Consider the hours of contact.	An unanswered call during the day would not necessarily imply that the resident was at work.
b. 1. Consider the newspaper circulation.	b. Selecting a sample from a circulated newspaper excludes those who do not have access to the paper.
2. Consider the urge to respond.	In emotive issues such as these, only those with strong views will bother to respond, so the sample will represent extreme points of view.

Exercise 11.4 Sources of bias

1. **WE9** a. Calculate the lower and upper margins for a 5% margin of error if 11% of the batteries in a sample were found to be batteries with defects.
    b. If the confidence level is 95%, what is the risk of the results falling outside the margins calculated in part a?

2. a. Calculate the lower and upper margins for a 4% margin of error if 96% of the members of a sample of school students were found to be Year 12 students.
    b. If the confidence level is 92%, what is the risk of the results falling outside the margins calculated in part a?

3. **WE10** In the same survey discussed in Worked example 10, the authors of the study state that 'The petroleum industry — through oil and gas companies, related industrial services, and consulting services — is the largest employer, either directly or indirectly, of professional engineers and geoscientists in Alberta... These professionals and their organizations are regulated by a single professional self-regulatory authority — APEGGA.' This fact is not stated in the article. Why do you think that the author of the article did not mention this fact?

4. The manager of a local supermarket asked 10 customers 'Do you like to shop at our supermarket?'. Eight customers replied 'Yes'. Once he collected this data, the manager placed a poster stating that '80% of people like to do their shopping in our supermarket.' Is this correct?

5. Calculate the confidence level of a survey if 2.5% of its results do not fall within a given margin of error.

6. Calculate the lower and upper margins for
    a. a 3% margin of error if 20% of the members of a sample were found to be people over 40 years old.
    b. a 5% margin of error if 70% of the cars in a sample were found to be sedans.

7. If the confidence level for the samples in question 6 is 96%, calculate the percentage of results that will not fall within the margins of error calculated.

8. Rewrite the following questions, removing any elements or words that might contribute to bias in responses.
    a. The poor homeless people, through no fault of their own, experience great hardship during the freezing winter months. Would you contribute to a fund to build a shelter to house our homeless?
    b. Most people think that, since we've developed as a nation in our own right and broken many ties with Great Britain, we should adopt our own national flag. You'd agree with this, wouldn't you?
    c. You'd know that our Australian 50 cent coin is in the shape of a dodecagon, wouldn't you?
    d. Many in the workforce toil long hours for low wages. By comparison, politicians seem to get life pretty easy when you take into account that they only work for part of the year and they receive all those perks and allowances. You'd agree, wouldn't you?

9. Rewrite parts a to d in question 8 so that the expected response is reversed.

11.5 Review: exam practice

11.5.1 Data collection: summary

Census and surveys

- The Australian Bureau of Statistics (ABS) is the official body which conducts the census in Australia.
- A census is the process of collecting, analysing and presenting information collected from the entire population.
- Census night is the night when information is collected for the census. In Australia, census night occurs every five years.
- Data collection is a process in which data is collected to obtain information and draw conclusions about issues of concern regarding a given population.
- A population consists of a complete group of people, objects, events, etc. with at least one common characteristic.
- In an Excel worksheet, the RAND( ) command generates a random number between 0 and 1.
- RANDBETWEEN(a,b) is an Excel worksheet command which generates a random number from **a** to **b**, where **a** is the lowest number required and **b** is the highest number required.
- A random number generator is a calculator or computer software function which produces random numbers.
- A sample is any subset of a population. Any sample of a population has to maintain all the characteristics of the entire population.
- Sampling is the process of selecting a sample of a population to provide an estimate of the entire population.
- Simple random sampling is the basic method of selecting a sample. This type of sampling ensures that each individual of a population has an equal chance of being selected for the sample.
- Generally, the required sample size is $\sqrt{N}$, where N is the size of the population.
- Systematic sampling or systematic random sampling requires a starting point chosen at random with members of the population chosen at regular intervals.
- Maximum starting point $= N - I(s - 1)$,
  - where N is the size of the population

    s is the size of the sample

    I is the whole number of $\dfrac{N}{s}$
- Self-selected sampling is used when the members of a population are given the choice to participate in a research or not.
- Stratified sampling is used when there are variations in the characteristics of a population. This method requires the population to be divided into sub-groups called strata.

Simple survey procedures

- The target population is the entire population from which a sample is chosen to be surveyed.
- A survey population are those from the target population that are able to be selected in the sample.
- A questionnaire is a list of questions used to collect data from the survey's participants.
- In open-ended questions the participants have the opportunity of stating their responses in their own words.
- **Closed** questions provide a list of answers from which the participants have to choose one or more options.
- Partially closed questions provide both answer choices and an option of creating own responses.

Sources of bias

- Bias is a distortion that occurs during the data collection process which produces results that are not representative of the population.
- Sampling errors occur due to the fact that surveys are conducted on one sample out of the large number of possible survey samples.
- The margin of error is the range of values that result when using a sample of a population.

  Lower margin $= x\% - \dfrac{e\%}{2}$ and Upper margin $= x\% + \dfrac{e\%}{2}$

  where $x\%$ is the percentage of the sample considered, $e\%$ is the percentage error
- The confidence level is used to find the percentage of results that will not fall within the margin of error stated.
  - Risk $= 100\% -$ Confidence level
- A measurement error is, in general, the difference between the measured value and the real value.
- A misrepresentation of the results of a survey is an untrue statement that does not reflect the characteristics of the population.

Exercise 11.5 Review: exam practice

Simple familiar

1. **MC** For which of the following data collection situations is a census required?
   - **A.** The employment status of a group of 1200 people over 18.
   - **B.** The employment status of all people living in Australia at a given time.
   - **C.** The employment status of a group of females aged between 30 and 40 years.
   - **D.** Temperatures in Australia from June 2017 to June 2018.

2. **MC** Which of the following represents a population?
   - **A.** A group of 25 members of a fitness club.
   - **B.** Every 10th cereal box from a batch of 500 boxes.
   - **C.** Every second male listener of a radio show.
   - **D.** All the girls in Year 11.

3. **MC** A publishing company of mathematics textbooks is checking an Essential Mathematics textbook for printing errors. Every 5th page is checked in one book and every 23rd page is checked in a second book. This type of sampling is called:
   - **A.** stratified sampling.
   - **B.** simple random sampling.
   - **C.** systematic sampling.
   - **D.** random sampling.

4. **MC** An animal farm is selecting a sample of 20 animals to try a new vaccine. The farm has 140 cows and 60 horses. The number of cows and horses required to be part of this sample is respectively
   - **A.** 14 and 6.
   - **B.** 6 and 14.
   - **C.** 7 and 3.
   - **D.** 3 and 7.

5. **MC** One advantage of basic random sampling over self-selected sampling is
   A. participants are more committed to the study.
   B. it allows participants to choose whether to take part in the study or not.
   C. it provides random selection.
   D. it requires a smaller sample.

6. **MC** A random number generator is used to generate 40 numbers between 0 and 1 to select a sample of 40 from a population of 100.
   A. The command RANDBETWEEN(0, 40) in an Excel worksheet was used.
   B. All generated numbers have to be multiplied by 100.
   C. Choose the first two non-zero digits in each number.
   D. All generated numbers have to be multiplied by 40.

7. **MC** The lower and upper margins for a 7% margin of error if 12% of the batteries of a sample were found to be defective are respectively:
   A. 5% and 19%.   B. 13% and 1%.   C. 0% and 19%.   D. 8.5% and 15.5%.

8. **MC** If the confidence level is 90%, the risk of the results to fall outside the margins calculated in question **7** is:
   A. 90%.   B. 10%.   C. 19%.   D. 12%.

9. **MC** Which of the following statements is correct?
   A. Closed questions are questions that only some people can answer.
   B. A bias is when a person has a different opinion than the rest of the sample.
   C. The target population is the part of the population to be surveyed.
   D. Open-ended questions are designed to allow people to write their own answer.

10. **MC** Blanca, the local veterinarian, records the name of every dog owner. This type of data collection is a
    A. survey because only the names of dog owners are recorded.
    B. census because the names of all dog owners are recorded.
    C. survey because only the dogs are taken into account.
    D. survey because the dogs are a sample of the animal population.

11. **MC** Toby is conducting a poll to determine the likely winner of an election. Which of the following methods of selecting his sample contains sampling bias?
    A. Using a computer to generate a page number, column number and position number from the telephone book.
    B. Randomly going up to people in a shopping centre on a Monday morning.
    C. Selecting the first person on every page of the telephone book.
    D. Randomly dialling telephone numbers and surveying whoever answers the telephone.

12. **MC** A random number generator is used to select 5 students from a class of 30. The first 10 numbers generated in a list are:
    87, 49, 28, 07, 16, 58, 10, 21, 19, 45, ...
    Reading the random numbers from left to right, the first student number selected would be:
    A. 28.   B. 87.   C. 7.   D. 10.

Complex familiar

13. When we obtain data from the whole population, we conduct a _____; however, a survey obtains data from a _____ of the population.

---

14. For each of the following, state whether a census or a survey has been used.
    a. Two hundred people in a shopping centre are asked to nominate the supermarket where they do most of their grocery shopping.
    b. To find the most popular new car on the road, 500 new car buyers are asked what make and model car they purchased.
    c. To find the most popular new car on the road, the make and model of every new car registered is recorded.
    d. To find the average mark in the Mathematics half-yearly examination, every student's mark is recorded.
    e. To test the quality of tyres on a production line, every 100th tyre is road tested.
15. State whether the following questions are open-ended, closed or partially closed.
    a. What is your preferred Mathematics topic?
    b. What is your preferred Mathematics topic?
       1. Algebra   2. Geometry   3. Statistics   4. Trigonometry
    c. What is your preferred Mathematics topic?
       1. Algebra   2. Geometry   3. Statistics   4. Trigonometry   5. Other
16. Select a sample of 5 students from your class
    a. 'by hand'.
    b. using a random number generator.
    c. Compare your sample with the samples of other classmates.

Complex unfamiliar

17. Calculate the lower and upper margins for a
    a. 7% margin of error if 10% of the members of a sample were found to be small dogs.
    b. 4% margin of error if 85% of the cars in a sample were found to be red cars.
    c. 8% margin of error if 92% of the electronic devices of a sample were found to be laptops.
18. Surveys are conducted on samples to determine the characteristics of the population. Discuss whether the samples selected would provide a reliable indication of the population's characteristics.

	Sample	Population
a	Year 11 students	Student drivers
b	Year 12 students	Students with part-time jobs
c	Residents attending a Neighbourhood Watch meeting	Residents of a suburb
d	Students in the school choir	Music students in the school
e	Cars in a shopping centre car park	Cars on the road
f	Males at a football match	People who watch TV
g	Users of the local library	Teenagers

19. A manufacturing company of tablets is testing two batches of 150 tablets each for defects. In the first batch every 6th tablet is checked while in the second batch every 10th tablet is checked.
    a. What method of sampling is used for this testing?
    b. How many tablets are being tested for defects in the two batches?
    c. What is the maximum starting point for each batch?
    d. Calculate the lower and upper margin for each batch for a 3% margin of error if 4% of the members of a sample were found to be defective.

20. It is important that a sample is chosen randomly to avoid bias. Consider the following situation.

The government wants to improve sporting facilities in Brisbane. They decide to survey 1000 people about what facilities they would like to see improved. To do this, they choose the first 1000 people through the gate at a football match at Suncorp Stadium.

In this situation it is likely that the results will be biased towards improving facilities for football. It is also unlikely that the survey will be representative of the whole population in terms of equality between men and women, age of the participants and ethnic backgrounds.

Questions can also create bias. Consider asking the question, 'Is football your favourite sport?' The question invites the response that football is the favourite sport rather than allowing a free choice from a variety of sports by the respondent.

Consider each of the following surveys and discuss:
   a. any advantages, disadvantages and possible causes of bias
   b. a way in which a truly representative sample could be obtained.

1. Surveying food product choices by interviewing customers of a large supermarket chain as they emerge from the store between 9:00 am and 2:00 pm on a Wednesday

2. Researching the popularity of a government decision by stopping people at random in a central city mall

3. Using a telephone survey of 500 people selected at random from the phone book to find if all Australian states should have Daylight Saving Time in summer

4. A bookseller uses a public library database to survey for the most popular novels over the last three months

5. An interview survey about violence in sport taken at a rugby league football venue as spectators leave

Answers

Chapter 11 Data collection

Exercise 11.2 Census and surveys

1. a. Census. The entire population of the online business.
    b. Survey. 50% of the Year 12 student population is a sample of the population.
    c. Census. The entire population of the hospital's patients.

2. a. Survey for the Year 12 students and census for the entire student population.
    b. Survey for the passenger airplanes and census for the entire number of airplanes.

3. a. Survey; a sample of 150 suburbs.
    b. Census; all the population of Australia.
    c. Census; all the secondary school population.
    d. Survey; sample of the population who viewed the play.
    e. Census; all the customer population.

4. a, b Any sample with 5 different numbers between 1 and 26.

5. a, b Any sample with 8 different numbers between 1 and 29.

6. 153 people

7. a. Any sample with 10 different numbers between 1 and 100.
    b. Any sample with 10 different numbers between 1 and 100.

8. 1 to 8 if selecting every sixth car, or 1 to 4 if selecting every seventh car.

9. From 1 to 6 selecting every fifth product.

10. a. 5 boys and 6 girls
    b. 8 adults and 2 children

11. Publicising at the local library, local radio station, letter drop in the surrounding households, advertisements at the local shopping centres.

    All applicants could be accepted as there are no restrictions set on the study.

12. Publicising using advertisements in the local paper, or at the local gym.

13. Systematic sampling

14. a. Stratified sampling.
    b. 6 black sheep and 4 white sheep

15. a. Systematic sampling.
    b. 20 phones in the first batch, 14 phones in the second batch.
    c. Maximum starting point is 5 in the first batch when selecting every fifth phone, and 9 in the second batch when selecting every seventh phone.

16. 19 males and 14 females

17. 7 white cars and 8 black cars

18. a. 15 students.
    b. Either using a random number generator or using systematic sampling.

19. a. Any sample with 12 different numbers between 1 and 140.
    b. Any sample with 12 different numbers different numbers between 1 and 140.

20. a. 28.
    b. Any sample with 28 different numbers different numbers between 1 and 800.

Exercise 11.3 Simple survey procedures

1. a. All Year 12 students.
    b. All students in the school.
    c. All students in Australia.

2. a. All the milk bottles in the dairy farm.
    b. All the customers of the Internet provider.

3. a. The year 11 students.
    b. The years 11 and 12 students.
    c. All students in the school.

4. a. Assumes all people have mobile phones. A possible replacement question is 'Do you own a mobile phone? If yes, what brand do you have?'
    b. Double-barrelled question. It should be separated into two questions: 'On a scale of 1 (strongly agree) to 5 (strongly disagree), should drinking alcohol be permitted at the beach? Should smoking be banned at the beach?'
    c. Assumes that all people go to church on Sundays and all people are of a certain religion. A possible replacement question is 'Do you attend a place of prayer?
        • Never
        • Rarely
        • Sometimes
        • Often
        • Regularly'

5. a. Double-barrelled question. 'Circle the preferred meal: pizza, pasta, both.'
    b. Double negative question. 'Are your parents' expectations of you clear?'

6. a. closed
    b. open-ended
    c. closed
    d. partially closed

7. a. Various answers are possible; see the worked solutions
    b. Various answers are possible; see the worked solutions
    c. Various answers are possible; see the worked solutions

8. Possible answers:
    Open-ended question: 'What is your preferred search engine when surfing the Internet?'
    Partially closed question: 'The Internet search engine I prefer to use is:
    ☐ Firefox ☐ Google ☐ Safari ☐ Other ....'
    Closed question: 'The Internet search engine I prefer to use is:
    ☐ Firefox ☐ Google ☐ Safari ☐ None of these ☐ I don't use the Internet'

9. a. Double barrelled question. Possible replacement: 'What mode of transport do you use to go to work?
        ☐ Bus ☐ Tram ☐ Other ....'
    b. Some people might have other opinions such as smoking or lack of social skills or too much technology.
        Possible replacement: 'In your opinion what is the most important issue facing teenagers today?'
    c. Leading question suggesting the desired answer. Possible replacement: 'How do you feel now after stopping smoking?'

d. Too personal. Possible replacement: Your annual income is
☐ less than $30 000
☐ $30 000-$50 000
☐ $50 000-$70 000
☐ $70 000-$90 000
☐ more than $90 000

10. a. partially closed
    b. closed
    c. open-ended

11. See sample responses in the worked solutions in eBookPLUS.

12. See sample responses in the worked solutions in eBookPLUS.

Exercise 11.4 Sources of bias

1. a. Lower margin $= 8.5\%$, upper margin $= 13.5\%$
   b. 5%

2. a. Lower margin $= 94\%$, upper margin $= 98\%$
   b. 8%

3. Could be to hide the fact that the engineers and geoscientists surveyed are, in some way, dependent on or paid by the petroleum industry. The reader might not trust the results as much or might think that this connection makes the geoscientists and the engineers somewhat biased.

4. This is a biased sample as the people surveyed are already shopping at the given supermarket.

5. 97.5%

6. a. Lower margin $= 18.5\%$, upper margin $= 21.5\%$
   b. Lower margin $= 67.5\%$, upper margin $= 72.5\%$

7. 4%

8. See sample responses in the worked solutions in eBookPLUS.

9. See sample responses in the worked solutions in eBookPLUS.

11.5 Review: exam practice

Simple familiar

1. B	2. D	3. C	4. A
5. C	6. D	7. D	8. B
9. D	10. D	11. B	12. A

Complex familiar

13. census, sample

14. a. Survey   b. Survey
    c. Census   d. Census
    e. Survey

15. a. Open-ended   b. Closed
    c. Partially closed

16. a. Any set of 5 students.
    b. Any set of 5 students.
    c. The sets would most likely be different.

Complex unfamiliar

17. a. Lower margin $= 6.5\%$ and upper margin $= 13.5\%$.
    b. Lower margin $= 83\%$ and upper margin $= 87\%$.
    c. Lower margin $= 88\%$ and upper margin $= 96\%$.

18. a. There would be many more student drivers in Year 12 than in Year 11 — perhaps also some in Year 10.
    b. Students with part-time jobs are in lower year levels as well.
    c. Residents not at the Neighbourhood Watch meeting have been ignored.
    d. Other music students who play instruments and don't belong to the choir have been excluded.
    e. The composition of cars in a shopping centre car park is not representative of the cars on the road.
    f. Females have been excluded.
    g. Users of the local library would not reflect the views of teenagers.

19. a. Systematic sampling.
    b. 25 tablets, 15 tablets.
    c. 6 and 10.
    d. Lower limit $= 1.5\%$, upper limit $= 5.5\%$.

20. See fully worked solutions

1. a. This is not a representative sample of the shopping public. Only those emerging from the store at that time have an opportunity to voice their opinion. The advantage of surveying by this method is that it's convenient and cheap.
   b. A large supermarket chain has a huge database, and they would be able to analyse the product choices of food items.

2. a. This method again is convenient and cheap. It, however, captures the opinions of only those who visit the central city mall.
   b. Choosing a random sample from the electoral roll would select a representative sample from the voting public.

3. a. Many people have unlisted numbers and mobile numbers which are not listed in the telephone book. These people would not have a chance of being selected. Also, it should be mentioned that the sample should represent all the states in proportion to their population.
   b. Select people at random from the electoral roll, making sure there is representation in population proportion from all states and from all areas in the state.

4. a. This method is cheap and efficient, and would give a fairly reliable indication of the most popular novels. If this is restricted to just one library, however, it may cause bias because of demographic influences.
   b. The databases of state and council libraries in the areas of interest should yield reliable results.

5. a. This sample would represent only those at one particular football match. There are many other areas of sport which experiences acts of violence. The avid TV sports watchers have not been included in the sample.
   b. It is difficult to get a truly random sample on this topic. It is very emotive, and only those with strong opinions tend to respond.

PRACTICE ASSESSMENT 4

General Mathematics: Unit 2 examination

Unit
Unit 2: Money, travel and data

Topic
Topic 1: Managing money
Topic 2: Time and motion
Topic 3: Data collection

Conditions

Technique	Response type	Duration	Reading
Paper 1: Simple (24 marks) Paper 2: Complex (16 marks)	Short response	60 minutes	5 minutes

Resources	Instructions
• QCAA formula sheet: • Notes not permitted • Scientific calculator permitted	• Show all working. • Write responses using a black or blue pen. • Unless otherwise instructed, give answers to **two decimal places.**

Criterion	Marks allocated	Result
Foundational knowledge and problem-solving	40	

A detailed breakdown of examination marks summary can be found in the PDF version of this assessment instrument in your eBookPLUS.

Question 1 (3 marks)

Jack is an electrician. He claims tax deductions of $1600 on tools, $180 on overalls, $10 per week on dry cleaning and $15 per week on work-related phone expenses. Calculate Jack's total tax deductions.

_____

_____

_____

Question 2 (3 marks)

Ashlee earns a gross annual salary of $87 000. She also earns $3740 per year from investments. If Ashlee has tax deductions totalling $7250, calculate her taxable income.

_____

_____

_____

Question 3 (4 marks)

Zoe left work at 5:20 pm and walked for 15 minutes to the train station. She had to wait 8 minutes for the train, and the train trip was 40 minutes. Zoe then had to walk for another 16 minutes to her house. What time did she get home:

a. in 12-hour time?

b. in 24-hour time?

_____

_____

_____

Question 4 (4 marks)

Twenty-five students were asked their opinion on what they thought was bad about online gaming. They provided the following responses:

1. It's addictive.
2. The time I spend on it.
3. I don't want to stop playing.
4. A lot of guns.
5. Not good for my eyes.
6. A lot of violence.
7. Too much killing.
8. It stops me from doing other things.
9. A lot of fighting.
10. Gory graphics.
    a. Classify the responses into two main categories.
    b. Which response or responses don't fit into your two categories?

_____

_____

_____

Question 5 (6 marks)

Calculate the following.

a. The average speed (in m/s) when covering 180m in 25 seconds
b. The distance covered (in km) when travelling at an average speed of 60 km/h for 45 seconds
c. The time (in seconds) it takes to travel 20 km travelling at an average speed of 12 m/s
d. The average speed (in km/h) when covering 782 m in 75 seconds.

Question 6 (4 marks)

The graph shows the money spent by a school on staff professional development in 2008, 2012 and 2016. Draw another bar graph that minimises the appearance of the decline in money spent on staff professional development.

Staff professional development spending

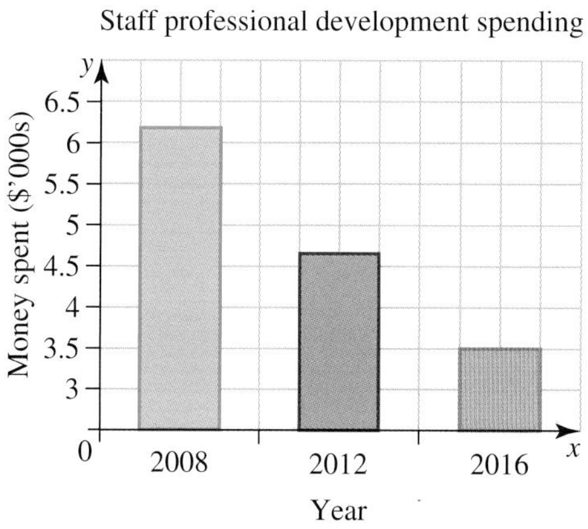

Part B: Complex — total marks: 16

Question 7 (6 marks)

Ross earns a weekly wage of $2100 per week. The table shows the tax to be paid on a range of taxable incomes.

Taxable income	Tax on this income
0–$18 200	Nil
$18 201–$37 000	19c for each $1 over $18 200
$37 001–$87 000	$3572 plus 32.5c for each $1 over $37 000
$87 001–$180 000	$19 822 plus 37c for each $1 over $87 000
$180 001 and over	$54 232 plus 45c for each $1 over $180 000

a. Calculate Ross's gross annual pay.
b. If he has a total of $6200 of deductions, determine his taxable income.
c. Calculate the total annual tax payable.
d. Calculate the weekly PAYG tax that would be deducted from Ross's wages.

Question 8 (6 marks)

Paul and Charlotte sell life insurance. Paul is paid a commission of 8.5% and Charlotte is paid $350 plus 4.5% commission.

a. Calculate how much Paul earns for a week in which his sales are $7000.
b. Calculate how much Charlotte earns for a week in which her sales are $7000.
c. In another week Charlotte earns $800. Determine the value of Charlotte's sales.
d. Paul wishes to earn $800 in a week. How much should his sales be?

Question 9 (4 marks)

Naomi's monthly budget is shown below.

Income		**Expenses**	
Salary after tax	$2175	Rent	$500
Dividends from shares	$55	Electricity	$85
		Phone	$70
		Health insurance	$30
		House contents insurance	$15
		Car registration	$60
		Car insurance	$100
		Petrol	$50
		Food	$155
		Clothing	$150
		Sport	$120
		Miscellaneous	$125
Total	$2230	Total	$1460

a. Calculate the amount available for saving.

b. Naomi wants to have a holiday in Bali (estimated cost $3000). Determine how long she needs to save.

c. Suggest **three** areas and **explain** how Naomi could reduce her expenses to save for her Bali holiday. Make sure she has spending money to take with her.

APPENDIX
Maths skills workbook

Introduction

We use mathematical skills every day, quite often without even realising it. This book contains activities that help to develop and refine some more commonly used mathematical processes.

There are many different ways to approach questions, and we have all learned different ways of getting to the same answer. If you have your own methods that are mathematically correct, you may wish to stay with these more familiar steps. Check with your teacher that your processes are acceptable.

KEY SKILL 1 Fractions

Every day of our lives we make decisions that involve fractions. We might be cutting a cake into portions or dividing money among a group of friends. Changing a recipe to cater for more or fewer people requires an understanding of fractions. Many occupations require a knowledge of how to calculate fractions of quantities.

Language of fractions

Numerator: the top number
Denominator: the bottom number
Simplest form: when there are no numbers other than 1 that can divide exactly into the numerator and denominator
Proper fraction: a fraction with a numerator smaller than the denominator
Improper fraction: a fraction with a numerator larger than the denominator
Mixed number: a whole number and a proper fraction

Questions

1. Write the fraction for the amount shown.

 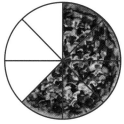 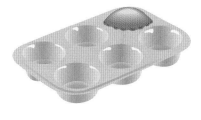

   a.   =   b.   =   c.   =

2. Write out the fraction that is described, then cancel down this fraction into its simplest form.

 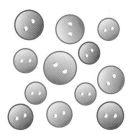 

   a. White squares   b. Blue buttons   c. Orange fish

3. Cancel each of these fractions into their simplest form by finding the largest number that divides into both the numerator and denominator.

   a. $\dfrac{10}{15} = $ ——   b. $\dfrac{20}{40} = $ c. $\dfrac{21}{35} = $

   d. $\dfrac{40}{25} = $ e. $\dfrac{25}{75} = $ f. $\dfrac{160}{400} = $

KEY SKILL 2 Equivalent fractions

Equivalent fractions are equal to each other: they are worth the **same** amount. Equivalent fractions can be found by either multiplying or dividing both numerator and denominator by the same amount.

 $\dfrac{1}{2}$

 $\dfrac{2}{4}$

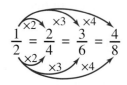

$$\frac{1}{2} = \frac{2}{4} = \frac{3}{6} = \frac{4}{8}$$

WORKED EXAMPLE

Fill in the missing values: $\dfrac{2}{5} = \dfrac{4}{} = \dfrac{6}{15} = \dfrac{}{30}$.

THINK

1. To make a 2 into a 4 you need to × 2.

2. To make a 5 into a 30 you need to × 6.

WRITE

$$\overset{\times 2}{\underset{\times 2}{\frac{2}{5} = \frac{4}{10}}}$$

$$\overset{\times 6}{\underset{\times 6}{\frac{2}{5} = \frac{12}{30}}}$$

Questions

Draw in your own links and find the missing numbers.

1. $\dfrac{4}{5} = \dfrac{}{10} = \dfrac{}{20} = \dfrac{24}{}$

2. $\dfrac{2}{3} = \dfrac{4}{} = \dfrac{}{12} = \dfrac{20}{}$

3. $\dfrac{2}{5} = \dfrac{}{10} = \dfrac{}{25} = \dfrac{12}{}$

4. $\dfrac{2}{7} = \dfrac{4}{} = \dfrac{16}{} = \dfrac{20}{}$

5. $\dfrac{5}{9} = \dfrac{}{18} = \dfrac{}{45} = \dfrac{50}{}$

6. $\dfrac{7}{11} = \dfrac{14}{} = \dfrac{35}{} = \dfrac{}{121}$

7. Ben ate 15 chocolates from a box of 25 chocolates.
   a. Write this as a fraction.

   _____

   b. Write the fraction in its simplest form.

   _____

   c. What fraction is left?

   _____

KEY SKILL 3 Multiplying fractions

Multiplying fractions is an important skill to have as it will not just be used for fractions but also when calculating **percentages.**

Being able to cancel down is important because it minimises the size of the numbers that are being used.

Remember that you can cancel the **numerator** and the **denominator** from one fraction or cancel **diagonally** across two fractions.

Remember the saying: **cancel one from the top with one from the bottom**.

WORKED EXAMPLE

a. Multiply $\dfrac{3}{4} \times \dfrac{5}{7}$. **b.** Multiply $\dfrac{6}{7} \times \dfrac{5}{12}$. **c.** Multiply $\dfrac{4}{5} \times 100$.

THINK

WRITE

a. 1. Can any numbers on the top lines cancel down with numbers on the bottom? No.

$\dfrac{3}{4} \times \dfrac{5}{7}$

2. Write out the two numbers on the top lines together on one single top line, and write out the two bottom numbers together on a single bottom line.

$= \dfrac{3 \times 5}{4 \times 7}$

3. Multiply the top numbers, and multiply the bottom numbers.

$= \dfrac{15}{28}$

b. 1. Can any numbers on the top lines cancel down with numbers on the bottom? Yes.

$\dfrac{6}{7} \times \dfrac{5}{12}$

2. Cancel down by finding what number will divide exactly into the top and bottom numbers.

$= \dfrac{{}^{1}\cancel{6}}{7} \times \dfrac{5}{{}^{2}\cancel{12}}$

3. Multiply what remains after cancelling down.

$= \dfrac{1}{7} \times \dfrac{5}{2}$

4. Multiply the top numbers, and multiply the bottom numbers.

$= \dfrac{5}{14}$

c. 1. Make 100 into a fraction by putting it over 1.

$\dfrac{4}{5} \times \dfrac{100}{1}$

2. Check to see if there is any cancelling to be done.

$= \dfrac{4}{{}_{1}\cancel{5}} \times \dfrac{\cancel{100}^{20}}{1}$

3. Multiply the top numbers, and multiply the bottom numbers.

$= \dfrac{4}{1} \times \dfrac{20}{1}$

$= \dfrac{80}{1}$

$= 80$

Questions

1. Check to see what can be cancelled down, then multiply these fractions.

   a. $\dfrac{5}{6} \times \dfrac{3}{10}$

   b. $\dfrac{7}{10} \times \dfrac{5}{21}$

   c. $\dfrac{2}{15} \times \dfrac{10}{11}$

   d. $\dfrac{3}{8} \times \dfrac{12}{13}$

   e. $\dfrac{25}{63} \times \dfrac{9}{35}$

   f. $\dfrac{15}{42} \times \dfrac{14}{25}$

   g. $\dfrac{20}{50} \times \dfrac{100}{1}$

   h. $\dfrac{25}{75} \times \dfrac{6}{7}$

2. Multiply each of the fractions with the whole number.

   a. $\dfrac{3}{5} \times 100$

   b. $\dfrac{25}{75} \times 100$

   c. $\dfrac{1}{5} \times 100$

   d. $\dfrac{49}{50} \times 100$

   e. $\dfrac{6}{9} \times 45$

   f. $\dfrac{2}{3} \times 90$

   g. $\dfrac{1}{6} \times 54$

   h. $\dfrac{20}{55} \times 110$

3. a. In a box of 50 matches, only $\dfrac{4}{5}$ of them would strike. How many matches is this?

   b. In a school, $\dfrac{6}{7}$ of a group of 84 Year 11 students have a part-time job. How many is this?

   c. In a bag of 720 lollies, $\dfrac{2}{9}$ of them are jelly beans. How many jelly beans are there?

   d. In a box of 300 nails, $\dfrac{1}{20}$ are faulty and cannot be used. How many nails can be used?

KEY SKILL 4 Decimal numbers

We work with decimal numbers (decimals) every day. Money is given in decimal numbers. Many measurements are given in decimals. When using a calculator, often the answers are decimal numbers, sometimes with many numbers after the decimal point. It is vital that we can accurately reduce these large numbers to something more useful by **rounding off**. Probably the most common amount to **round off** to is 2 decimal places. All money is **rounded off** to 2 places, giving dollars before the decimal point and cents after it.

WORKED EXAMPLE

Round off the following numbers to 2 decimal places.
a. **2.47423**   b. **76.4285**

THINK	WRITE
a. 1. Underline the first two numbers after the decimal point.	2.47423
2. Draw a line to cut through the number immediately after the underlined numbers.	2.47|423
3. Look at the number that is immediately after the line that has cut through the original decimal number.	4
4. If it is **between 0** and **4**, all numbers after the cutting line can be removed.	2.47|423 = 2.47
b. 1. Underline the first two numbers after the decimal point.	76.4285
2. Draw a line to cut through the number immediately after the underlined numbers.	76.42|85
3. Look at the number that is immediately after the line that has cut through the original decimal number.	8
4. If it is **between 5 and 9**, add 1 to the last number before the cutting line and then remove all of the numbers after the cutting line.	76.42|85 = 76.43

Questions

1. Round off these numbers to 2 decimal places.

   a.   234.023674   b.   48.5822233   c.   0.165201
        = 234.02\|3674   = 48.58\|22233   = 0.16\|5201

        = _____   = _____   = _____

   d.   200.171436   e.   0.9878900   f.   0.00982315
        = 200.17\|1436   = 0.98\|78900   = 0.00\|982315

        = _____   = _____   = _____

2. Using the same method, round the following numbers off to 3 decimal places.

   a.   0.5962017   b.   48.5822233   c.   0.165801
        = 0.596\|2017   = _____   = _____

        = _____   = _____   − _____

KEY SKILL 5 Fractions to decimals

The easiest way to convert fractions to decimals is to use a calculator.

a. Write $\dfrac{4}{5}$ as a decimal number.

b. Write $\dfrac{20}{16}$ as a decimal number.

THINK

a. 1. $\dfrac{4}{5}$ means '4 divided by 5'.

   2. Use the calculator to convert $\dfrac{4}{5}$ into a decimal.

b. 1. $\dfrac{20}{16}$ means '20 divided by 16'.

   2. Use the calculator to convert $\dfrac{20}{16}$ into a decimal.

WRITE

$\dfrac{4}{5}$ is $4 \div 5$.

$4 \div 5 = 0.8$

$\dfrac{20}{16}$ is $20 \div 16$.

$20 \div 16 = 1.25$

Questions

1. Write each of the following as decimals, rounding to 2 decimal places where necessary.

   a. $\dfrac{1}{8}$ b. $\dfrac{2}{3}$ c. $\dfrac{75}{100}$ d. $\dfrac{1}{4}$

   e. $\dfrac{3}{4}$ f. $\dfrac{90}{100}$ g. $\dfrac{1}{3}$ h. $\dfrac{25}{75}$

   i. $\dfrac{25}{100}$ j. $\dfrac{1}{2}$ k. $\dfrac{50}{100}$ l. $\dfrac{30}{40}$

2. Convert the following improper fractions into decimal numbers. Round to 2 decimal places where necessary.

   a. $\dfrac{15}{4}$ b. $\dfrac{9}{2}$ c. $\dfrac{56}{20}$ d. $\dfrac{35}{2}$

   e. $\dfrac{150}{6}$ f. $\dfrac{31}{5}$ g. $\dfrac{226}{5}$ h. $\dfrac{93}{5}$

   i. $\dfrac{221}{10}$ j. $\dfrac{37}{4}$ k. $\dfrac{450}{8}$ l. $\dfrac{59}{3}$

KEY SKILL 6 Percentage skills

The three topics of decimals, fractions and percentages are very closely related and the skills used in one topic may be useful for others.

It is useful to remember that when you think of **percentages**, think of the number **100**. It's always the best place to start. For example:

$$50\% \text{ means:}$$
$$50 \text{ out of } 100$$
$$\text{or}$$
$$\frac{50}{100}$$
$$\text{or}$$
$$50 \div 100$$
$$\text{or}$$
$$0.50$$

Questions

Complete the following table. The first two have been done for you.

	Percentage	Fraction	Will it cancel down?	Calculator steps	Decimal number
1	10%	$\frac{10}{100}$	$\frac{1}{10}$	$10 \div 100$	0.1
2	20%	$\frac{20}{100}$	$\frac{1}{5}$	$20 \div 100$	0.2
3	25%				
4	30%				
5	50%				
6	75%				
7	80%				
8	90%				
9	12.5%				
10	150%				
11	33.33%				
12	66.67%				

13. Mara collected shells at the beach. She collected 100 shells, of which 70 were in perfect condition.
    a. What percentage was in perfect condition?
    b. Write as a fraction the number of perfect shells out of the total number of shells.
    c. Write as a decimal the number of perfect shells out of the total number of shells.

KEY SKILL 7 Percentages

Quantities are often expressed as a percentage of an amount; for example, 2% of pet owners have a rabbit. This statement gives a proportion; for example, 2 out of 100 pet owners have a rabbit.

a. Find 40% of 620.

b. Find 12.5% of 180.

THINK

WRITE

a 1. Write 40% as a fraction.

$$\frac{40}{100}$$

2. Multiply this by 620.

$$\frac{40}{100} \times 620$$

3. Use the calculator for this calculation.

$$40 \div 100 \times 620 = 248$$

b 1. Write 12.5% as a fraction.

$$\frac{12.5}{100}$$

2. Multiply this by 180.

$$\frac{12.5}{100} \times 180$$

3. Use the calculator for this calculation.

$$12.5 \div 100 \times 180 = 22.5$$

Questions

1. Fill in the missing numbers and solve these problems.
   a. Find 20% of 380.

   $= \qquad \times$

   $=$

   b. Find 15% of 175.

   $= \qquad \times$

   $=$

2. Now create your own working out for the following.
   a. 60% of 80

   b. 12.5% of 204

   c. 250% of 84

   d. 18% of 44

   e. 15.5% of 360

   f. 33.33% of 180

   g. $55\frac{3}{4}$% of 96

   h. $18\frac{1}{4}$% of 688

KEY SKILL 8 Fractions into percentages

A percentage is another way of writing a fraction, but in this case the denominator is always 100. It is a useful way to make comparisons.

As an example, we can say that 30% is the same as $\dfrac{30}{100}$ or 90% is the same as $\dfrac{90}{100}$.

WORKED EXAMPLE

What percentage is 15 out of 50?

THINK	WRITE
1. 15 out of 50 means 15 divided by 50.	$15 \div 50$
2. Now multiply by 100 to make a percentage.	$\dfrac{15}{50} \times 100$
3. Use the calculator for this calculation.	$5 \div 50 \times 100 = 30\%$

Questions

1. What percentage is:
   a. 45 out of 90

   _____

   b. 5 out of 8

   _____

   c. 12 out of 60?

   _____

2. The table below shows the results for two Maths tests. Turn each score into a percentage, then shade in or circle the *best* test score for each student. Round off to 2 decimal places if required.

Student	Test 1 (out of 80)	Working out	Score (%)	Test 2 (out of 60)	Working out	Score (%)
Michelle	75	$75 \div 80 \times 100 =$	93.75%	54		
Yang	15			15		
Simon	40			32		
Alba	49			45		
Christina	65			55		

3. Convert the following amounts into percentages. (Round off to 2 decimal places if required.)
   a. 12 seconds out of 60 seconds

   _____

   b. 15 kg out of 75 kg

   _____

   c. 35 cents out of $2.00

   _____

KEY SKILL 9 Ratios

A ratio is a comparison between two or more amounts (or values). The symbol : is used to separate the values. A ratio could be written as 5 : 4. This ratio means 'five **compared** to four', or we might say 'five to four'. The ratio 7 : 10 : 9 could be described as 'seven to ten to nine'.

WORKED EXAMPLE

The Essential Mathematics class has 9 boys and 13 girls in it. Write this as a ratio.

THINK	WRITE
1. The values, in the order of the question, are 9 and 13.	9 boys and 13 girls
2. Use the ratio symbol to separate the values.	9 : 13

Questions

1. Write the following as ratios.
   a. 10 compared to 13 = _____ : _____
   b. Five compared to seven = _____ : _____
   c. 100 compared to 1 = _____ : _____
   d. 11 to 3 = _____ : _____
   e. 6 to 30 = _____ : _____
   f. 12 to 7 = _____ : _____

2. Write the following as ratios.
   a. A farmer had 350 cows and 7 bulls. _____ : _____
   b. A car travelled 150 km in 2 hours. _____ : _____
   c. Peter spent $3 on 2 ice-creams. _____ : _____
   d. A cake requires 3 cups of flour and 2 eggs. _____ : _____
   e. Cordial needs 4 parts of water and 1 part of cordial syrup. _____ : _____
   f. A jet travelled 640 km in one hour. _____ : _____
   g. A bag of lollies had 10 jelly beans, 15 snakes and 9 mint leaves in it.

   _____ : _____ : _____
   h. A bowl of fruit contained 4 bananas, 2 apples, 35 grapes and 2 oranges.

   _____ : _____ : _____

KEY SKILL 10 Using ratios

A common example of a ratio is the **cement to water** ratio of 4:1, which is used in making concrete. This means that the amount of cement used is four times greater than the amount of water used. The actual amounts of cement and water used or the amount of cement being made is not stated, because the ratio is a comparison of the two quantities.

WORKED EXAMPLE

Shade these bricks into the ratio of 3 : 2.

Think/Write

1. You could shade them to look like this:

2. Or you could re-organise them to look like this:

Questions

1. Shade in these rows of bricks in the given ratios.

   a. 4 : 1

   b. 3 : 4

   c. 5 : 1

   d. 2 : 3

2. On the pictures below, draw a dividing line to separate them in the ratio given.

   a. 7 : 2

   b. 3 : 2

   c. 1 : 1

   d. 2 : 1

KEY SKILL 11 Simplifying ratios

Ratios are like fractions — sometimes they can be cancelled down into smaller numbers (or simplest form).

WORKED EXAMPLE

Cancel down 12 : 2 into its simplest form.

THINK

1. Find the highest common factor (the biggest number that divides into all parts of the ratio).

2. Divide all parts of the ratio by this number.

WRITE

12 and 2 are both divisible by 2.

$$\frac{12}{2} : \frac{2}{2}$$

$$= 6 : 1$$

Questions

1. Cancel these down into their simplest form.

   a. $6 : 2$ b. $15 : 10$ c. $9 : 3$

   d. $30 : 12$ e. $9 : 6$ f. $20 : 16 : 12$

2. On a necklace there are 15 black beads and 10 white beads. What is the ratio of black to white beads in simplest form?

3. In a fruit bowl there are 6 mandarins and 2 bananas. What is the ratio of mandarins to bananas in simplest form?

4. A school has 320 boys and 360 girls. What is the ratio of boys to girls in simplest form?

PROJECT In the kitchen

Use the skills you have been practising on the previous pages to answer these questions and solve these problems.

1. When cooking for a large number of people, recipes need to be scaled up to make sure enough is made for everyone.

   A basic chocolate cake requires: 2 cups of flour, $\frac{2}{9}$ cup of cocoa, 60 g of butter, $1\frac{1}{2}$ cups of milk (375 mL), 1 cup of sugar and 1 egg.

   How much of each ingredient would you need for 2, 3, 4, 5 or 10 cakes?

	1	2	3	4	5	10
Flour (cup)	2					
Cocoa (cup)	$\frac{2}{9}$					
Sugar (cup)	1					
Butter (g)	60					
Egg	1					
Milk (cup)	$1\frac{1}{2}$					

2. Each one of these chocolate cakes is cut into 8 slices for serving. How many cakes would you need to make for 120 guests at a function to all have one slice of chocolate cake?

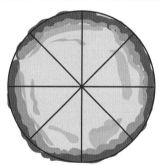

_____

_____

_____

3. Write out the quantities of each of the chocolate cake ingredients that you will be required to buy in order to make the cakes for the 120 guests described in question 2.

_____

_____

4. The bakery that is chosen to make these chocolate cakes charges $5.00 for each cake. What is the total cost of the cakes required in question 2?

5. The baker has calculated that the cost of electricity to heat the ovens and cook the cakes is 10% of the cost of a cake.
   a. What is the cost of the electricity to cook one cake? _____
   b. What is the total cost of electricity to cook all of the cakes that have been ordered? _____

6. The baker has offered to fill the chocolate cakes with fresh cream for an increase of 15% to the cost.

   a. If the original cost of a single cake were $5.00, by how much would the cost of one cake increase?

   _____

   _____

   b. What is the new cost of a single cake? _____

   c. What is the total cost of chocolate cakes, with cream, to serve 120 guests (from question 2)?

   _____

7. You are helping to organise a child's party and your first job is to make up the lolly bags for the guests to take home. The instructions you are given are:
   • each bag must contain 20 lollies
   • the lollies are to be snakes, jelly babies, mint leaves and black jelly beans, in the ratio of $4:6:7:3$.

   a. How many of each of the lollies will you have to place in a bag?

   Snakes _____

   Jelly babies _____

   Mint leaves _____

   Black jelly beans _____

   b. If 30 guests will receive a lolly bag, how many of each of the four types of lollies will you need to purchase? _____

8. Your next task is to purchase the hot pastries (party pies, mini sausage rolls and mini pasties) for lunch. You are assuming that each guest will eat four items.

   a. What is the total number of pastries you will have to buy? _____

   b. The mix of pastries is to be $\frac{1}{3}$ party pies, $\frac{1}{2}$ sausage rolls and $\frac{1}{6}$ pasties. How many of each type will you need to buy? _____

9. Your final task is to order the soft drinks. Allow 800 mL for each child.

   a. How much soft drink is required (in mL)? _____

   b. Change this into litres. _____

   c. You decide to order 20% more to cover any spillage.

      i. What is the extra amount you will need to purchase? _____

      ii. What is the total amount of soft drink you will need to buy? _____

Answers

Appendix Maths skills workbook

KEY SKILL 1 Fractions

1. a. $\dfrac{7}{10}$ b. $\dfrac{5}{8}$ c. $\dfrac{1}{6}$

2. a. $\dfrac{1}{2}$ b. $\dfrac{1}{3}$ c. $\dfrac{3}{4}$

3. a. $\dfrac{2}{3}$ b. $\dfrac{1}{2}$ c. $\dfrac{3}{5}$

   d. $\dfrac{8}{5}$ e. $\dfrac{1}{3}$ f. $\dfrac{2}{5}$

KEY SKILL 2 Equivalent fractions

1. $\dfrac{8}{10}, \dfrac{16}{20}, \dfrac{24}{30}$ 2. $\dfrac{4}{6}, \dfrac{8}{12}, \dfrac{20}{30}$ 3. $\dfrac{4}{10}, \dfrac{10}{25}, \dfrac{12}{30}$

4. $\dfrac{4}{14}, \dfrac{16}{56}, \dfrac{20}{70}$ 5. $\dfrac{10}{18}, \dfrac{25}{45}, \dfrac{50}{90}$ 6. $\dfrac{14}{22}, \dfrac{35}{55}, \dfrac{77}{121}$

7. a. $\dfrac{15}{25}$ b. $\dfrac{3}{5}$ c. $\dfrac{2}{5}$

KEY SKILL 3 Multiplying fractions

1. a. $\dfrac{1}{4}$ b. $\dfrac{1}{6}$ c. $\dfrac{4}{33}$

   d. $\dfrac{9}{26}$ e. $\dfrac{5}{49}$ f. $\dfrac{1}{5}$

   g. 40 h. $\dfrac{2}{7}$

2. a. 60 b. $\dfrac{100}{3}$ c. 20

   d. 98 e. 30 f. 60

   g. 9 h. 40

3. a. 40 b. 72

   c. 160 d. 285

KEY SKILL 4 Decimal numbers

1. a. 234.02 b. 48.58 c. 0.17

   d. 200.17 e. 0.99 f. 0.01

2. a. 0.596 b. 48.582 c. 0.166

KEY SKILL 5 Fractions to decimals

1. a. 0.125 b. 0.67 c. 0.75

   d. 0.25 e. 0.75 f. 0.9

   g. 0.33 h. 0.33 i. 0.25

   j. 0.5 k. 0.5 l. 0.75

2. a. 3.75 b. 4.5 c. 2.8

   d. 17.5 e. 25 f. 6.2

   g. 45.2 h. 18.6 i. 22.1

   j. 9.25 k. 56.25 l. 19.67

KEY SKILL 6 Percentage skills

3. $\dfrac{25}{100}, \dfrac{1}{4}, 1 \div 4, 0.25$ 4. $\dfrac{30}{100}, \dfrac{3}{10}, 3 \div 10, 0.3$

5. $\dfrac{50}{100}, \dfrac{1}{2}, 1 \div 2, 0.5$ 6. $\dfrac{75}{100}, \dfrac{3}{4}, 3 \div 4, 0.75$

7. $\dfrac{80}{100}, \dfrac{4}{5}, 4 \div 5, 0.8$ 8. $\dfrac{90}{100}, \dfrac{9}{10}, 9 \div 10, 0.9$

9. $\dfrac{12.5}{100}, \dfrac{1}{8}, 1 \div 8, 0.125$ 10. $\dfrac{150}{100}, \dfrac{3}{2}, 3 \div 2, 1.5$

11. $\dfrac{33.33}{100}, \dfrac{1}{3}, 1 \div 3, 0.33$ 12. $\dfrac{66.67}{100}, \dfrac{2}{3}, 2 \div 3, 0.67,$

13. a. 70% b. $\dfrac{7}{10}$ c. 0.7

KEY SKILL 7 Percentages

1. a. 76 b. 26.25

2. a. 48 b. 25.5

   c. 210 d. 7.92

   e. 55.8 f. 60

   g. 53.52 h. 125.56

KEY SKILL 8 Fractions into percentages

1. a. 50% b. 62.5% c. 20%

2.

	Test 1	Test 2
Michelle	93.75%	90%
Yang	18.75%	25%
Simon	50%	53.33%
Alba	61.25%	75%
Christina	81.25%	91.67%

3. a. 20% b. 20% c. 17.5%

KEY SKILL 9 Ratios

1. a. 10 : 13 b. 5 : 7

   c. 100 : 1 d. 11 : 3

   e. 6 : 30 f. 12 : 7

2. a. 350 : 7 b. 150 : 2

   c. 3 : 2 d. 3 : 2

   e. 4 : 1 f. 640 : 1

   g. 10 : 15 : 9 h. 4 : 2 : 35 : 2

KEY SKILL 10 Using ratios

1. a.
   b.
   c.
   d.

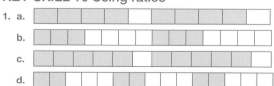

2. See sample responses in the worked solutions in eBookPLUS.

KEY SKILL 11 Simplifying ratios

1. a. $3:1$ b. $3:2$ c. $3:1$
   d. $5:2$ e. $3:2$ f. $5:4:3$

2. $3:2$

3. $3:1$

4. $8:9$

PROJECT In the kitchen

1.

	1	2	3	4	5	10
Flour (cup)	2	4	6	8	10	20
Cocoa (cup)	$\frac{2}{9}$	$\frac{4}{9}$	$\frac{2}{3}$	$\frac{8}{9}$	$\frac{10}{9}$	$\frac{20}{9}$
Sugar (cup)	1	2	3	4	5	10
Butter g	60	120	180	240	300	600
Egg	1	2	3	4	5	10
Milk (cup)	$1\frac{1}{2}$	3	$4\frac{1}{2}$	6	$7\frac{1}{2}$	15

2. 15

3. Flour = 30 cups, cocoa = 3.3 cups, sugar = 15 cups, butter = 900 g, eggs = 15, milk = 22.5 cups

4. $75

5. a. 50c   b. $7.50

6. a. 75c   b. $5.75   c. $86.25

7. a. Snakes = 4, jelly babies = 6, mint leaves = 7, black jelly beans = 3
   b. Snakes = 120, jelly babies = 180, mint leaves = 210, black jelly beans = 90

8. a. 120
   b. Party pies = 40, mini sausage rolls = 60, pasties = 20

9. a. 24 000 mL
   b. 24 L
   c. i. 4.8 L   ii. 28.8 L

GLOSSARY

allowance an additional amount of money given to workers to allow them to complete certain tasks

annual leave loading an additional payment received on top of the 44-week annual-leave pay

area the amount of flat surface enclosed by a two-dimensional shape. It is measured in square units, such as square metres, m^2, or square kilometres, km^2.

arithmetic mean one measure of the centre of a set of data. It is given by the formula:

$$\text{mean} = \frac{\text{sum of all data values}}{\text{total number of data values}}$$

When data are presented in a frequency distribution table, it is given by the formula:

$$\bar{x} = \frac{\sum (f \times x)}{n}$$

Australian Bureau of Statistics (ABS) the statistical agency of the federal government. The ABS collects data and publishes a wide range of reports for use by the governments of Australia and the community.

average speed the total distance of a trip divided by the total time taken; average rate of change of distance with respect to time

back-to-back stem-and-leaf plot a stem plot used to compare two different sets of data. Back-to-back stem plots share the same stem, with one data set appearing on the left of the stem and the other data set appearing on the right.

bias occurs when some individuals are more likely to be selected for study than others, which may result in a sample that is not representative of the whole population

BIDMAS the order in which calculations are performed. The order is: brackets; index (or power); division, multiplication from left to right; addition and subtraction from left to right.

budget a list of all planned income and costs

capacity the maximum amount of fluid that can be contained in an object. It is usually applied to the measurement of liquids and is measured in units such as millilitres (mL), litres (L) and kilolitres (kL).

categorical data data that involve grouping or classifying things, not numbers; for example, grouping hair colour

census collection of data from a population (e.g. all Year 10 students) rather than a sample

census night held every five years, the night all people who are in Australia fill in the census form

central tendency a single value that represents the middle of a frequency distribution

closed describes questions to which the participant is provided with a list of answers from which the participant must choose one or more options

column graph graph in which equal width columns are used to represent the frequencies (numbers) of different categories

commission a percentage of the sale price given to a salesperson when a sale is made

comparison rate an interest rate that gives a more accurate indication of the total cost of a loan. It is a comparison rate that reduces to a single percentage the interest rate and all the additional fees

comprehensive motor vehicle insurance that covers damage to your own vehicle and other people's property, as well as theft and some other risks, plus legal costs

compulsory third-party (CTP) mandatory motor vehicle insurance required by each state and territory. CTP covers any person that you might injure while you are driving.

concentration the ratio of the amount of solute to the amount of solvent

confidence level the level of confidence that a population parameter will lie in an interval. For example, a confidence level of 95% means there is a 5% chance that the results will not lie in the interval.

continuous data numerical data that can take any value that lies within an interval. Continuous data values are subject to the accuracy of the measuring device being used.

conversion graphs line graphs that convert one quantity into another

council rates a fee home owners pay to support councils in providing parklands, libraries, rubbish collection and road maintenance

data collection the process by which information is collected about a given population

density the ratio of mass to volume

dependent describes the variable whose value changes because of a change in the independent variable

discount a price reduction on an item

discrete data numerical data that is counted in exact values, with the values often being whole numbers

discretionary spending spending that occurs over a given time interval that is not a fixed amount

dot plot a plot in which every data value is represented by a dot, used to identify the most common values

equivalent ratios ratios that are equal in value, e.g. $1 : 2 = 3 : 6$

estimate an approximate answer when a precise answer is not required

fixed spending spending that occurs over a given time interval that is a fixed amount

frequency the number of times a score occurs in a set of data

frequency distributions *see* **frequency distribution tables**

frequency distribution table table used to organise data by recording the number of times each data value occurs

frequency tables *see* **frequency distribution tables**

fuel consumption the amount of fuel used per 100 km

gradient a measure of the slope of a line at any point

grouped column graphs graphs that display the data for two or more categories, allowing for easy comparison

grouped data numerical data that is arranged in groups to allow a clearer picture of the distribution and make it easier to work with

histogram a display of continuous numerical data similar to a bar chart, in which the width of each column represents a range of data values and the height of the column represents that range's frequency

hourly wage a set amount of money earned each hour

income tax a tax levied on people's financial income and based on an income tax table

independent describes the variable that is changed to produce a change in a dependent variable

interest payment earned for having money stored in a bank or financial institution

interest rate the percentage of the principal that is paid out in a given time period as interest

kilowatt-hour (kWh) a standard unit of electrical energy consumption

line graph graph containing points joined with line segments

margin of error the range of values that result when using a sample of a population

mean commonly referred to as the average; a measure of the centre of a set of data. The mean is calculated by dividing the sum of the data values by the number of data values.

measurement error the difference between the measured value and the real value

measures of central tendency mean, median and mode

measures of spread statistical values that indicate how far data values are spread either from the centre, or from each other

median the middle value(s) of a data set when the values are placed in numerical order

Medicare levy a portion of taxpayer's funds used to pay for Medicare (healthcare for Australian residents)

megajoules (MJ) a standard unit of energy

misrepresentation an untrue statement that does not reflect the characteristics of the population as determined by a survey or census

mode the category or data value(s) with the highest frequency. It is the most frequently occurring value in a data set.

monthly income the amount of income that a person receives in one month

nominal data categorical data that has no natural order or ranking

numerical data data that can be counted or measured

one-way tables *see* **frequency distribution tables**

open-ended describes questions to which participants can state their responses in their own words

ordinal data categorical data that can be placed into a natural order or ranking

outlier an extreme value or unusual reading in the data set, generally considered to be any value beyond the lower or upper fences

overtime any additional hours a person works after a certain number of hours a week. This is usually paid at a higher rate than the normal hourly wage

partially closed questions that provide a compromise between open-ended and closed questions Participants can choose options from a list of answers and have the option of creating their own responses if the options aren't suitable

Pay As You Go (PAYG) tax a withholding tax system administered by the Australian Taxation Office

penalty rate the pay rate paid when workers in some jobs are required to work on weekends and public holidays

pension government financial support given to some people after they have retired

per for each. For example, the car travels at 60 km per hour means that the car travels 60 km for each hour driven.

per cent the amount out of 100, or per hundred; for example, 50 per cent (or 50%) means 50 out of 100 or $\frac{50}{100}$

percentage relative frequency or **% relative frequency** the frequency of a score as a proportion of the total number of scores, expressed as a percentage

picture graphs or **pictographs** graphs that use pictures to display categorical data

piecework a fixed method of payment typically for production of individual items

pie graphs type of graph mostly used to represent categorical data. A circle is used to represent all the data, with each category being represented by a sector of the circle, whose size is proportional to the size of that category compared to the total.

population the whole group from which a sample is drawn

principal the amount that is borrowed or invested

quantitative data information collected as numerical values, for example, the number of children, time spent shopping

questionnaire a list of questions used to collect data from a population

random number generator a device or program that generates random numbers between two given values

rate a measure of how one quantity is changing compared to another

rate in the dollar the rate at which council taxes are charged

ratio the relationship between two or more values commonly expressed as $\frac{a}{b} \Leftrightarrow a : b$

recurring decimals decimals that have one or more digits repeated continuously; for example, 0.999... They can be expressed exactly by placing a dot or horizontal line over the repeating digits as in this $8.343\,434 = 8.3\dot{4} = 8.\overline{34}$.

registration fee a combination of administration fees, taxes and charges paid to legally drive a vehicle on the road

relative frequency the frequency of a particular score divided by the total sum of the frequencies

representative a sample that represents the larger population well

royalty a payment made to authors, composers or creators for each copy of the work or invention sold Royalties are typically calculated as a percentage of the total sales.

salary a fixed amount of money earned in one year, but usually paid a portion of this each fortnight or month

sample part of a population chosen to give information about the population as a whole

sample size the number of participants in a sample

sampling the process of selecting a sample of a population to provide an estimate of the entire population

sampling errors errors that occur in sampling that reduce the accuracy of the estimates the sample can provide of the entire population

scale a series of marks indicating measurement increasing in equal quantities

scale factor the ratio of the corresponding sides in similar figures, where the enlarged (or reduced) figure is referred to as the image and the original figure is called the object

scientific notation used to express very large or very small numbers. To express a number in scientific notation, write it as a number between 1 and 10 multiplied by a power of 10; for example, 100 is written as 1×10^2

self-selected sampling a voluntary sample made up of people who self-select into a survey

side-by-side bar chart bar charts that contain multiple sets of categorical data presented as multiple bars in the same chart

simple random sampling a type of probability sampling method in which each member of the population has an equal chance of selection

stamp duty a tax levied by all Australian territories and states on property purchases, such as homes, land and vehicles

stem-and-leaf plot an arrangement used for numerical data in which data points are grouped according to their numerical place values (the 'stem') and then displayed horizontally as single digits (the 'leaf')

step graphs discontinuous graphs formed by two or more linear graphs that have zero gradients

strata sub-groups into which a population is divided

stratified sampling a sampling method where groups within a population have a similar representation in the sample

superannuation a percentage of annual salary that is set aside for retirement

survey collection of data from a sample of a population

survey population the number of participants in a sample selected from a population

systematic random sampling sampling in a way that ensures that each member of the population has an equal chance of being chosen

systematic sampling a sampling method where the data values chosen to be in the sample are selected at regular intervals

tally a mark made to record the occurrence of a score

target population the population from which data is collected

taxable income the amount of income remaining after tax deductions have been subtracted from the total income

tax deductions work-related expenses that are subtracted from taxable income, which lowers the amount of money earned and amount of tax paid

third-party fire and theft third-party property insurance with some extra features that cover your vehicle

third-party property motor vehicle insurance that covers damage to other people's property and legal costs, but not damage to your own vehicle

time measurement used to work out how long we have been doing things or how long something has been happening

timeline a graphical way of displaying a series of events in the order that they occur

timetable a list of events that are scheduled to occur

total income the sum of all money earned by an individual

two-way frequency table a table that displays two categorical variables according to the frequencies of predetermined groupings

two-way tables tables that list all the possible outcomes of a probability experiment in a logical manner

ungrouped data numerical data that is not arranged in groups to enable exact analysis

vehicle insurance covers the costs associated with vehicle accidents, such as repairs, replacement vehicle, injuries to people

volume the amount of space a 3-dimensional object occupies. The units used are cubic units, such as cubic centimetres (cm^3) and cubic metres (m^3).

INDEX